D SCIENTIFIC SERIES IN COMPUTER SCIENCE

ned

ontrol Theory of Robotic Systems (*J M Skowronski*)

n Introduction to Chinese, Japanese, and Korean Computing
K T Huang & T D Huang)

athematical Logic for Computer Science (*Z W Lu*)

omputer Vision and Shape Recognition (Eds. *A Krzyzak, T Kasvand & C Y Suen*)

ochastic Complexity in Statistical Inquiry (*J Rissanen*)

Perspective in Theoretical Computer Science — Commemorative Volume
Gift Siromoney (Ed. *R Narasimhan*)

mputer Transformation of Digital Images and Patterns (*Z C Li, T D Bui,*
Y Tang & C Y Suen)

ay Grammars, Patterns and Recognizers (Ed. *P S P Wang*)

uctural Pattern Analysis (Eds. *R Mohr, Th Pavlidis & A Sanfeliu*)

Computational Model of First Language Acquisition (*N Satake*)

Design and Implementation of ConcurrentSmalltalk (*Y Yokote*)

m Humans to Computers — Cognition Through Visual Perception
V Alexandrov & N D Gorsky)

oduction to Theoretical Computer Science (*Ma Xiwen*)

igital Optical Cellular Image Processor — Theory, Architecture and
ementation (*K-S Huang*)

puter Epistemology — A Treatise on the Feasibility of the Unfeasible or Old
s Brewed New (*T Vámos*)

ications of Learning and Planning Methods (Ed. *N G Bourbakis*)

nces in Artificial Intelligence — Applications and Theory (Ed. *J Bezdek*)

duction to Database and Knowledge-Base Systems (*S Krishna*)

rn Recognition: Architectures, Algorithms and Applications
R Plamondon & H D Cheng)

acter and Handwriting Recognition — Expanding Frontiers (Ed. *P S P Wang*)

are Science and Engineering — Selected Papers from the Kyoto Symposia
I Nakata & M Hagiya)

nces in Machine Vision — Strategies and Applications
C Archibald & E Petriu)

g

matical Foundations of Parallel Computing (*V V Voevodin*)

age Architectures and Programming Environments (Eds. *T Ichikawa &*
otani)

ete list of published titles in the series, please write in to the publisher.

Advances in
MACHINE VISIC

Strategies and Applicat

WORL

Publis

11: C
12: A
(J
13: M
14: C
15: St
16: A
fo
17: Co
Y
18: Ar
19: Str
20: A C
21: The
22: Fro
(*V*
23: Intr
24: A D
Imp
25: Com
Idea
26: Appl
27: Adva
28: Introd
29: Patte
(Eds.
30: Char
31: Softw
(Eds.
32: Advar
(Eds.

Forthcomin

33: Mathe
34: Langu
H Tsu

For a compl

World Scientific Series in Computer Science – Vol. 32

Advances in MACHINE VISION

Strategies and Applications

Edited by

Colin Archibald

Autonomous Systems Laboratory
National Research Council of Canada
Ottawa, Ontario K1A 0R6
Canada

Emil Petriu

Department of Electrical Engineering
University of Ottawa
Ottawa, Ontario K1N 6N5
Canada

World Scientific
Singapore • New Jersey • London • Hong Kong

Published by

World Scientific Publishing Co. Pte. Ltd.

P O Box 128, Farrer Road, Singapore 9128

USA office: Suite 1B, 1060 Main Street, River Edge, NJ 07661

UK office: 73 Lynton Mead, Totteridge, London N20 8DH

Library of Congress Cataloging-in-Publication data is available.

ADVANCES IN MACHINE VISION: STRATEGIES AND APPLICATIONS

ISBN 981-02-0976-2

Printed in Singapore by JBW Printers & Binders Pte. Ltd.

Preface

This book is about new ideas for extracting information from sensor data. Each paper was invited, and reviewed by other experts in the field, *and* by the editors. We have achieved an interesting cross-section of machine vision. There are 20 papers presented from eight countries. The desire to force the book to contain only the ideas in our fields of interest has been resisted, and we have been satisfied to edit for technical and English correctness.

The text is organized around five section titles which will help direct those who are browsing. It may be interesting to note, however, that interesting theoretical developments can be found in the section called 'Applications,' and interesting applications are apparent to the avid reader, even in the most theoretical papers.

Typesetting of this book was done using LaTeX. We thank Gerhard Roth for helping us get started with this. The communication between the authors and the editors was done by e-mail, including the submission of the final manuscript. It is still amazing that world wide communication can occur with such ease.

Colin would like to specifically thank the authors for their efficiency, and friendliness during the preparation of this book. It was a pleasure to be working with you.

– *Colin Archibald and Emil Petriu*

The Cover

The Discrete Fourier Transform (DFT) may be formulated as a multiplication of a matrix and the time-series vector. The corresponding complex-valued matrix of numbers appears on the cover as a colour-coded image, where the crosshairs denote the time and frequency axes. Appropriate 2-D windows into this image lead to variants of the Short Time Fourier Transform (STFT) and the wavelet transform. Coordinate transformations (rotations, shears, etc.) applied to the image further extend these well known time-frequency methods into new time-frequency "perspectives." (See *Wavelets and "Chirplets": Time-Frequency "Perspectives" with Applications* on page 99.)

– *Steve Mann*

Table of Contents

IV. Perceptual and Geometric Model Building

V. Applications

Minimal Subset Random Sampling for Pose Determination and Refinement

Gerhard Roth
Institute for Information Technology
National Research Council of Canada
Ottawa, Canada K1A OR6

Martin D. Levine
McGill Research Center for
Intelligent Machines
Montreal, Canada H3A 2A7

Abstract

Pose determination is the process of finding the pose (position and orientation) of a part with a known geometry in a scene using sensor data. No prior estimate of the pose is assumed to be available. If an estimate of the pose is available then pose determination becomes pose refinement. It is shown that both can be modelled as optimizations of a cost function, which in the worst case has many local minima. This optimization model makes it clear that the most important issue is how to efficiently find the global minimum of the cost function, since this will produce the best pose. The minimal subset principle states that a minimal subset of correspondences between model and sensor data can be used to hypothesize a pose. If enough minimal subsets are chosen randomly then the correct pose is likely to be found. This principle is used to produce a general algorithm for pose determination and refinement which is capable of dealing with many different kinds of part models. Elimination theory makes it possible to create a hypothetical pose from a subset of correspondences whenever the model primitives are described by a polynomial equation. The algorithm can be easily parallelized, and a number of possible approaches to parallelization are discussed.

1 Introduction

To estimate the pose of an object is to use sensor data to compute the position and orientation of that object with respect to a known reference frame. The pose is information which is essential for manipulation and planning purposes in robot vision. It will be shown that finding the pose is equivalent to solving a particular kind of optimization problem, and that all previous approaches can be modelled in this fashion. This optimization model highlights the fact that the most important issue is how to quickly find the correct pose, which means finding the global minimum of a particular cost function. The model is also applied to the pose refinement problem, which is simply a restricted version of pose determination.

A pose determination and refinement algorithm based on the minimal subset principle is proposed. A minimal subset of correspondences is the smallest number necessary to produce a unique pose. The algorithm functions by repeatedly choosing a minimal subset of correspondences between model and data primitives and using this subset to produce a hypothetical pose. This principle has been previously applied to certain pose determination problems [1, 2], but its generality and relation to optimization have not been noted or understood. It will be shown that any pose determination problem can be solved in this fashion, as long as the ability to compute a hypothetical pose from a minimal subset of correspondences exists. That this ability exists is demonstrated for a wide variety of geometric primitives. The resulting algorithm has a simplicity and generality which has not heretofore been obtained in a single approach for pose determination and refinement. Some examples are used to illustrate the concepts.

2 Previous Approaches

There are many previous approaches to pose determination and refinement. One of the earliest techniques for pose determination uses pose clustering [3, 4]. In this work, each possible pairing of image and model features votes in a parameter space of transformations. The best cluster in this parameter space is the computed pose. This approach is closely related to the methods based on the Hough Transform [5, 6].

A completely different technique is based on the use of interpretation trees [7]. Here a tree of assignments of image to model primitives is pruned using simple rules based on local geometric properties. This process continues until the largest consistent set of assignments is found, and this set is used to compute the pose. The difficulty with this approach is that the execution time is an exponential function of the length of the assignment tree.

One way to achieve an execution time which is not exponential in the number of features is the hypothesize and test paradigm. This is the basis for a number of approaches to pose determination [8, 2], of which the most recent is the alignment method [9]. The idea is to use a small number of assignments of model to image features to hypothesize a pose, and then to verify whether this pose is correct. This is also the same principle used in geometric hashing, in which the verification process is made faster by a hash table which is built off-line. This hash table is used online to find all the model primitives which match a hypothesized pose, from which the best pose is then determined [10].

In pose determination no estimate of the pose is available beforehand, while in pose refinement the assumption is made that such an estimate already exists, and that this estimate is a good one. This initial estimate is used to select some subset of correspondences between model and image features, and once made, these correspondences are assumed to be correct. Given such correspondences, pose refinement has been shown to be an optimization problem, and nonlinear optimization algorithms have been applied to obtain the solutions [11]. The assumption is that the initial pose estimate is close enough to the final estimate to guarantee convergence of the optimization algorithm. There has also been some work in pose refinement using robust statistics [12]. The use of robust methods in the optimization algorithm makes it possible to ignore some percentage of incorrect correspondences between model and image features. We will discuss the differences between pose determination and refinement in more detail in the next section. Minimal subset random sampling can be applied to both problems, since both are described by our optimization model.

3 Optimization Model

There are so many pose determination and refinement algorithms that it seems difficult to find any commonality between them. In this section we will show that all such algorithms can be modelled in a simple way. This model shows that the process of finding the best pose is equivalent to computing the global minimum of a cost function. The latter measures the conformity of the model to the sensor data, and the best pose maximizes this conformity.

To explain these concepts more clearly, a number of formal definitions are necessary. The assumption is made that the raw sensor data consist of an unordered set of points in two-or three-dimensional Euclidian space. Such sensor data are called geometric sensor data [13], and can be obtained directly from active sensors such as laser rangefinders [14], by processing passive sensor data using methods such as stereo vision, or by edge detection on intensity images [15]. A geometric primitive is a curve or surface that can be described by a number of free parameters. The object whose pose is to be determined is modelled as a union of k such geometric primitives, labelled $m_1, m_2, \dots m_k$, and it is assumed to be a rigid object. An individual geometric primitive can be whatever is appropriate to the task at hand. Examples of geometric primitives are points, lines, planes, or more complex shapes such as cylinders or cones. Similarly, a set of l geometric primitives labelled $s_1, s_2, \dots s_l$ are extracted from the sensor data. Only a subset of the sensor primitives need belong to the object whose pose is to be determined.

The pose is defined by a transformation T, which when applied to the sensor primitives, should align them with the model primitives. The type of transformation depends on the particular geometric primitives and on the dimension of the geometric sensor data. When applied to the sensor primitives, the transformation T produces a set of transformed primitives labelled $s'_1, s'_2, \ldots, s'_l$ where $s'_j = T(s_j)$. Each transformed sensor primitive s'_j can be placed in correspondence with any member of a set q_j of model primitives. The only restriction on the primitive types is that the sensor primitive must be completely contained within the corresponding model primitive. This means for example, that the model primitive could be a line, and the sensor primitive a point, but not vice versa.

For pose determination, the set q_j consists of all the model primitives, while for pose refinement it is usually assumed that the set consists of a single model primitive. This is because a prior pose estimate of T is used to associate each transformed sensor primitive with a single model primitive. Therefore the cardinality of this set ($||q_j||$) is at least one, and at most k. For every such correspondence there must be some way of measuring the conformity of a sensor to a model primitive. This is done by a distance function d, which measures the similarity between two primitives. Similarity in this case is not necessarily the Euclidian distance but can be any appropriate measure. Assume the transformed sensor primitive s'_j is placed in correspondence with model primitive m_i. Then $d(m_i, s'_j)$ is a function which returns a scalar, and this scalar indicates the degree of match between the transformed sensor primitive and the model primitive. The particular form of d is not important, but it must return zero when the two primitives match perfectly. It is clear that d is an indirect function of T, the transformation between the model and the sensor data, since $s'_j = T(s_j)$.

The final component of our model is the cost function h, which is defined as $h(d_1, d_2, \ldots, d_r)$, where r is the total number of possible correspondences. From the preceding definition it is clear that $r = \sum_{j=1}^{l} ||q_j||$ since each of the l primitives can correspond to any model primitive in the associated correspondence set. The cost function h returns a scalar which is the degree of match between the model and the sensor primitives under the transformation T. As the transformation T changes, the value of h changes since it is a function of the distance measurements, which are themselves a function of T. The objective is to find the value of T which produces the minimum h since this is the pose which maximizes the match between the model and sensor data. This model is quite general, and we will show that by using different h and d functions it is possible to account for all known pose determination and pose refinement algorithms.

The first use of this model is to understand the difference between pose determination and refinement. For pose refinement, a prior pose estimate of T is used to associate each transformed sensor primitive with a single model primitive. This is done by applying the initial estimate of T to each sensor primitive and then making a correspondence between this primitive and the closest model primitive. In this case the total number of correspondences r is equal to l, since $||q_j|| = 1$ for $j = 1, \ldots, l$. If each member of the set of initial correspondences is correct, a nonlinear optimization algorithm can be used to find the best pose T [11]. For our model this means that the h function has a fixed number of local minima. Each local minimum indicates a different solution to the underlying system of nonlinear equations that is used in the optimization process [16]. The number of such local minima is small, since the total number of solutions is limited. Thus if the initial correspondences are correct, pose refinement is equivalent to minimizing a cost function which has only a small number of local minima.

If no prior estimate of the pose is available, then it must be determined which model primitive corresponds to each sensor primitive. With no prior estimate, each transformed sensor primitive can be put into correspondence with every model primitive. Thus r is equal to lk, since $||q_j|| = k$ for $j = 1, \ldots, l$. Note that only a small subset of the initial correspondences is correct since each sensor primitive can only correspond to a single model primitive. Somehow the h function must select the correct correspondence for each sensor primitive and ignore the rest. (We will use the robust statistics term "inlier", for a correct correspondence, and "outlier" for an incorrect correspondence [17].) This ability implies that h has many local minima, since each local minimum implicitly partitions all correspondences into inliers and outliers, and there are many such partitions. The local minimum which produces the smallest value of the cost function is the global minimum, and this is the one with the best pose transformation T. Since it cannot be assumed that all the initial correspondences are correct, pose determination is equivalent to the optimization of a cost function which may have many local minima.

A simple example will clarify these concepts. Consider the case where the model and sensor primitives consist of a finite set of points in the plane. Define $d(m_i, s'_j) = |m_i - s'_j|^2$ so that d is simply the square of the distance between the model and sensor points. The transformed sensor point s'_j is defined as $s'_j = Rs_j + M$ where R is a 2D rotation matrix, M is a 2D translation vector and s_j is the initial sensor point. The simplest possible h function would sum all the d values; that is $h(T) = \sum_{j=1}^{l} d(m_i, s'_j)$, where model point m_i has been placed in correspondence with each sensor point s'_j. This h function assumes that every one of the l sensor points is placed in correct correspondence with a single model point. This is the classical pose refinement algorithm, since to correctly make such a correspondence there must be some prior estimate of the pose which is reasonably accurate. The global minimum can be found by a standard nonlinear optimization algorithm, and this is clear if the definition for d is inserted directly into the h function.

A single incorrect correspondence can arbitrarily bias the pose if a non-robust optimization method is used to compute the pose transformation T. An improvement would be to limit the effect of an incorrect correspondence by changing the distance function d. If the value of this function did not increase as the distance between the transformed sensor and model primitive increased, then the influence of a single bad correspondence would be limited. This idea of limiting the influence of a single outlier is called the influence function approach in the field of robust statistics [17]. In this case, $h(T) = \sum_{j=1}^{l} \rho(d(m_i, s'_j))$ where ρ is the influence function. This approach has been used in the pose refinement literature [12]. This modification makes the pose refinement more robust, since the h function is now capable of discarding a certain percentage of bad correspondences. However, as noted previously, this means that the h function has potentially many more local minima since the number of local minima increases dramatically as the percentage of bad correspondences increases. For pose refinement, an initial estimate of the pose exists, so the total number of bad correspondences is a much smaller fraction of all the correspondences than for a pose determination algorithm, and therefore the number of local minima in h is also smaller. According to the robust statistics literature, the percentage of bad correspondences that can be tolerated is 50% of the total. This limitation will be discussed later in the paper, and it will be shown that in a practical sense it does not exist.

The clustering and Hough transform approaches to pose determination can be shown to be equivalent to template matching, where a template is a band of fixed size around the model. The template size is defined by the cell (or cluster) size in parameter space, while the template shape is defined by the part model and the hypothesized pose T. For the 2D point example, the template is a circle of radius r around each model point; the template match counts the number of sensor points, which under the transformation T, are within a distance r of a model point. The appropriate cost function is $h(T) = \sum_{i=1}^{k} \sum_{j=1}^{l} c(m_i, s'_j)$, where $c(m_i, s'_j)$ is zero if $d(m_i, s'_j) < r$ and is one otherwise. Note that the input to the cost function h consists of each of the kl possible correspondences, thus while the previous example dealt with pose refinement, this example deals with pose determination.

In a companion paper [18] we described a similar optimization model for primitive extraction, which is the problem of finding geometric primitives in sensor data. The ability to extract primitives is assumed to exist in this paper, since the sensor primitives $s_1, \ldots, s_l$ are necessary for pose determination and refinement. The analogy between the two approaches is shown in Table 1. In this paper, an outlier is an incorrect correspondence, and in both papers the ability to tolerate outliers means that the associated cost function h may have a large number of local minima. Any algorithm that can deal with outliers must be able to find the global minimum from among the many local minima. This ability is essential for both pose determination, and in practice, for pose refinement as well, since a single bad correspondence can arbitrarily bias the computed pose.

4 Minimal Subset Random Sampling

The optimization model shows that in some sense pose determination and refinement are mathematically solved problems. The main issue is how to efficiently find a good solution. This means determining the global minimum of a cost function which may have many local minima. There is a number of known methods for solving such global optimization problems [19]. The difficulty with these is their computational cost; they are very slow and are impractical for robot vision without massively parallel hardware such as a connection machine [20]. The minimal subset concept is based on

Primitive Extraction and Fitting
Input = Points Goal = Find Optimum Parameter Vector A Minimize $h(A) = h(e_1, \ldots, e_r)$ where e_i is Residual Error for point i Inlier = point on primitive Outlier = point not on primitive
Pose Determination and Refinement
Input = Correspondences Goal = Find Optimum Pose T Minimize $h(T) = h(d_1, \ldots, d_r)$ where d_{ij} is Distance Measure for correspondence ij Inlier = correct correspondence Outlier = incorrect correspondence

Table 1: Relation between Primitive Extraction and Pose Determination

the observation that it is often possible to compute the characteristics of an entire set from a subset. For pose determination this means that computing the pose from a subset of correspondences often produces a pose nearly as good as the one computed using all the correspondences. A key issue is the size of the minimal subset. This is equal to the minimal number (which we label P) of correspondences between model and sensor primitives that produces a unique pose transformation T. The answer depends on the particular geometric primitives that have been chosen and on the dimension of the geometric data. The relationship of P to different types of model primitives will be discussed later in the paper.

In computer vision applications, a minimal subset of correspondences can be used to produce a hypothesis in the hypothesize and test paradigm [2]. In our case, the hypothesis is a particular pose T, and a randomly chosen minimal subset of correspondences is used to generate this value of T. The cost function h tests this hypothesis by measuring the degree of overlap between the model and sensor primitives under T. The idea of using a minimal random subset of correspondences to hypothesize a pose has been introduced previously [1, 21], but it was not generalized to deal with a wide variety of geometric primitives. The following pseudo-code shows how minimal subset random sampling can be applied to any pose determination or refinement problem. The input is the model defined by model primitives $m_1, \ldots, m_k$, the sensor primitives $s_1, \ldots, s_l$, and the correspondence subsets $q_1, \ldots, q_l$:

For Z randomly chosen sets of P correspondences from the correspondence subsets

1. *Compute the pose T that aligns the P model and sensor primitives.*
2. *Compute the distances $d_1, \ldots, d_r$ between model and sensor primitive of every correspondence with respect to pose T.*

3. Rank the goodness of the pose T by evaluating the cost function $h(d_1, \ldots, d_r)$.

4. Save the pose T with the smallest cost as the best pose.

The output consists of the best pose T and the associated value of the cost function h. This information, along with the initial definition of the h function, enables each of the possible correspondences in the correspondence sets to be marked as an inlier or an outlier.

4.1 Computing a Hypothetical Pose

We have previously discussed step 2 above, how to compute the distance measure between primitives. In this section, we will discuss step 1, how to compute the hypothesized pose. First consider how to choose Z, the number of randomly chosen minimal subsets. It is assumed that each of the P correspondences uses a different sensor primitive. Since there are at most k model primitives that could correspond to a single sensor primitive, the number of possible sets of random samples of size P is k^P. In practice this number is too large to use as a value of Z for any reasonably sized k; however, a much smaller value of Z can be used. Let the probability of a single correspondence of a sensor primitive to a model primitive in the correspondence subset being correct be ϵ. Then the probability of all P correspondences in a single one of the Z draws being correct is ϵ^P. Therefore the probability of at least one of the Z draws producing a correct set of P correspondences is $1-(1-\epsilon^P)^Z$. This formula is a simple application of combinatorial analysis and the same result is presented in [1]. Using this formula the value of Z can be computed for any given ϵ and P. Usually the probability of success is set to .95, and Z is set to the value necessary to achieve this probability. While the value of Z computed in this way is less than k^P, it is still quite large if ϵ is small. This analysis shows that pose determination is inherently a very difficult problem because $\epsilon = 1/k$, since there are k elements in each correspondence set. Therefore any prior estimate of the pose is potentially very useful since it increases ϵ by decreasing the number of possible correspondences in each correspondence subset.

Consider the case where the sensor primitives are points, and the model primitives are defined by a polynomial equation. This is one of the most natural ways to perform pose determination and refinement because using only points as sensor primitives means primitive extraction is not necessary. The model primitives are defined by a polynomial equation, which in 2D is $\sum_{i=0}^{m}\sum_{j=0}^{n} a_{ij}x^iy^j = 0$ and in 3D is $\sum_{i=0}^{l}\sum_{j=0}^{m}\sum_{k=0}^{n} a_{ijk}x^iy^jz^k = 0$. These polynomial equations define a curve in 2D and a surface in 3D. Using only polynomials for geometric primitives appears at first to be restrictive. However, it has been shown that all of the basic primitives in a computer aided design (CAD) system can be converted to such implicit polynomial equations [22]. Since the objects whose pose we wish to find usually have an associated definition on a CAD system, such a representation is not restrictive in practice.

Next we consider how to produce a hypothetical pose from P correspondences. Application of the pose transformation T should align the sensor primitives with the model primitives to which they correspond. If we have P correspondences, then substituting the transformed sensor primitive into the appropriate model equation will produce a system of polynomial equations. The unknown in the system is the pose transformation T, and solving the system will produce the hypothetical pose. The maximum number of such equations is six for 3D primitives and three for 2D primitives, since these are the degrees of freedom of objects in space. Thus P is at most three for the 2D case, and at most six for the 3D case.

The following example in which the model consists of a circle and a line, and the sensor data is a set of points in the plane, illustrates this technique. In Section 6 we will discuss an application to a practical pose determination problem that uses this model. In the case of the plane, the pose transformation T consists of a rotation and translation so the pose has three degrees of freedom and the number of correspondences from sensor points to model primitives necessary to define a unique transformation T is three. The pose transformation T applied to a point (x, y) in the plane produces new points (x', y'), which are defined as follows:

$$x' = (x+h)\cos(\theta) - (y+k)\sin(\theta) \tag{1}$$
$$y' = (x+h)\sin(\theta) + (y+k)\cos(\theta) \tag{2}$$

The pose of the model is represented by the transformation $T = (\theta, h, k)$ which defines the rotation and translation in the plane. In order to make the above equations polynomial functions of the unknowns, $u1$ is substituted for $\cos(\theta)$ and $u2$ for $\sin(\theta)$. If the equation $u1^2 + u2^2 - 1 = 0$ is added to encode the constraint that $\cos(\theta)^2 + \sin(\theta)^2 = 1$, then the transformation $T = (u1, u2, h, k)$ can be mapped one-to-one with $T = (\theta, h, k)$.

The model will be positioned at the origin, with the center of the circle being the origin and the line parallel to the y-axis. This makes the equation of the circle $x^2 + y^2 - r^2 = 0$, and the equation of the line $x - r - p = 0$, where r and p are known. The model is shown in Figure 1 part (a), and three sensor points are shown in part (b). If T is applied to a sensor point, then the transformed point must be on the corresponding model primitive. The assumption is that two sensor points are on the circle, with the third sensor point on the line. If the three sensor points are $(x0, y0), (x1, y1)$ and $(x2, y2)$, then substituting the transformed x and y values into the model equations for the circle and the line will produce the following polynomial equations.

$$((x0+h)u1 - (y0+k)u2)^2 + ((x0+h)u2 + (y0+k)u1)^2 - r^2 = 0 \tag{3}$$
$$((x1+h)u1 - (y1+k)u2)^2 + ((x1+h)u2 + (y1+k)u1)^2 - r^2 = 0 \tag{4}$$
$$((x1+h)u1 - (y1+k)u2) - r - p = 0 \tag{5}$$
$$u1^2 + u2^2 - 1 = 0 \tag{6}$$

Since the number of equations (four) equals the number of unknowns $(u1, u2, h, k)$, this polynomial system can be solved and has a fixed number of solutions. In this case there are four solutions, and these are shown in Figure 1 parts (c) through (f). Of course not all of these provide a good match to the model.

This method for producing the set of equations whose solution is the hypothesized pose T can be used whenever the sensor and model primitives are defined by polynomial equations, and not just for the special case where the sensor primitives are points. As has just been shown, however, when the primitives are points the equations are particularly simple. This same approach has also been used when the sensor primitives are 2D line segments and the model primitives are 3D line segments [16]. This situation occurs when polygonal objects are observed through a camera lens to produce a 2D image. Each 3D edge projects onto a corresponding 2D line segment. Another similar situation occurs when the model primitives are planes and the sensor primitives are 3D line segments [23]. In both cases a system of polynomial equations is produced by a minimal set of correspondences of sensor to model primitives.

In order to determine a hypothetical pose T from this minimal set of correspondences, it is necessary to solve the nonlinear system of polynomial equations created by substituting the transformed sensor primitives into the equations of the model primitives. The ability to automatically solve for any nonlinear polynomial system is necessary to generate a hypothetical pose T from P correspondences.

5 Solving Nonlinear Polynomial Systems

There are two approaches to solving such systems. These are numerical methods and symbolic methods. Traditionally numerical methods, such a Newton-Raphson, have been used, but these require an initial estimate which must be close to the actual solution. Even though numerically based methods that use homotopies [24, 25] find all the solutions, they are computationally very costly. Symbolic methods, on the other hand, are more efficient since the final result is a polynomial equation in a single unknown which can be quickly solved. From each solution, simple back-substitution can be used to solve for the other unknown variables. The difficulty is how to eliminate all the free variables in order to create a polynomial equation with a single unknown.

There are two symbolic methods which can be used for this purpose. They are the Gröbner basis technique [26] and multi-variable resultants [27], both of which belong to the field of elimination theory. The end result of these methods is the elimination of all but one of the free variables, to create a polynomial in a single unknown. The solution to a polynomial equation with a single unknown variable exists in a closed form if it is of low order, or can be found numerically using fast algorithms. The solution for this unknown is then used to solve for another unknown, and so on in a recursive fashion until all the unknowns have been found [28]. Even though it has been proven mathematically that both of these symbolic methods are always able to eliminate all the free variables except one, the problem is that the time required to do

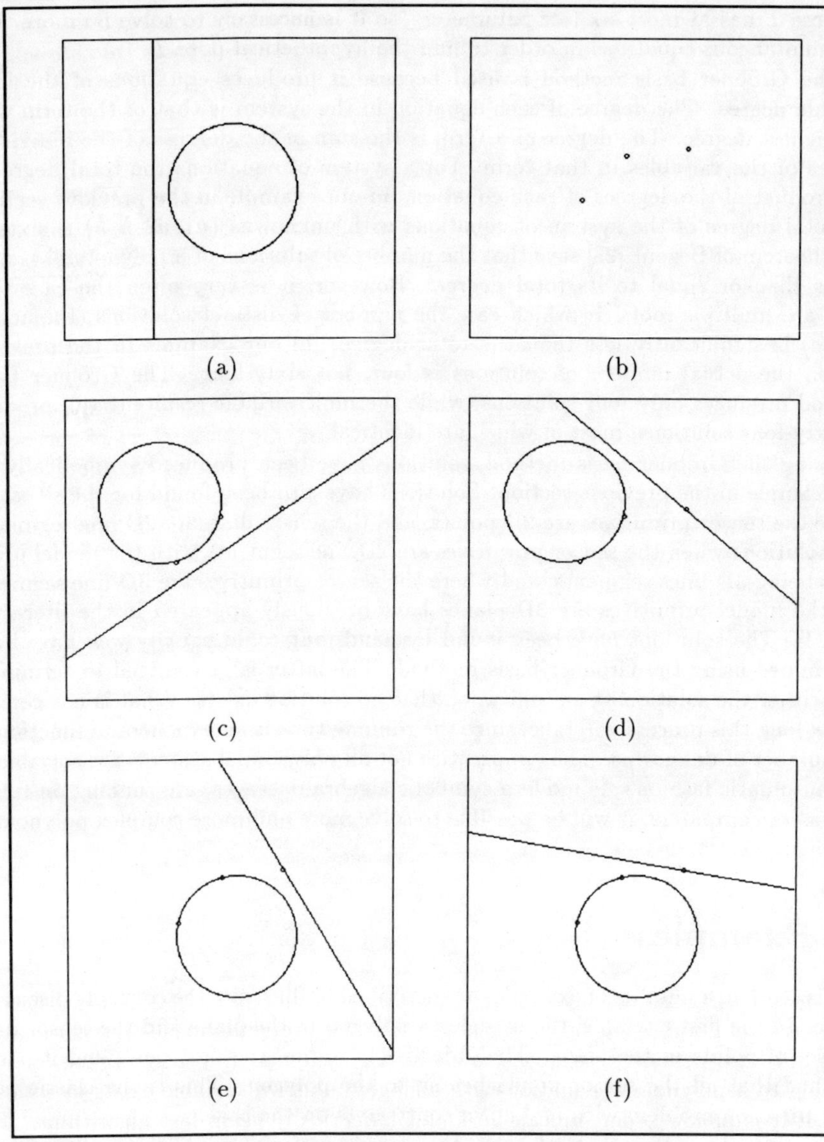

Figure 1: Circle and Line Model (a) Model at the origin (b) Three sensor points (c) (d) (e) (f) Four computed poses with model superimposed on sensor points

this is an exponential function of the number of equations and unknowns. Fortunately, the pose T has at most six free parameters, so it is necessary to solve no more than six simultaneous equations in order to find the hypothetical pose T.

The Gröbner basis method is used because it produces equations of the least possible degree. The degree of each equation in the system is that of the term with the highest degree. The degree of a term is the sum of the degrees of the individual powers of the variables in that term. For a system of equations the total degree is the product of the degrees of each equation. In our example in the previous section, the total degree of the system of equations with unknowns $(u1, u2, h, k)$ is sixteen. The theorem of Bezout [28] says that the number of solutions of a polynomial system is less than or equal to its total degree. However, it is very often the case that there are multiple roots, in which case the number of distinct solutions (the actual degree) is significantly less than the total degree. In our example in the previous section the actual number of solutions is four, not sixty-four. The Gröbner basis method produces only four solutions, while the multivariable resultant will produce all sixty-four solutions, most of which are identical.

Using the Gröbner basis method, solutions have been produced symbolically for the example in the previous section. Solutions have also been found for the situation where the sensor primitives are 2D points, and the sensor data are 2D line segments. The solutions when the sensor primitives are 2D line segments with the model primitives being 3D lines segments, and where the sensor primitives are 3D line segments and the model primitives are 3D planes have previously appeared in the literature [16, 23]. The solutions have been found by hand, but could equally well have been determined using the Gröbner basis method. The latter is guaranteed to terminate with either the solution or an indication that no solution exists. What is not certain is how long this process will take, since the running time is an exponential function of the number of unknowns. Thus in practice not all polynomial systems are solvable in this automatic fashion. As modern symbolic algebra packages, are running on larger and faster computers, it will be possible to solve more and more complex polynomial systems.

6 Examples

In this section, a number of examples are described to illustrate the concepts discussed above. In the first example, the model is a polygon in the plane and the sensor data is a set of points in the plane. The objective is to find the best pose, and it is not assumed that all the sensor points belong to the polygon. This is the classic pose determination problem with no known constraints on the pose.

To evaluate a hypothetical pose, the pose transformation T is applied to the sensor points and the number of sensor points within a small distance from each line segment of the model is counted. In practice, this distance must be set empirically from a careful study of the sensor and its noise characteristics. In this example, it

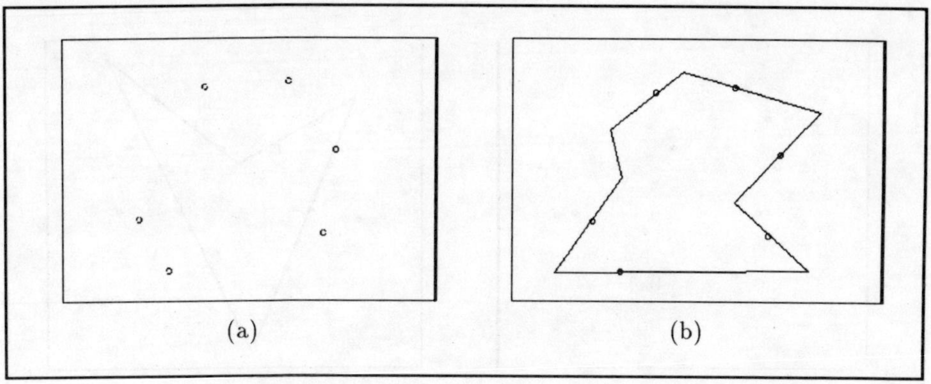

Figure 2: Two Dimensional Pose Determination with no Spurious Points (a) Initial sensor points (b) Computed pose with model superimposed on sensor points

is arbitrarily chosen. In Figures 2 and 3, two instances of the pose determination algorithm are shown. Part (a) of both figures shows some simulated sensor data, and part (b) shows the model with the best pose aligning the model and sensor data. In Figure 2 all the sensor data is consistent with the model, while in Figure 3 this is not the case. The pose was determined using minimal subset random sampling in which three sensor points were randomly selected, a hypothetical pose determined and the pose evaluated as described above. The hypothetical pose T was computed by solving a nonlinear system of equations defined by these correspondences. The hypothesize and test strategy using random sampling makes it possible to avoid searching the entire interpretation tree to find the correct pose, as was done in [29]. This means that as the number of model primitives increases, this approach will become faster than a method which must search the entire interpretation tree [9].

The following example applies random sampling to the pose refinement problem. The assumption in pose refinement is that an estimate of the pose is available, along with some indication of its accuracy. With the same model primitives as the previous example, such a situation would occur when the polygon describes the walls of a room, and the points consist of sensor data taken from a laser rangefinder [30]. If the rangefinder is mounted on an autonomous vehicle, then the vehicle itself can provide the initial pose estimate from its internal navigation system [31]. However, the accuracy of this estimate decreases as the vehicle travels, so it must be periodically corrected by a pose refinement process. A correction is made by adjusting the vehicle pose till the sensor data has the best alignment with the 2D polygon environment model. This has been accomplished previously in the literature using a pose refinement algorithm based on numerical optimization [31]. While effective, the difficulty with this approach, and in fact with any local optimization procedure, is that there is no way of predicting the range of convergence, even for noise-free data.

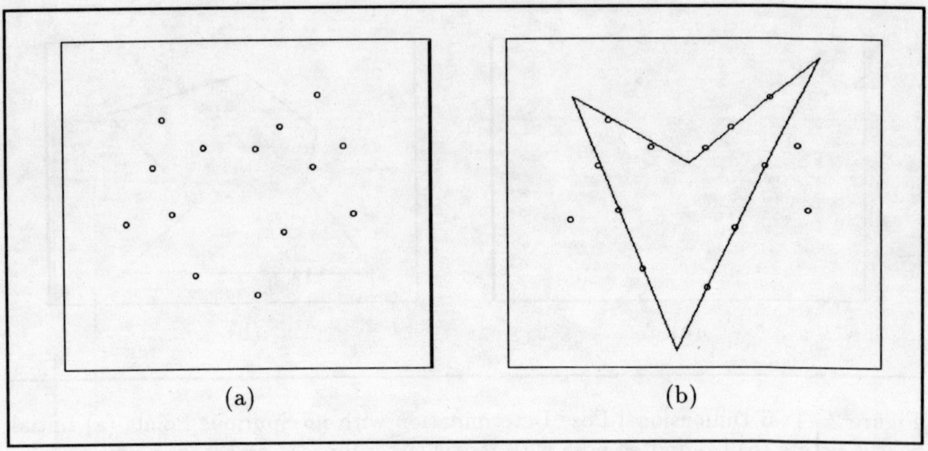

Figure 3: Two Dimensional Pose Determination with Spurious Points (a) Initial sensor points (b) Computed pose with model superimposed on sensor points

In other words, as the error in the initial estimate of the vehicle position increases, the optimization procedure will at some point fail to converge to the correct answer. In terms of the optimization model presented in Section 3, the process converges to a local instead of the global minimum. As we have discussed previously, this is due to the fact that the optimization approach to pose refinement assumes that the correspondences between the sensor and model primitives are correct. By contrast, the random sampling algorithm does not make this assumption, and is guaranteed to converge (at least probabilistically) given enough random samples. It is not difficult to adjust the number of random samples according to the accuracy of the initial pose estimate. If it is very accurate, few random samples are needed, and as this estimate worsens more random samples are required.

In the worst case when the initial estimate is arbitrarily bad, pose refinement becomes pose determination. Fig. 4(a) show some simulated sensor data, along with the initial pose estimate that might be provided by the autonomous vehicle superimposed on the sensor data. Fig. 4(b) shows the refined pose obtained using the random sampling algorithm. Even though the same basic algorithm that was used in pose determination is used in pose refinement, the result is obtained much faster because the initial vehicle position is used to limit the number of possible correspondences during the random sampling process. We plan to apply this approach to adjusting the position of a mobile vehicle using actual range data.

The final example has as input a number of three-dimensional points produced by a laser rangefinder mounted on a robot wrist [32]. The objective is to find the pose of an object called an "H-Fixture", which is the approximate shape and size of one of the grapple fixtures for the space station. This rangefinder collects parallel profiles of the

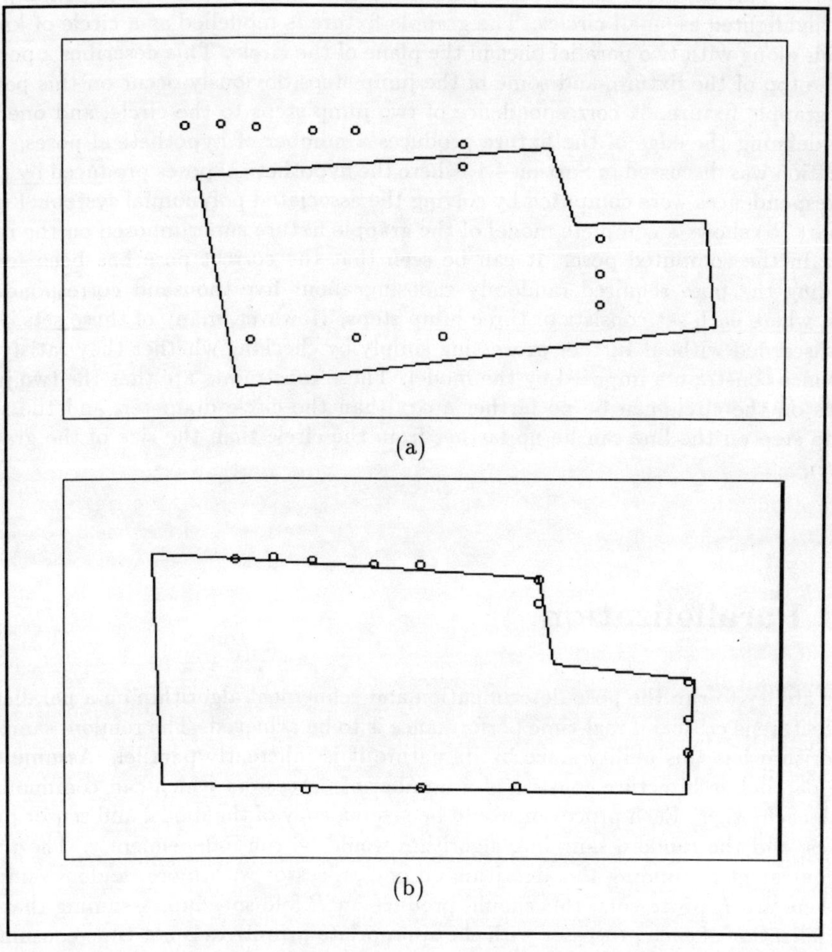

Figure 4: Two Dimensional Pose Refinement superimposed on sensor points (a) Initial sensor points and initial pose superimposed on sensor points (b) Final pose with model superimposed on sensor points

object, and from these profiles is extracted a subset of the input called jump points. Thse are points which have a significant difference in depth from their neighbours. These jump points define the 3D outline of the object as seen from the rangefinder. Figure 5, part (a) shows a number of parallel profiles along with the jump points which are highlighted as small circles. The grapple fixture is modelled as a circle of known radius along with two parallel lines in the plane of the circle. This describes a portion of the top of the fixture, and some of the jump steps obviously occur on this part of the grapple fixture. A correspondence of two jump steps to the circle, and one to a line defining the edge of the fixture produces a number of hypothetical poses. This situation was discussed in Section 4.1, where the hypothetical poses produced by three correspondences were computed by solving the associated polynomial system. Figure 5, part (b) shows a complete model of the grapple fixture superimposed on the range data in the computed pose. It can be seen that the correct pose has been found. Finding the pose required randomly choosing about five thousand correspondence sets, where each set consists of three jump steps. However, many of these sets could be discarded without further processing simply by checking whether they satisfy the distance constraints imposed by the model. These constraints are that the two jump steps on the circle can be no farther apart than the circle diameter, and the third jump step on the line can be no farther from the circle than the size of the grapple fixture.

7 Parallelization

The ability to run the pose determination and refinement algorithm on a parallel architecture is critical if real-time performance is to be achieved. The random sampling algorithm has this ability since by its nature it is inherently parallel. Assume that the parallel architecture consists of a number of processors which can communicate with each other. Each processor would be given a copy of the model and sensor primitives, and the random sampling algorithm would be run independently. The result is equivalent to running the algorithm on one processor with more random samples. If there are R processors, this should produce an R fold speedup, assuming that the initialization of each processor with the appropriate primitives is not time-consuming. This is a reasonable assumption, since for pose determination and refinement there is usually a small number of model primitives.

This basic approach will work on a variety of different parallel architectures. All that is necessary is the ability to transfer data between processors at a reasonable rate. The ability to parallelize the pose determination and refinement process is not evident in other approaches in which the hypothesis generation is done in a more deterministic fashion [11].

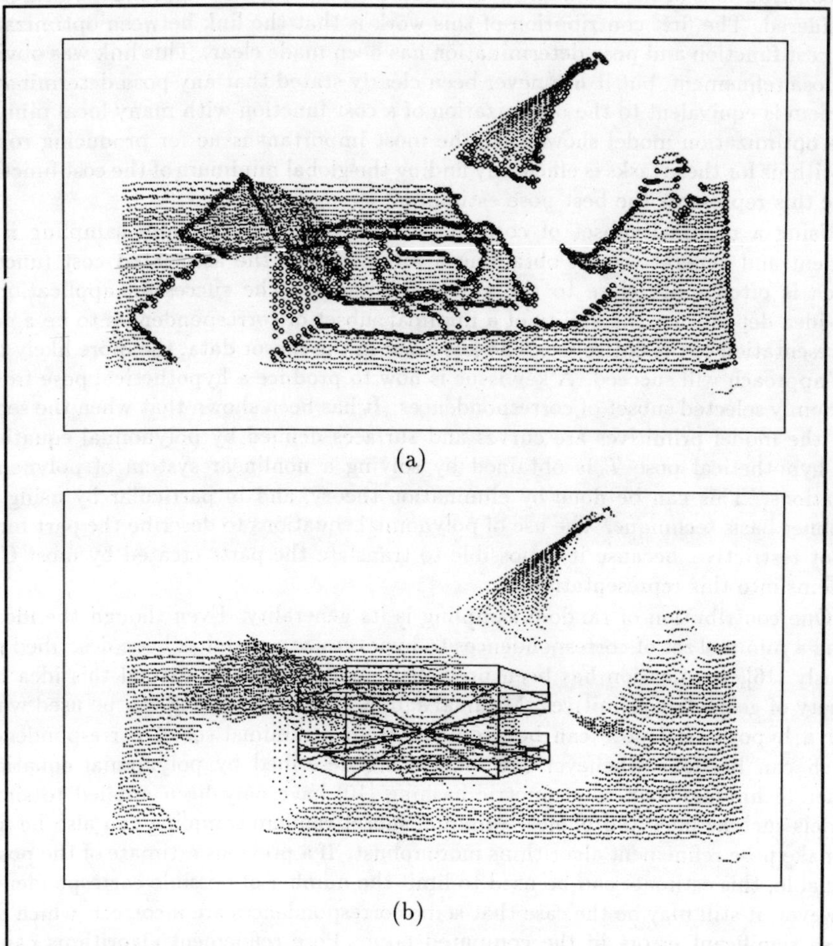

Figure 5: Grapple fixture pose determination (a) Jump steps with range profiles (b) Final pose of grapple fixture drawn on range profiles

8 Discussion and Conclusions

Since there has already been so much work done in pose determination and refinement, the contributions of the minimal subset random sampling paradigm must be carefully considered. The first contribution of this work is that the link between optimization of a cost function and pose determination has been made clear. This link was obvious for pose refinement, but it has never been clearly stated that any pose determination problem is equivalent to the optimization of a cost function with many local minima. This optimization model shows that the most important issue for producing robust algorithms for these tasks is efficiently finding the global minimum of the cost function, since this represents the best pose estimate.

Using a minimal subset of correspondences chosen by random sampling is an efficient and general way of obtaining a minimum for the associated cost function which is often quite close to the global minimum. The successful application of this idea depends on the ability of a minimal subset of correspondences to be a good representation of the entire set. The less noise in the sensor data, the more likely that this approach will succeed. A key issue is how to produce a hypothetical pose from a randomly selected subset of correspondences. It has been shown that when the sensor and the model primitives are curves and surfaces defined by polynomial equations, the hypothetical pose T is obtained by solving a nonlinear system of polynomial equations. This can be done by elimination theory, and in particular by using the Gröbner basis technique. The use of polynomial equations to describe the part model is not restrictive, because it is possible to translate the parts created by most CAD systems into this representation.

One contribution of random sampling is its generality. Even though the idea of using a minimal set of correspondences to hypothesize a pose has been described previously [16], no mention has been previously made of how to extend this idea to a variety of geometric primitives. Minimal subset random sampling can be used whenever a hypothetical pose can be produced from a minimal set of correspondences, which can be done whenever the models are described by polynomial equations. Other techniques such as geometric hashing [10] have only been applied to simple models such as lines and planes. Minimal subset random sampling can also be used to make pose refinement algorithms more robust. If a previous estimate of the pose is available, this estimate can be used to limit the number of possible correspondences. However, it still may be the case that some correspondences are incorrect, which may cause significant errors in the computed pose. Pose refinement algorithms can be made more robust using the influence function approach [12] from robust statistics, but even for noise-free data, convergence depends on a good initial starting estimate. This is not the case for random sampling since for noise-free data convergence is guaranteed if enough samples are taken. It is also the case that using random sampling, pose refinement algorithms can deal with more than 50% bad correspondences. The difficulties that previous robust pose refinement methods using influence functions have had with a large percentage of bad correspondences are due to the heuristic way in which the starting position of the associated optimization algorithm was chosen [12]. With random sampling, the global minimum can be found even when there are more than 50% incorrect correspondences. In a number of the presented examples the percentage of bad correspondences (outliers) is considerably more than 50%.

The random sampling algorithm can also be easily parallelized on a variety of different parallel architectures. This ability is essential if real-time performance is to be achieved for pose determination and refinement. The conclusion is that random sampling is a powerful approach to solving pose determination and refinement problems, and has not been previously considered in its full generality.

Acknowledgements

Gerhard Roth would like to thank the National Research Council of Canada (NRCC) for providing education leave for graduate studies. Martin D. Levine would like to thank the Canadian Institute for Advanced Research and PRECARN for its support. This work was partially supported by the National Science and Engineering Council of Canada. The authors would like to thank C. Archibald of NRCC for calibrating the laser rangefinder used to provide the data in one of the examples.

References

[1] M. Fischler and R. C. Bolles, "Random sample consensus," *Communications of the ACM*, vol. 24, pp. 381–395, June 1981.

[2] C. Archibald and C. Merritt, "Pose determination of objects from sparse range images," in *Proceedings of 2nd Conference on Intelligent Autonomous Systems*, (Amsterdam, The Netherlands), Dec. 1989.

[3] G. Stockman, "Matching images to models for registration and object detection via clustering," *IEEE Trans. Pattern Analysis and Machine Intelligence*, vol. 4, no. 3, pp. 229–241, 1982.

[4] G. Stockman, "Object recognition and localization via pose clustering," *Computer Vision, Graphics and Image Processing*, vol. 40, pp. 361–387, 1987.

[5] D. Thompson and J. Mundy, "Three dimensional model matching from an unconstrained viewpoint," in *IEEE International Conference on Robotics and Automation*, (Raleigh, North Carolina), pp. 208–220, March 1987.

[6] T. Silverberg and D. Anderson, "An iterated hough procedure for three-dimensional object recognition," *Pattern Recognition*, vol. 17, no. 6, pp. 621–629, 1984.

[7] W. Grimson and T. Lozano-Perez, "Model-based recognition and localization from sparse range and tactile data," *The International Journal of Robotics Research*, vol. 3, no. 3, 1984.

[8] N. Ayache and O. Faugeras, "Hyper: A new approach to the recognition and positioning of two dimensional objects," *IEEE Trans. Pattern Analysis and Machine Intelligence*, vol. 8, no. 1, pp. 44–53, 1986.

[9] D. Huttenlocher and S. Ulman, "Object recognition using alignment," in *First Inernational Conference on Computer VIsion*, (London, England), pp. 102–111, 1987.

[10] H. Wolfson, "Model based object recognition by geometric hashing," in *European Conference on Computer Vison 90*, (526-536), April 1990.

[11] D. Lowe, "Fitting parameterized three-dimensional models to images," *IEEE Trans. Pattern Analysis and Machine Intelligence*, vol. 13, pp. 441–450, May 1991.

[12] M. Haralick, H. Joo, C. Lee, X. Zhuang, V. Vaidya, and M. Kim, "Pose estimation from corresponding point data," *IEEE Transactions on Systems, Man and Cybernetics*, vol. 19, pp. 1426–1439, Nov. 1989.

[13] P. J. Besl, "Describing geometric signals," in *workshop on range image processing, vision interface 89*, (London, Ontario, Canada), Canadian image processing & pattern recognition society, June 1989.

[14] M. Rioux, "Laser rangefinders based on synchronized scanning," *Applied Optics*, vol. 23, pp. 3837–3844, 1985.

[15] M. D. Levine, *Vision in man and machine.* New York: McGraw-Hill, 1985.

[16] M. Dhome, M. Richetin, J. Lapreste, and G. Rives, "Determination of the attitude of 3-d objects from a single perspective view," *IEEE Trans. Pattern Analysis and Machine Intelligence*, vol. 11, pp. 1265–1278, Dec. 1989.

[17] F. Hampel, E. Ronchetti, P. Rousseeuw, and W. Stahel, *Robust statistics: the approach based on influence functions.* New York: Wiley, 1986.

[18] G. Roth and M. Levine, "Segmentation of geometric signals using robust fitting," in *10th International Conference on Pattern Recognition*, (Atlantic City), pp. 826–831, June 1990.

[19] A. Torn, *Global Optimization.* Berlin: Springer-Verlag, 1989.

[20] W. D. Hillis, *The connection machine.* ACM Distinguished Dissertation, Cambridge, Massachusetts: The MIT Press, 1985.

[21] R. C. Bolles and M. A. Fischler, "A ransac-based approach to model fitting and its application to finding cylinders in range data," in *Seventh International Joint Conference on Artificial Intelligence*, (Vancouver, British Colombia, Canada), pp. 637–643, 1981.

[22] T. Sederberg and D. Anderson, "Implicit representation of parametric curves and surfaces," *Computer Vision, Graphics and Image Processing*, vol. 28, pp. 72–84, 1984.

[23] H. H. Chen, "Pose determination from line-to-plane correspondences: existence condition and closed-form solutions," *IEEE Trans. Pattern Analysis and Machine Intelligence*, vol. 13, pp. 530–541, June 1991.

[24] A. Morgan, *Solving polynomial systems using continuation for science and engineering.* Prentice Hall, Englewoord Cliffs, 1987.

[25] J. Renegar, "One the efficiency of newton's method in approximating all zeros of a system of a complex polynomials," *Mathematics of Operations Research*, vol. 12, pp. 121–148, Feb. 1987.

[26] B. Buchberger, "Gröbner basis: an algorithmic method in polynomial ideal theory," in *Recent trends in multidimensional system theory* (N. Bose, ed.), Hingham, Mass.: Reidel, 1985.

[27] J. Canny, *The complexity of robot motion planning.* Cambridge: The MIT Press, 1987.

[28] C. Hoffman, *Geometric and solid modelling, an introduction.* San Mateo, California: Morgan Kaufman, 1989.

[29] P. Gaston and T. Lozano-Perez, "Tactile recognition and localization using object models: the case of polyhedra in the plane," *IEEE Trans. Pattern Analysis and Machine Intelligence*, vol. 6, no. 3, pp. 257–266, 1984.

[30] F. Blais, M. Rioux, J. Domey, and J. Beraldin, "A very compact real time 3-d range sensor for mobile robot applications," in *Mobile Robotics 3*, vol. 1007, (Cambridge, Mass.), pp. 330–338, SPIE, 1988.

[31] I. Cox and J. Kruskal, "On the convergence of noisy images to line segment models," in *Second International Conference on Computer Vision*, (Tampa, Florida), pp. 252–258, 1988.

[32] M. Rioux and F. Blais, "Compact three dimensional camera for robotic applications," *Journal of the Optical Society of America*, vol. 3, pp. 1518–1520, 1986.

Robust High Breakdown Estimation and Consensus

Peter Meer
Department of Electrical and Computer Engineering
Rutgers University
Piscataway, NJ 08855-0909

Abstract

Application of robust regression methods to computer vision problems becomes increasingly popular. These methods make use of a priori available information through the assumed model from which, however, deviations are allowed. Thus robust methods are more powerful than nonparametric techniques which, in order to avoid erroneous assumptions, do not involve models at all. We describe the least median of squares estimator and show that for computer vision data its straightforward use may not yield the desired results. Most of the artifacts can be avoided by using a somewhat different, consensus based approach. We present the principle behind the consensus paradigm.

1 Introduction

In the last two decades, to avoid the pitfalls of least squares, robust estimation techniques were developed by statisticians. Recently these robust techniques have become popular in computer vision. See Meer *et al.* (1990b) for a survey of robust techniques as applied in computer vision. Some of the robust operators can tolerate a discontinuity (edge) and recover the model corresponding to the absolute majority of the data in the processing window. The least median of squares (LMedS) estimator, proposed by Rousseeuw(1984), is one of this class and is the most frequently used in computer vision. However, the LMedS estimator was developed for statistical applications and its utility in solving computer vision problems cannot be taken for granted. In Section 3 we describe LMedS and show its limitations for noisy data. These limitations can be avoided by a *consensus* based technique which retains the desirable properties of both the least squares and the least median of squares estimators. In Section 4 we present the consensus principle.

2 The Weighted Least Squares Estimator

The weighted least squares (WLS) estimator is a frequently used tool in computer vision. We only mention here the properties used later in the paper. For proofs and/or more detailed discussions see a textbook on linear estimation. The noise corrupting the linear model used in WLS is assumed to be of zero mean. In this case the estimated parameters are unbiased, i.e., their expected value is correct. The WLS estimates become minimum-variance estimates by proper choice of the weights. However, it can be shown that for the small window operators usually employed in computer vision the achieved accuracy does not suffice (Meer *et al.*, 1990a).

The implicit assumption behind WLS is that the entire data set can be characterized by only *one* parameter vector β. For piecewise data, a case frequently met in computer vision but less common in statistics, this is not true. In piecewise data a model discontinuity is also present and (at least) two parameter vectors are required to describe the entire data set. For example, changes in reflectance or depth yield piecewise data.

It is well known that least squares estimation cannot handle piecewise data. This is caused by the violation of the zero mean noise assumption. The data not represented by the model has to be regarded as corrupted by an additional non-zero mean noise process. It is shown in the next section that model estimation for piecewise data can be achieved only by high breakdown point robust techniques which tolerate non-zero mean noise.

Least squares polynomial surface estimation of piecewise constant data (i.e., containing a step edge) has a property of importance in this regard. Let the step edge in the data be of height h and corrupted with noise having standard deviation σ. The surface is estimated in a window centered on the discontinuity. The simplest, degree-0 model is that of a constant plane (the number of unknown parameters $p = 1$). The least squares estimated value yields a plane lying between the two original horizontal surfaces that form the step edge. When a planar, degree-1 model ($p = 3$) is used, the estimated plane lies across the edge and intersects both surfaces.

Both fits are erroneous and do not correspond to either of the original surfaces. However, up to moderate signal-to-noise (SNR) ratios, when the standard deviation of the residuals (the noise's standard deviation estimate) is computed the degree-0 model yields an increase relative to σ while the degree-1 model may not. Thus, when a window operator performing degree-0 least squares estimation (averaging) slides along piecewise constant noisy data, the standard deviation estimate increases in the neighborhoods of step edges. The increase is significant enough to be used for coarse detection of the presence of a model discontinuity (edge). Thus a reliable indication of the presence of a discontinuity is obtained. This result is widely applied for adaptive filtering in the signal processing literature. For an application in computer vision see Park and Meer (1990).

To conclude this section two observations can be emphasized:

- When all the data belongs to one model corrupted with zero mean noise the least squares estimates are unbiased but may have a large standard deviation.
- At moderate signal-to-noise ratios the degree-0 least squares estimate of the noise's standard deviation can be used as warning about the presence of a step edge in the data.

3 The Least Median of Squares Estimator

In this section we discuss a robust estimator which can handle weakly corrupted piecewise data. For convenience we describe the one-dimensional case; extension to multi-dimensions is immediate.

The ability of an estimator to handle severely corrupted data is captured by its *breakdown point* ε^*. This is the smallest fraction of the data which can yield arbitrary estimate values. For example, the degree-0 least squares estimate (the *mean* of n data points, z_i) has $\varepsilon^* = 1/n$, since one large valued erroneous point already compromises the result. The asymptotical breakdown point of the degree-0 LS estimator thus is 0. The zero breakdown point property is common for all least squares estimators (Hampel *et al.*, 1986, p.328).

Two types of severely corrupted data can be distinguished. *Outliers* are data points with values $z_{i,j}$ which cannot be represented by the assumed model. Thus noiseless piecewise data can always be described by one model and a set of outliers. For example, one of the planes forming a roof edge is taken as the model and the points belonging to the other are then regarded as outliers relative to that model. Once zero mean noise is also present, however, the dichotomy between the model and the outliers may not be easy to establish. (This subject is discussed later in this section.)

Leverage points are data points corresponding to outlying regressor variable values $x_k(i,j)$. These points are not necessarily "bad" since their values may be close to the value predicted by the model. However, if a data point is far away from the rest and cannot be accounted for by the model, it may exert an increased influence on the estimate. The sparse data in some computer vision applications (for example, stereo) may contain leverage points which compromise the result in spite of the fact that their value is within the range of the data. Detection of leverage points requires special techniques. See for example Rousseeuw and van Zomeren (1990).

The breakdown point of an estimator cannot exceed 0.5 without making use of a priori information about the data. Indeed, at least half of the data should be represented by the model to be the unique solution of the estimation. The *median*, the degree-0 least absolute deviations estimator (belonging to the L_1 family of estimators), tolerates close to half the data being severely corrupted. At the limit, the median has $\varepsilon^* = 0.5$. It can be shown, however, that higher degree surface fitting in L_1 is sensitive to leverage points and it has $\varepsilon^* = 0$ (Hampel *et al.*, 1986, p.328).

Another degree-0 estimator is the *mode* of the data. For a continuous probability distribution function (p.d.f.) the mode is the most probable variable value, i.e., the maximum of the p.d.f. (Without loss of generality we can consider a unimodal p.d.f. for the moment.) Assume that n outcomes were obtained from the p.d.f. and they are ordered by increasing values. First the half-length of a window containing $N \leq n/2$ points is computed in all the $n - N + 1$ positions along the ordered sequence. (The half-length is defined as half the difference between the data point values at locations i and $i + N - 1$ in the ordered sequence.) The midpoint of the shortest window is taken as the mode since the maximum of the p.d.f. implies the most outcomes in an interval.

Let the window size equal to half the data size, i.e., $N = \lfloor n/2 \rfloor$, where $\lfloor \cdot \rfloor$ is the floor function. Then it can be shown that the mode $\hat{\beta}_0$ minimizes the median of the squared residuals, i.e., satisfies

$$\min_{\hat{\beta}_0} \; \operatorname*{med}_{i} (z_i - \hat{\beta}_0)^2 \tag{1}$$

and that the value of (1) is the squared half-length of the window yielding the mode (Rousseeuw and Leroy, 1987, p. 166). The use of squares instead of absolute values assures the uniqueness of the solution for n an even number (*ibid*, p. 170). The presence of the median in (1) makes the mode a robust degree-0 estimator with breakdown point close to 0.5.

The minimization problem (1) can be generalized to an arbitrary model $\hat{\beta}$ generating the residuals r_i

$$\min_{\hat{\beta}_0} \; \operatorname*{med}_{i} \; {r_i}^2. \tag{2}$$

The expression (2) defines the *least median of squares* (LMedS) estimator introduced in statistics by Rousseeuw (1984). The already mentioned book of Rousseeuw and Leroy (1987) gives an excellent practical analysis of the estimator. The finite sample breakdown point of the LMedS estimator is

$$\varepsilon^* = \frac{\lfloor n/2 \rfloor - p + 2}{n} \tag{3}$$

and asymptotically yields 0.5.

Since (2) has no analytical solution, to find the LMedS estimates a numerical technique, projection pursuit, is employed.

3.1 The LMedS Algorithm

Assumptions: The model is defined by the parameters β_k, $k = 0, \ldots, (p-1)$. The data contains n points with a fraction $\varepsilon < 0.5$ of outliers.

Step 1. Randomly select a p-tuple Π from the data.

Step 2. For this p-tuple compute the parameter values $\hat{\beta}_k(\Pi)$ by solving the p equations. Note that if a linear model is assumed the p-tuple must yield a full rank coefficient matrix for the system.

Step 3. Retain all parameters but $\hat{\beta}_0(\Pi)$. Project the data into the β_0 subspace by computing

$$\alpha_i(\Pi) \triangleq z_i - \sum_{k=1}^{p-1} \hat{\beta}_k(\Pi) x_k(i) \quad i = 1, \ldots, n. \tag{4}$$

Step 4. Find the mode of the $\alpha_i(\Pi)$ sequence and allocate it to $\hat{\beta}_0(\Pi)$. Store the corresponding shortest half-window size

$$\delta(\Pi) = [\min_{\hat{\beta}(\Pi)} \operatorname{med}_i r_i{}^2]^{1/2} \tag{5}$$

Step 5. Repeat Steps 1 to 4 q times. The final LMedS estimates are from the p-tuple yielding

$$\delta = \min_{\Pi} \delta(\Pi)$$

The search procedure (Step 5) has the goal of finding at least one p-tuple containing only data points represented by the model, i.e., *inliers*. Projection of such a p-tuple into the β_0 space (Step 3) can yield at most εn outliers. The mode detection procedure (Step 4) then assures the recovery of $\hat{\beta}_0(\Pi)$, and that the entire parameter vector satisfies (2).

The random sampling (Step 1) reduces the complexity of the LMedS algorithm to a manageable amount of computation. Mode computation per p-tuple requires $O(n \log n)$ operations and there are $O(n^p)$ possible p-tuples. Let the tolerated probability of error in random sampling be $Q \ll 1$. Then the probability that for q independently selected p-tuples at least one contains only inliers is

$$1 - [1 - (1 - \varepsilon)^p]^q \leq 1 - Q \tag{6}$$

From (6) q can easily be computed. For example, when a degree-1 model ($p = 3$) is fit to piecewise data with $\varepsilon = 0.45$, and the tolerated probability of error is $Q = 0.01$, choosing $q = 26$ different 3-tuples suffices to find the three LMedS estimates. Note that q does not depend on the data size. Since the LMedS estimates correspond to one of the p-tuples, the only characteristic of interest about the structure of the data is ε.

A "hidden" assumption behind the LMedS estimator is that the inliers are only weakly corrupted by independent identically distributed Gaussian noise. That is, the parameter values obtained from a p-tuple containing only inliers are close to the correct ones. The LMedS algorithm also returns a robust estimate of the noise's standard deviation:

$$\hat{\sigma} = 1.4826 \left[1 + \frac{5}{n-p} \right] \delta \tag{7}$$

where the term $5/(n-p)$ is the finite sample size correction (*ibid*, p. 202) and 1.4826 is the correction factor for median based Gaussian noise standard deviation estimates. Once the noise corrupting the model is defined (zero mean normal with variance $\hat{\sigma}^2$) the data can be classified into inliers and outliers. The weights

$$w_i = 1 \quad \text{if } |r_i| \leq 2.5\hat{\sigma} \tag{8}$$

mark the inliers with about a 0.95 confidence level (the uncertainty is due to the finite data size). If the nature of the noise is known a priori other correction factors and thresholds can be computed. The LMedS algorithm and the inlier/outlier dichotomy procedure, however, become unreliable when significant noise corrupts the data. This problem is discussed below.

3.2 Performance Analysis

An optimization principle similar to the one used in the LMedS algorithm was already proposed for computer vision applications by Fischler and Bolles (1981) as the Random Sample Consensus (RANSAC) paradigm. In RANSAC the model derived from a p-tuple determines the set of data points agreeing with it within a given tolerance limit. If the number of points in the set exceeds a threshold, no more p-tuples are drawn. Thus RANSAC requires the a priori information of a tolerance limit and a cardinality threshold. LMedS automatically selects the model representing at least half the data by minimizing a quality measure of the fit, the median of the squared residuals. For a more detailed comparison of the two techniques see Meer *et al.* (1990b).

The high breakdown point of the LMedS estimator makes it an ideal operator for handling piecewise data. The estimator identifies the model corresponding to the majority of the points in the set and discriminates the points not belonging to it, i.e., the outliers. The LMedS algorithm (as presented above) was applied to several computer vision problems. Kim *et al.* (1989), Roth and Levine (1990), Sinha and Schunck (1990) have used it in surface reconstruction; Kumar and Hanson (1989) in camera position estimation; Tirumalai *et al.* (1990) in dynamic stereo.

We have already emphasized the necessary condition for obtaining "good" LMedS estimates, the existence of at least one p-tuple from which reliable model parameters can be extracted. When significant zero mean noise corrupts *all* the data points such a p-tuple may not exist. In this case a p-tuple containing only noisy inliers yields, unbiased estimates (since we are using a least square type technique for the computation) but with large variance. The noise can also erase the clear dichotomy between data points belonging to the model and outliers.

Consider again piecewise constant data containing a step edge of amplitude h (ε close to 0.5) corrupted with zero mean noise taking values in the range $(-a, a)$. This definition is for convenience only; for Gaussian noise a could be, say, 3σ. Suppose also for the moment that $a \leq h/2$.

First we use the degree-0 LMedS estimator, i.e., compute the mode of the data. Assume that we have succeeded in recovering the uncorrupted value of the horizontal surface corresponding to the majority of the pixels. Then the squared residuals are distributed as follows: slightly more than half in the interval $(0, a^2)$ and the rest in the interval $[(h-a)^2, (h+a)^2]$. The estimate of $\hat{\sigma}$, the model noise standard deviation, is proportional to the square root of the median of squared residuals, and thus it will exceed a. But a is the range of the noise and therefore $\hat{\sigma}$ is a strong overestimate. Large $\hat{\sigma}$ reduces the weights and outliers may be classified as inliers.

When a degree-1 LMedS estimator is applied to the noisy piecewise constant data the large variance of the 3-tuple based estimates makes any plane orientation possible. Assume that we have chosen two 3-tuples, one generating the correct horizontal fit, the other yielding a plane tilted at about 45 degrees across the edge. The latter fit is similar to what a least squares estimator would have recovered from the window. If significant noise is present a is close to $h/2$), most of the squared residuals for the tilted plane fit are between 0 and a^2. The median of this residual sequence is significantly lower than the median of the sequence obtained when the horizontal fit is employed (see above). The LMedS criterion (2) seeks the minimum and therefore the tilted plane is preferred. Note that tilted planes are always obtained from some 3-tuples since any sample containing pixels from both surfaces can yield one. The estimator will choose the correct, horizontal plane only when the amplitude of the corrupting noise is very small relative to the amplitude of the step, that is, the SNR is very high. The correct fit, however, implies an overestimate of σ as was shown above.

For noisy data with a discontinuity the mode detection procedure can also fail. Such data yields a bimodal distribution. The significant noise level spreads the values across the entire range of the data and no clear bimodality can be seen. The mode detection procedure finds that the shortest window corresponds to a value between the two surfaces forming the original step edge.

The severity of the artifacts of the LMedS estimator for piecewise constant data were investigated by a simulation study (Mintz *et al.*, 1990). We have shown that application of LMedS to noisy piecewise constant data with a large fraction of outliers can result in failure.

- The high breakdown point property of the LMedS estimator no longer is satisfied and performance similar to least squares is often obtained.

- The LMedS estimator cannot recover the correct model when this is misspecified, i.e., non-essential parameters are introduced.

- The inlier/outlier dichotomy can become erroneous because of overestimation of the noise's standard deviation.

In the next section we propose a new approach toward computer vision problems based on seeking consensus among local processes.

4 The Consensus Paradigm

The principle behind the LMedS estimator (and the RANSAC paradigm) can be described as follows:

> Compute a candidate model based on a randomly chosen small subset of the data. Apply this model to all the data. Compute a global quality measure for the model. Optimize the quality measure by repeating the procedure several times.

To have reliable LMedS estimates it is necessary that at least one candidate model carries the correct parameter values. We have shown that when this condition is not satisfied the LMedS estimator loses most of its attractive properties. The high breakdown point of an estimator can be interpreted as seeking the consensus in the data about the assumed model. Optimality in LMedS is achieved by minimizing the criterion (2) and in RANSAC by maximizing the number of inliers. We can regard this optimality also as a *consensus* about the model representing the majority of data points.

However, consensus can be achieved in other ways. Assume that the majority of the data obey a model of known structure but unknown parameters. The agreement with the model is characterized by the quantity γ.

Step 1. Define a simple local process which provides unbiased estimates of γ whenever the data agrees with the assumed model.

Step 2. Build a global distribution of γ.

Step 3. Find by a robust procedure the expected value of γ for agreement with the model, Γ.

Step 4. Use Γ to select the reliable local processes.

Step 5. Estimate the model parameters with these processes.

The local estimates are obtained by a least squares based procedure which has the advantage of being computationally inexpensive and providing unbiased estimates. The large variance of the estimates is no longer of concern since we use their global distribution. The robust procedure analyzing this global distribution is based on the mode seeking algorithm described in the previous section. The mode is a high breakdown point estimator and thus close to half the local estimates can be erroneous while the value of Γ indicating agreement with the model is still correctly recovered. The spread of the γ estimates around Γ is charaterized by the robust standard deviation (7). The "inliers" of this distribution thus mark the local processes where agreement with the model was obtained. The applicability of the described paradigm is very large. We describe succinctly two applications.

In *multiresolution adaptive least squares* image smoothing (Park and Meer, 1990) local fits for a constant plane are computed at every pixel. Several neighborhood sizes 7×7, 5×5, 3×3 are used. For given neighborhood size the standard deviation of the residuals is used as γ. Then $\gamma \leq \Gamma$ indicates the pixels whose smoothed value can be taken from the given neighborhood since the neighborhood falls on a homogeneous region not containing an edge. The analysis proceeds from large to small neighborhoods.

In the *consensus by decomposition* (CBD) estimators (Mintz *et al.* 1990) a candidate model is computed for a small neighborhood of every data point. Each model parameter is mapped into a separate parameter-image based on the same sampling structure as the input. Note that, these parameter-images are that of a noisy step-edge. The consensus technique is applied for each parameter-image separately. The standard deviation of the residuals from a constant are used as γ. The points around a discontinuity (if exist) are eliminated by mapping the outliers of the γ distribution into the parameter-image. A second mode seeking procedure recovers the robust parameter estimate. The CBD estimation proceeds from the highest degree coefficients of a polynomial surface model toward the intercept. The necessary condition for a reliable CBD estimate is that the majority of the candidate models carry unbiased parameter estimates. This condition is much weaker than the one required by LMedS and makes possible the elimination of the latter's artifacts.

5 Conclusion

We have described a new principle to approaching computer vision problems. Whenever no clear dichotomy between "good" and "bad" data exist, the robust high breakdown methods from statistics may fail. The decomposition of the problem into non-robust local estimation procedures yields unbiased distributions for the sought estimates. When these distributions are analyzed with a simple mode seeking algorithm all the desired properties of high breakdown estimation are preserved.

References

[1] M.A. Fischler and R.C. Bolles (1981): "Random sample consensus: A paradigm for model fitting with applications to image analysis and automated cartography," *Communications of the ACM* **24**, 381–395.

[2] F.R. Hampel, E.M. Ronchetti, P.J. Rousseeuw and W.A. Stahel (1986): "Robust Statistics: An Approach Based on Influence Functions", John Wiley & Sons: N.Y.

[3] D.Y. Kim, J.J. Kim, P. Meer, D. Mintz and A. Rosenfeld (1989): "Robust computer vision: A least median of squares based approach," In *Proceedings of the DARPA Image Understanding Workshop*, Palo Alto, CA, May 1989, 1117–1134.

[4] R. Kumar and A.R. Hanson (1989): "Robust estimation of camera location and orientation from noisy data having outliers," In *Proceedings of the IEEE Workshop on Interpretation of 3D Scenes*, Austin, TX, November 1989, 52–60.

[5] P. Meer, D. Mintz and A. Rosenfeld (1990a): "Least median of squares based robust analysis of image structure," In *Proceedings of the DARPA Image Understanding Workshop*, Pittsburgh, PA, September 1990, 231–254.

[6] P. Meer, D. Mintz, D.Y. Kim and A. Rosenfeld (1990b): "Robust regression methods in computer vision: A review," *International Journal of Computer Vision* **6**, 59–70.

[7] D. Mintz, P. Meer and A. Rosenfeld (1990): "Consensus by decomposition: A paradigm for fast high breakdown point robust estimation," Computer Vision Laboratory. University of Maryland, College Park, CAR-TR-525, December 1990.

[8] R.-H. Park and P. Meer (1990): "Multiresolution adaptive least squares smoothing of images," Computer Vision Laboratory, University of Maryland, College Park, CAR-TR-513.

[9] G. Roth and M.D. Levine (1990): "Segmentation of geometric signals using robust filtering," In *Proceedings of the 10th International Conference on Pattern Recognition*, Atlantic City, NJ, June 1990, 826–831.

[10] P.J. Rousseeuw (1984): "Least median of square regression," *J. Amer. Stat. Assoc* **79**, 871–880.

[11] P.J. Rousseeuw and A.M. Leroy (1987): *Robust Regression & Outlier Detection*, John Wiley & Sons: N.Y.

[12] P.J. Rousseeuw and B.C. van Zomeren (1990): "Unmasking multivariate outliers and leverage points," With comments by R.D. Cook and D.H. Hawkins; D. Ruppert and D.G. Simpson; P.J. Kempthorne and M.B. Mendel; and rejoinder. *Journal of the American Statistical Association* **85**, 633–651.

[13] S.S. Sinha and B.G. Schunck (1990): "A robust method for surface reconstruction," In *Proceedings of International Workshop on Robust Computer Vision*, Seattle, WA, October 1990, 183–199.

[14] A.P. Tirumalai, B.G. Schunck and R.C. Jain (1990): "Robust dynamic stereo for incremental disparity map refinement," In *Proceedings of International Workshop on Robust Computer Vision*, Seattle, WA, October 1990, 412–434.

Boundary Feature Extraction and Structural Verification of Objects

Q.M. Wu and M.G. Rodd
Department of Electrical & Electronic Engineering
University College of Swansea
SWANSEA SA2 8PP, United Kingdom

Abstract

Several algorithms are developed for verification of regular-shaped objects by breaking boundaries into fundamental primitives such as lines and circular arcs. After discussing the algorithms for detection of extreme points having high curvatures, an algorithm is proposed for detecting knot-points formed by basic primitive segments. For verification of object structures, the existing flat industrial objects are classified into four categories. For each category, corresponding effective verification algorithms are developed, according to the characteristics of the objects in each class.

1 Introduction

With the rapid progress in computer vision research, machine-vision inspection systems can be expected to play a more and more important role in industrial automation. To achieve this goal, many techniques and systems have been developed over the past decade. However, due to the inherent diversity of inspection tasks, it is difficult to develop generic techniques applicable to a wide range of applications. Very often, an *ad hoc* approach is adopted for each individual application task which is time- and labor-consuming.

In this paper, a set of generic algorithms are presented for structural verification of *man-made, flat industrial parts.* Such shape inspection is a common and critical issue in many inspection tasks — as diverse as the monitoring of the quality of punched sheet metal, the inspection of products (e.g., biscuits) in the food industry, or in the monitoring of mechanical parts (e.g., washers) in manufacturing. Therefore it is important to develop a generic methodology for solving these inspection problems.

Most structural verification tasks can be performed based on the boundaries of objects. Thus the extraction of the desired boundaries of objects in an industrial environment is of great importance to the high-level verification processes. As has been discussed in [?], the successful extraction of these boundaries requires both effective algorithms and a priori knowledge. In this paper, it is assumed that the boundaries of objects are available and issues of feature extraction and verification are discussed — all based on the object boundaries. The rest of the paper is organized into three sections. In Section 2, the boundary dominant point detection problem is discussed. Based on this, a set of verification algorithms are developed in Section 3. Finally, some conclusions are drawn in Section 4.

2 Feature Extraction by Curvature

Feature extraction is a key step towards the successful recognition, and subsequent verification, of industrial objects. Since the objective of this paper is to develop effective vision algorithms for verifying the structure of industrial objects, it is crucial to analyze what critical features can be derived from object boundaries, and which of these are of value in the recognition and verification processes. Having determined which features can be important, algorithms can then be examined for the efficient and reliable detection of such features. This section starts by investigating the dominant features of real-world man-made objects, and then moves towards proposing techniques, based (where appropriate) on well-substantiated theory, and algorithms to handle the extraction of such features. The validity of such algorithms is analyzed by experimentation in practical situations.

2.1 Dominant Features

Features derived from object boundaries can include both global features and local features. The use of global shape features, however, (such as an object's perimeter or area) is limited for inspection purposes to particular environments, mainly because such properties do not normally reflect *local* variations in object shapes. In most real situations, however, it is the *local* variations which require close inspection.

It is evident in the literature [?, ?]. that points having high curvatures and zero-crossing points make up an important set of features on object boundaries. However, despite much success achieved by the direct use of these important points in object recognition, there is also evidence that the geometric primitive elements between these points are important features as well. This point has been fully demonstrated in a paper by Fischler and Bolles [?].

The question which arises is: "If it is recognized that the idea of using primitive elements is important in the development of machine-vision systems, what primitives should be used?" Studying the scenes around us in the real world, it is clear that while natural scenes seem to be described by many primitives (some of them complicated), most industrial objects can be described by relatively few simple geometric primitives — mainly because most industrial parts are designed via CAD systems in which only limited, well-defined primitive elements are used. Among the small number of primitives often used in CAD systems, lines and circular arcs are the most popular. This paper therefore concentrates on feature extraction of such well-structured objects.

The first key step in segmenting a boundary into primitives is to highlight the *joint-points* (which are referred to as *knot-points* in this paper). These are the points which denote changes in primitive segments. The planar object boundary is then partitioned into primitive segments — typically lines and circular arcs. For example, in inspecting a regular shape consisting of lines and circular arcs, if the knot-points between different primitive segments can be located, geometric features can then be easily transformed into primitives, and their spatial relations transformed into a simple symbolic string. The inspection task can then take place by matching the string with a model — maybe compiled directly from a CAD system [?].

It is therefore clear that the dominant features on object boundaries will include points with high curvatures, the knot-points between different geometrical primitives and, of course, the primitives themselves. It must be pointed out that most points on an object boundary having high curvatures are also knot-points. Since there are obvious changes in curvature at knot-points, they can be detected by analysing the changes in curvature on object boundaries.

However, to estimate the curvature of a digitized curve is not a trivial task. In the remaining part of this section, these problems are studied, with a particular focus on efficient and deterministic implementation. Based on the outcome of this study, a flexible and robust way of segmenting curves, based on curvature analysis of the boundary, is proposed. It will be shown later how these techniques can be used in the development of efficient industrial inspection algorithms.

2.2 Curvature Estimation

In Euclidean space, the curvature k of a planar curve at a point $P(x, y)$ is defined as the instantaneous rate of change of the slope angle of the tangent at point P with respect to arc length. If $\Delta\alpha$ is used to denote the change of the slope when moving along the curve from point P to point P', where Δs is the change of the arc length, the curvature at point P can then be computed using

$$k = \frac{d\alpha}{ds} = \lim_{\Delta s \to 0} \frac{\Delta\alpha}{\Delta s}$$

For a continuous curve $y = f(x)$, this can be expressed in terms of derivatives as (see [?])

$$k = \frac{\ddot{y}}{(1 + \dot{y}^2)^{3/2}} \tag{1}$$

Curvature at any point of a curve can be accurately calculated using this formula, and appropriate information can be extracted by analysing changes in curvature.

However, when dealing with digitized curves, it is not immediately clear how to define the *discrete* analog of curvature Assuming that the boundary of an object is traced, say, based on *8-connectedness*, the change of slope takes discrete values which are multiples of 45^0, so that small changes in slope are impossible to detect. One solution to extracting "noise-free" curvatures from a contour is to "smooth" the curve. There are two categories of approaches which are correctly being employed for this purpose: the Gaussian-filter based approach [?] and the angle-detection schemes [?].

A). Gaussian-Filter Approach

This approach represents a planar curve by introducing a parameter l (a variable related to path length). Thus x and y coordinates can be represented as functions of l. To achieve the smoothing effect, each coordinate function is first smoothed before being used for computing the curvature. In order to compute the curvature at varying levels of detail, functions $x(l)$ and $y(l)$ are convolved with a one-dimensional Gaussian $g(l, \sigma)$ of standard deviation σ.

Another typical approach in this category initially finds the bounding contour of a shape using an edge detector (such as those derived by Hueckel [?] and by Canny [?]), which generates the orientation of each edge point. As the boundary is tracked, the orientation at each point is estimated. The orientation function, denoted as $\phi(l)$, is thus established. The first Gaussian derivative filtered response $g'(l, \sigma) * \phi(l)$ and the second Gaussian derivative filtered response $g''(l, \sigma) * \phi(l)$ are then computed.

According to the definition mentioned earlier, the curvature at a point is the instantaneous change of orientation (slope) over the change of the arc length. Thus the change of the orientation over a short arc length between points P and P' is an approximation of the curvature at the point P. Therefore, the filtered response $g'(l, \sigma) * \phi(l)$ reflects the curvature at point $P(x_l, y_l)$. Compared to the approach that computes curvature using filtered coordinate functions, the physical meaning of this method is relatively clear.

B). Angle-Detection Schemes

Angle-detection schemes provide another approach for estimating the curvature of a digitized curve. The fundamental principles of angle-detection schemes for curvature computation are derived from the curvature definition for continuous curves. It is clear that the direction change between two successive link elements can not be used to compute the curvature directly, as was mentioned previously. However, if the angular difference between two successive vectors to calculate the curvature is used, a *reasonably* accurate estimate of the curvature can be obtained. A representative procedure for several angle-detection schemes is described in [?].

The curvature calculated using this procedure is customarily called *incremental curvature*. The incremental curvature is a smoothed measure of curvature; the greater m is (often referred to as a *smoothing factor or smoothing parameter*), the heavier the smoothing. The smoothing effect in the incremental curvature method is achieved by using a straight-line segment with a length of m boundary points to approximate the digital segment. This has solved the problem that the slope change between any two successive discrete boundary points is only the multiple of 45^0 mentioned earlier, and leads to the estimate of the slope at a point P being closer to that computed by using a continuous curve to fit the digital one. It is clear that if a longer line segment is used in the approximation, the curvature estimate thus derived may not reflect the original fine features of the digital curve. On the other hand, if a small smoothing factor m is used, noise injected by digitizing the curve may be reflected in the estimate. Therefore, an appropriate smoothing factor has to be chosen. The "optimal" smoothing factor is dependent on the specific application. In practice, the factor appropriate to the class of curves to be processed can be chosen only by experimentation or by examining the class of curvature features that exist in the existing object categories. For a curve extracted from an image with a reasonably high resolution, The smoothing factor will range normally from a minimum of 5 to a maximum of 13 [?].

2.3 Extreme-Point Detection

There are several methods to locate peak curvature points and zero-crossings from curvature estimation. Assuming that the curvature estimate is computed using one of the approaches described in the preceding subsection, the location of curvature maxima (minima), or zero-crossings, simply involves the detection of peaks (or valleys) on a curvature plot. To locate these points accurately normally requires a two-stage processing procedure. In the first stage, a global curvature threshold T_c (for the complete curvature plot) is first of all used to select those boundary points with curvature values greater than T_c. For each corner on an object boundary, a point will be found which has a curvature value greater than T_c; often a small number of points, close to the corner point, have curvature values greater than the threshold. In the second stage, therefore, an input local threshold (for one peak only) is applied to the curvature estimates of the segment close to the peak to eliminate those points whose curvature estimates are not local maxima in a sufficiently large segment of the curve.

From the above discussion, a few observations can be made:

- *Computational Aspect:* Both of the algorithms in Section 2.2 (i.e., that which estimates the curvature by convolving the orientation function with a Gaussian kernel, and that which smoothes the coordinate functions) use Gaussian filtering to eliminate the noise injected by digitization and other sources. Due to the time-consuming convolution operation involved in Gaussian filtering, both algorithms are computationally expensive. In contrast, angle-detection schemes are conceptually, and computationally, simpler.

- *Level of Smoothing:* Both Gaussian-filtering approaches and angle-detection schemes estimate the curvature of a quantized curve by first undertaking a smoothing operation — the degree of smoothing depending on various parameters. In the Gaussian-filtering approach, the parameter is the standard deviation σ, and the change of σ from small to large will allow a coarse-to-fine detection scheme to be established. This will make it possible to find instances of curvature change primitives at one scale, even if they cannot be found at some other scales. In many cases, a useful parameter space (scale-space) can be established by plotting the zero-crossing points against the parameter l. In the angle-detection scheme, using different values of m will allow different degrees of smoothing. The larger m is, the heavier the smoothing. Therefore, in practical applications, an appropriate value of m must be chosen.

- *Threshold Selection:* The selection of the threshold value is important, since a high threshold could miss some critical points, but at the same time, would get rid of false points. A low threshold value, on the other hand, would give more-reliable critical-point detection, but would increase the number of false detections. Therefore the threshold selection requires intensive experimental verification.

- *Detection Accuracy:* The use of angle-detection schemes is accurate for detecting sharp corners because these methods explicitly measure the changes of slopes between two successive vectors, and hence points with high curvatures can be reliably detected. However, they do not produce satisfactory results when used to detect relatively smooth joints (which are normally difficult to perceive) because the linear fit destroys subtle curvature. The use of the Gaussian-filter-based approach enables the detection of these points since a peak (or valley), which corresponds to the position of the smooth joint, clearly shows in the second Gaussian derivative filtered response. This attribute will be discussed in greater detail later.

The review undertaken in this section has illustrated that each category of algorithms has some advantages and disadvantages. For industrial applications, however, better, more-generic, solutions should be sought.

2.4 Knot-Point Detection

In this subsection a new algorithm is described which resolves the problems inherent in the two categories of algorithms previously discussed. This algorithm achieves this primarily by *combining* their advantages for detecting critical points. It is shown how this can lead to an approach which is reliable and computationally relatively inexpensive. Importantly, the proposed algorithm is capable of locating knot-points between basic primitives consisting of straight lines and/or circular arcs (i.e., regions of constant curvature) — hence efficiently segmenting contours into a number of desired primitives for use in a syntactic inspection system.

All the possible types of knot-points, which can exist between primitives of lines and circular arcs, can be classified into two categories, in terms of the complexity of detecting them in a digitized curve. In the first category all the knot-points are sharp corners; in the second category the knot-points are either smooth joins (smooth corners) or the knot-points formed by short segments. The knot-points in the first category are easy to detect, while those in the second category are extremely difficult to extract.

As was mentioned in the last section, angle-detection operators are good for locating sharp corners, particularly in terms of their computationally simple implementation, their reliability, and their accuracy in locating sharp corners. Gaussian-filter based approaches are computationally more expensive but are reliable in locating smooth joins, etc. The new knot-point detector is thus developed by making use of the advantages of both. It can be described as follows:

The angle-detection scheme is first of all used to locate the sharp corners, and then a set of criteria are applied to distinguish if the segment between a pair of knots is a straight line, a circular arc, or a combination (i.e., a succession of primitive segments without knot points between them, which is referred to as a "combination segment"). If it is a combination segment, a priori knowledge about the primitive type (i.e. a smooth join, a nick or a bump) of a segment can be obtained from the criteria that have been satisfied by the segment. Thus, the scale-space approach can be applied to this particular segment to detect the primitive, guided by the a priori knowledge (which is not available when using the scale-space approach alone).

It will be shown that this scheme has advantages over several angle-detection schemes and the Gaussian-filter based approaches for detecting knot-points in the application classes referred to in this paper. Here, the criteria to be used is first of all described, and then the complete scheme is given.

2.4.1 Criteria for Linearity/Circularity

The following seven criteria are used to distinguish the three primitives — a straight line, a circular arc, and a combination segment (which may include a smooth join or a short segment).

Criterion A: Ratio of Segment Length Over the Distance Between Two Knots ($SLKD$ ratio)

It is again assumed that the boundary of an object is traced based on 8-connectedness, and has been coarsely divided by sharp corners detected using an angle-detection scheme. Let S represent the length of a segment, D the distance between the two knots, N_{od} the number of points on this segment with odd-valued chain codes, and N_{ev} the number of points with even-valued chain codes. The segment length and the distance between the two knots can be calculated, respectively, as

$$S = N_{ev} + \sqrt{2}N_{od} \tag{2}$$

(the sum of the number of step edge points, and number of the diagonal edge points multiplied by $\sqrt{2}$), and

$$D = \sqrt{(x_{i+1} - x_i)^2 + (y_{i+1} - y_i)^2} \tag{3}$$

where x_i, y_i, x_{i+1}, y_{i+1} are the coordinates of the two knot-points.

Thus the $SLKD$ ratio is S/D. For a straight-line segment, this ratio is approximately 1; for a circular arc segment and any other primitive segment, this ratio has a value greater than 1.

Criterion B: Ratio of Segment Length Over the Circular Arc Length ($SLAL$ ratio)

Assume that the segment between two knots is a circular arc. Its length can be calculated as

$$L = R\theta \quad , \tag{4}$$

where L is the circular arc length, R is the radius, and θ is the angle between two lines linking the two knots to the arc center. R and θ can be obtained using the following formulae (their derivation is given in Appendix).

$$R = \frac{(D/2)^2 + h^2}{2h} \tag{5}$$

$$\theta = 4arctan(\frac{h}{D/2}) \tag{6}$$

where h is the distance between the middle point (x_{cc}, y_{cc}) of the chord (linking the two knot-points) and the middle point (x_{cs}, y_{cs}) of the segment. h can be calculated using

$$h = \sqrt{(x_{cs} - x_{cc})^2 + (y_{cs} - y_{cc})^2} \tag{7}$$

The $SLAL$ ratio is defined as S/L. Since $SLAL$ is calculated by assuming the segment to be a circular arc, the $SLAL$ ratio of a real circular arc is approximately 1, and this ratio has a value of less than or greater than 1 for other segment types, depending on their smoothness.

Criterion C: Maximum Absolute Error

The maximum absolute error implies two different meanings in terms of segment types. For a line segment, this error is defined as the maximum distance of every boundary point in the segment from the line linking the two knot-points, i.e.,

$$E_{lmax} = \max_{i\in(1,N)} \left\{ \frac{|Ax_i + By_i + C|}{\sqrt{A^2 + B^2}} \right\} \tag{8}$$

where it is assumed that $Ax + By + C = 0$ is the line linking two knot-points. (Constants A, B and C can be derived using the two knot-points, (x_i, y_i) is a point on the segment, and N is the number of points in the segment.)

For a circular arc, the error is defined as the maximum distance of every boundary point in the segment from the circular arc defined by center (x_c, y_c) and radius R, i.e.,

$$E_{amax} = \max_{i \in (1,N)} \left\{ |\sqrt{(x_i - x_c)^2 + (y_i - y_c)^2} - R| \right\} \tag{9}$$

where (x_c, y_c) is the center, and R the radius of the circular arc, taking the line joining the two knot-points as a chord. (The derivation of this center and radius will be described in Appendix.)

It is clear that E_{lmax} is small if a segment is a straight line, and E_{amax} is small if a segment is a circular arc.

Criterion D: Accumulation Error

For computational efficiency (compared with calculating least-mean-square errors), the idea of *accumulation error* is introduced as one of our criteria. This has two definitions: respectively the sum of the distances of every boundary point in the segment from the line linking two knot-points, and the sum of the distances of every segment point from the circular arc defined in the previous criterion. These are

$$E_{lmod} = \sum_{i=1}^{N} \frac{|Ax_i + By_i + C|}{\sqrt{A^2 + B^2}} \tag{10}$$

where N is the number of points on the segment, A, B and C are constants for the line linking the two knot-points,
and

$$E_{amod} = \sum_{i=1}^{N} |\sqrt{(x_i - x_c)^2 + (y_i - y_c)^2} - R| \tag{11}$$

for line or combination segments (coarsely identified), and for circular arcs, respectively.

As with E_{lmax} and E_{amax}, E_{lmod} has a small value if a segment is a straight line, and E_{amod} has a small value if a segment is a circular arc. Compared with E_{lmax} and E_{amax}, E_{lmod} and E_{amod} are more accurate. This is because former are measures more sensitive to noise.

Criterion E: Sign Changes

The sign changes of the distances of every segment point from the line and the circular arc defined respectively in Criterion C can be recorded as a separate criterion. As will be seen later, this criterion is useful for testing primitive types. Depending on the segment type coarsely identified as a line or circular arc, this is a measure of the sign changes of

$$\frac{Ax_i + By_i + C}{\sqrt{A^2 + B^2}} \tag{12}$$

(where $i = 1, 2, ..., N$, A, B, and C are constants for the line linking the two knot-points)
and

$$\sqrt{(x_i - x_c)^2 + (y_i - y_c)^2} - R \tag{13}$$

($i = 1, 2, ..., N$), respectively referred to as $NSCL$ and $NSCA$.

Criterion F: Angle ψ Between the Chord and the Line Linking the Middle Points of the Chord and the Segment

If ψ is defined as the angle between the chord and the line linking the middle points of the chord and the segment, for a circular arc, ψ is equal to $\pi/2$. For other primitives, it usually deviates from $\pi/2$. Therefore this angle can be calculated and used to test whether a segment is a circular arc. Equation ?? gives the value of ψ.

$$\psi = arccos\ (\mathbf{V}_a \cdot \mathbf{V}_b / |\mathbf{V}_a||\mathbf{V}_b|) \tag{14}$$

where $\mathbf{V}_a$, $\mathbf{V}_b$ are the vector from the middle point of the chord to the middle point of the circular arc, and the vector from the middle point to one of the two knot-points, respectively.

Criterion G: Ratio of Accumulation Errors on Both Sides of the Middle Point of the Segment

Assume a segment has been coarsely identified as a circular arc and $P_1(x_1, y_1)$, ..., $P_i(x_i, y_i)$, ..., $P_N(x_N, y_N)$ are the points on the segment. As well as calculating the accumulation error (using Equation ??), the accumulation error can also be calculated using Equation ?? by taking the line linking two knots as $Ax + By + C = 0$ (where A, B and C are the constants derived from the two knot-points). If A_1 , A_2 are defined as

$$A_1 = \sum_{i=1}^{N/2} \frac{|Ax_i + By_i + C|}{\sqrt{a^2 + b^2}} \tag{15}$$

$$A_2 = \sum_{i=\frac{N}{2}+1}^{N} \frac{|Ax_i + By_i + C|}{\sqrt{a^2 + b^2}}, \tag{16}$$

A_1 should be equal to A_2 if a segment is a circular arc. Thus, another criterion for confirming a circular arc can be derived by checking whether $A_1 \approx A_2$.

2.4.2 Knot-Point Detection Algorithm

(i) Detecting Sharp Corners

The sharp corners can be detected by applying the angle-detection schemes. A corner is characterized by three incremental curvature regions — two for which the curvature δ_j^m fluctuates within relatively narrow limits due to digitization noise, separated by a third region consisting of a small number of nodes in which $\sum \delta_j^m$ equals a significant value. This characteristic enables us to detect sharp corners on the boundary of an object in a straightforward way. Extraction of sharp corner points simply involves (a) the estimation of discrete curvature, (b) global thresholding, and (c) local thresholding (or maybe local peak sharpening and thresholding), as mentioned in Section 2.3.

(ii) Testing Segment Types

Once the boundary has been segmented by sharp corners, various criteria can be applied to test the segment types. Criterion A is first applied to each segment. If the $SLKD$ ratio is within a given threshold, say greater than 0.9 and less than 1.1, this segment is regarded as a straight line, otherwise the other criteria is applied. The criteria used for identifying circular arc segments include criteria B, D, F and G. Criterion B is applied first. If this ratio is reasonably close to 1, criteria D, F and G are applied; the acceptability of a segment as a circular arc depends on:

- how close E_{amod} is to zero,
- how close angle ψ is to $\pi/2$, and
- how similar A_1 is to A_2.

The use of these criteria for segment identification will be discussed in detail in the experiments which follow. In many cases, just one of these criteria is enough to identify a circular arc segment. If a segment is accepted as neither a straight line nor a circular arc, it is regarded as a combination segment, which may include two or more primitive segments.

(iii) Detecting Knot-Points in Combination Segments

Two methods can be used for further processing combination segments. One is to split the segment at certain points [?], and the other is to apply a scale-space approach to these segments [?]. These types of segments almost all belong to the second category as defined in an earlier section, i.e., those with smooth joins which are referred to as the first class of combination segments, and those with short primitive segments which is referred to as the second class of combination segments.

First of all the detection of knot-points, on the first class of combination segments, is discussed. The common feature of this class of segments is that the change of slope is smooth, and the curvature discontinuity is therefore not easy to detect. However, there are indeed significant changes in curvature at these points. The detection of these discontinuities using an angle-detection scheme is however not reliable because of the "vague" features obtained to indicate the position of a point of discontinuity. As described in Section 2.3, several methods can be successfully applied for this purpose. To highlight the process, examples are given to show the location of knot points on several combination segments of the first class. Figure 1 shows several segments with smooth joins formed by circular arcs and/or line primitives and their Gaussian-filtered responses. It can be seen that detection of the knot-points on these segments involves simply the location of peaks or valleys in the filtered responses of orientation functions, with second Gaussian derivatives.

In our experiments, it was noted that the second method (estimating curvature using Gaussian-smoothed coordinate functions) performed better for locating the knot-points between two circular arcs with opposite signs for the flanking curvatures. The first method performed better in locating the knots on other smooth joins. Thus in some circumstances (e.g., when dealing with very complex shapes), one could use them, respectively, for the different types of smooth joins. A priori knowledge of the segment types can be obtained by applying the criterion test. For example, the use of Criterion E, to test the sign changes of the two expressions $(Ax_i + by_i + C)/(A^2 + B^2)^{1/2}$ and $(x_i - x_c)^2 + (y_i - y_c)^2)^{1/2} - R$ as i varies successively from 0 to N on a combination segment, can distinguish between the smooth join formed by two circular arcs with the same sign of flanking curvatures, and one with opposite signs. However, consideration of the noise effect, the sign changes, peaks, and zero-crossings close to the two ends are ignored. Once the knot-point(s) have been found, each subsegment can be confirmed to be a line or an arc by making appropriate criterion tests.

For the detection of knot-points on the second class of combination segments, again, one cannot rely on the angle-detection scheme which does not produce accurate estimation of curvature for a combination segment with short primitives. The scale-space approach is therefore employed. The detail of the approach is described in Asada and Brady's paper, "The Curvature Primal Sketch" [?]. To aid the interpretation of the "tree" established by "The Curvature Primal Sketch", the primitive types of a combination segment can be coarsely identified by making appropriate criterion tests before the scale-space approach is applied. For example, an "end" primitive (a primitive with two longer straight-line segments and one short line segment in the middle) normally has a large $SLKD$ ratio and no sign changes.

To demonstrate how this algorithm can be applied to effectively detect boundary knot-points, it is presented here an experiment. Figure 2 shows the boundary of an object and knot-points detected marked with small solid squares. The following describes how those knot-points are detected. In the first stage, the sharp corners P_0, P_1, ..., P_9 are located using the angle-detection scheme by choosing a threshold as 35^0. Each segment is then tested. To show how the criteria respond to different types

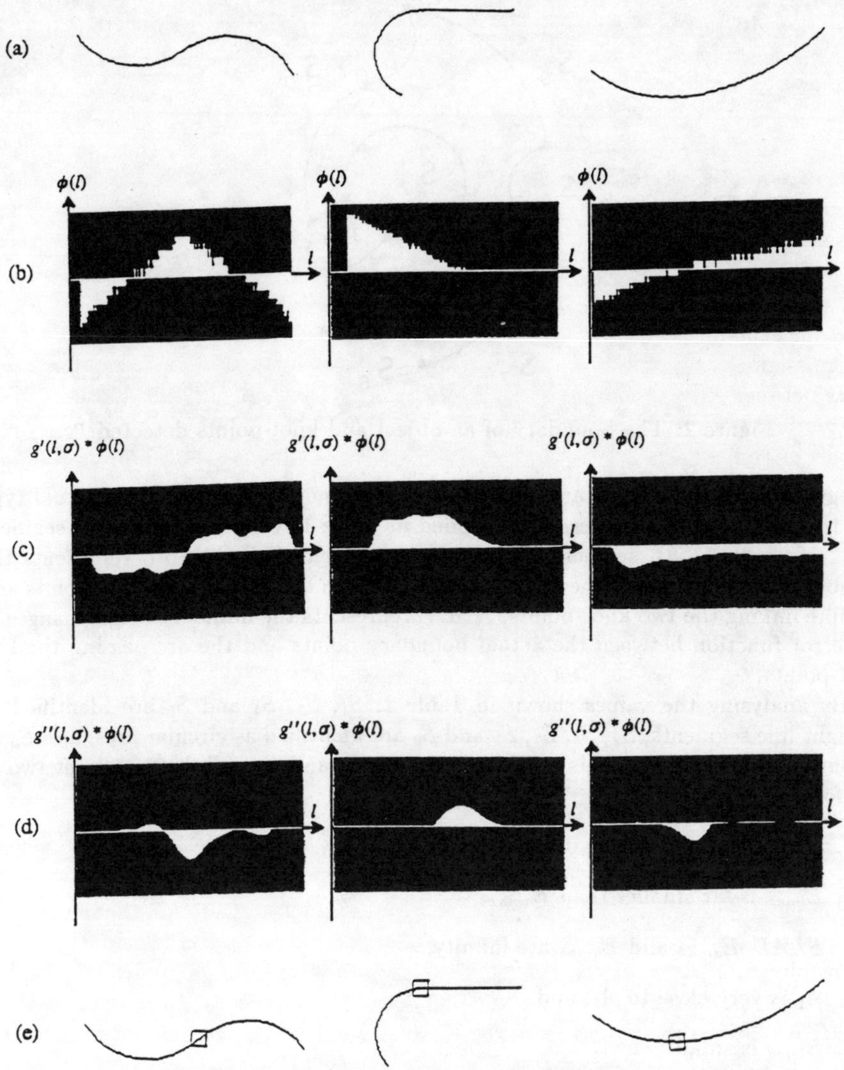

Figure 1: Knot-point detection on the first class of combination segments. (a) Segments with smooth joins. (b) The smooth joins in orientation space. (c) The first Gaussian-filtered response. (d) The second Gaussian-filtered response. (e) The knot-points detected.

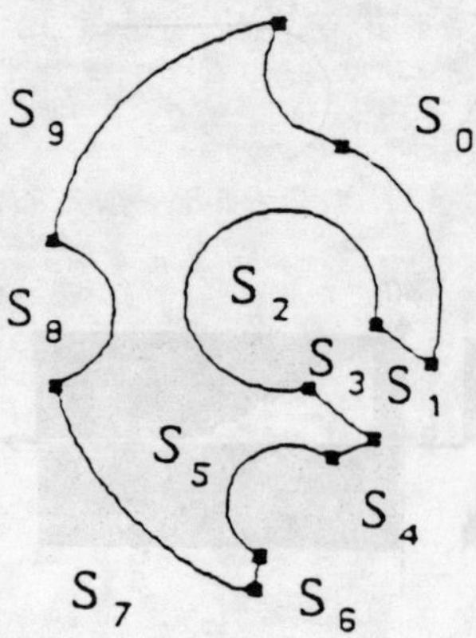

Figure 2: The boundary of an object and knot-points detected.

of segments, all the criteria are applied to each segment regardless of its actual type and the results of this process are included in Table 1, where S_i represents segment P_iP_{i+1} ($i = 0, \ldots, 8$), segment S_9 represents segment P_9P_0; $NSCL$ represents the number of sign changes of the error function between the actual boundary points and the line linking the two knot-points; $NSCA$ represents the number of sign changes of the error function between the actual boundary points and the arc passing the two knot-points.

By analysing the values shown in Table 1, S_1, S_3, S_4 and S_6 are identified as straight-line segments, S_2, S_5, S_7, S_8 and S_9 are identified as circular arcs, and S_0 as a combination segment. This is because S_1, S_3, S_4 and S_6 each have at least two of the following features:

1. $SLKD$ ratio is within 0.9 and 1.1,
2. E_{lmax} is far smaller than E_{amax},
3. $SLAL$, E_{amax} and E_{amod} are infinity,
4. A_1 is very close to A_2, and
5. E_{lmod} is small;

while S_2, S_5, S_7 and S_9 each have at least one of these features:

1. $SLAL$ ratio is close to 1,
2. angle ψ is very close to $\pi/2$,
3. E_{amax} is much smaller than E_{lmax},

	S_0	S_1	S_2	S_3	S_4	S_5	S_6	S_7	S_8	S_9
$SLKD$	1.32	1.03	6	0.98	1.32	1.9	1.26	1.12	1.44	1.52
$SLAL$	0.78	0.98	0.88	∞	0.77	0.89	∞	0.93	0.97	0.94
ψ	1.63	1.7	1.57	1.57	1.7	1.56	1.57	1.58	1.57	1.6
E_{lmax}	16	0.54	52	0.74	0.89	21	0.8	11	18.6	92
E_{amax}	16	10	1.5	∞	5.9	0.8	∞	2.4	1.1	3.3
E_{lmod}	1021	4.4	3864	6.2	3.6	84	3.9	44	617	1867
E_{amod}	602	87	63	∞	41	32.6	∞	93	21.3	153
$NSCL$	1	2	0	0	0	0	0	0	0	0
$NSCA$	3	1	8	0	1	1	0	3	5	4
A_1	353	2	1930	22.5	1.8	312	1.6	272	294	1825
A_2	667	2.37	1934	3.7	1.8	372	2.3	271	323	2041
Decision	combin.	line	arc	line	line	arc	line	arc	arc	arc

Table 1: **The results of the criterion tests on the segments obtained using the corner-detection scheme.**

4. A_1 is very close to A_2, and

5. $NSCL$ is 0.

A combination segment, such as S_0, does not have these features. To further segment the combination segment S_0 in the actual implementation, each coordinate function of the boundary points on S_0 was smoothed using a Gaussian filter with a standard deviation σ of 8, the curvatures were calculated and one zero-crossing point was located. In the same way as above, criterion tests were applied to the two subsegments, and they were identified as two circular arcs. The benefit of introducing new criteria can be seen in several areas. One criterion does not always give reliable classification of segments because of the noise. In contrast, the classification using several criteria gives correct results in most cases. This is because different segments of the same type still show different characteristics. However, not all segments need all of the criterion tests for a correct classification. For example, if a segment is a straight line, with parameter $h = 0$, and if this is not known, hence Equation ?? is used to calculate the radius of the "arc", the radius R of the arc will be infinity. This shows that the segment is likely to be a straight-line segment. In this case, there is no need to check other criteria, other than E_{lmod}, since a combination segment can also have the same effect when Equation ?? is used, and E_{lmod} will make the distinction.

3 Object-Structure Verification

Object inspection covers a broad area of topics, including the classification of objects, the verification of object structures and the detailed inspection of surfaces. The verification of object structures, which is studied in this section, includes measuring the various dimensions of the parts, as well as checking to see whether there are any flaws, such as breakages or cracks on the outer side of a casting, or defects such as nicks, bumps, or cut corners on the boundaries of flat parts.

Different classes of parts normally have different shapes, but these shapes can generally be classified into four categories according to the complexity of the primitives which combine together to form an object's contour. The four categories are: (1) circular objects, (2) rectangular objects (including squares), (3) objects composed of lines and circular arcs, and (4) objects composed of complex primitives, which means that the parameters of these primitives are difficult to estimate. This section shows that for each of those, efficient structural verification algorithms can be developed in terms of the common characteristics of objects within the same class. The concentration, however, is on the development of verification algorithms for objects in the first three classes (termed as *regular-shaped* objects), which are often encountered in many industrial inspection situations. The development of these algorithms will rely heavily on the techniques proposed in previous sections.

3.1 Rectangular Shapes

For generality, it is assumed that the position and orientation of objects to be inspected are not known a priori, but that the distance between the camera and the objects is known. The inspection thus involves verifying the size (length and width) of an object to determine if it is within the allowed tolerances, and locating defects such as nicks, bumps, and cut-corners on the boundary. The techniques proposed involve a sequence of four major steps.

- *Detection of local maxima on the curvature graph*: Since there are significant changes in edge orientation at the corners of an object, an analysis of the changes in curvature at boundary points provides useful information for locating the object's corners. The simple, but efficient, algorithm described in Section 2.2 is used here to calculate the curvature of a discrete curve. To find the major points of discontinuity, a threshold (its value is determined by experience, normally around $35^0 \sim 40^0$) is applied to the curvature graph, and those points with a curvature value greater than the overall threshold, and with a local maximum curvature value (since the curvature values of boundary points close to a corner point can also be greater than the overall threshold), are selected. As a result, only the real corner points, and any other points where there are possible defects causing high curvature values, are chosen and passed on to the next stage of processing.

- *Elimination of non-corner points*: To be able to estimate the size (length and width) of an object in the scene, it is necessary to know which points found in the first step are the "real" corners, and which are not. This can be achieved by the following simple procedure: Assuming that $P_1, ..., P_i, ..., P_M$ are the points selected by step (1), whether or not point P_i is a "real" corner depends on whether the angle between vectors P_iP_{i-1} and P_iP_{i+1} is equal, or close, to $\pi/2$.

- *Calculation of the dimensions of each object:* Having obtained the real corners of an object, its dimensions can then be determined by simply calculating the distance between each pair of adjacent corners. The average of the two longest distances is the object length, and the average of the two shortest is the object width. It should be noted that in circumstances where only three corners can be found, the object is simply rejected, since a cut corner exists.

- *Detection of ragged edges:* There are two ways of inspecting the boundary conditions. The first is to calculate the distance of each boundary point from the straight line formed by a pair of adjacent corners. The second is to calculate the distance between every boundary point and a straight line obtained by using the information on all boundary points between the two adjacent corners. It is clear that the former method is fast but not very accurate, and the latter is the converse. If an object fails either of these distance tests (i.e., a distance is greater than the allowed tolerance), it is rejected.

This procedure can be used in many practical applications. Figure 3(a) shows the boundaries of the objects in a good contrast image. Figure 3(b) shows a curvature plot for the fourth boundary (counted from left) shown in Figure 3(a). To find the major points of discontinuity, a threshold of $T = 35^0$ (chosen by experimental experience as mentioned earlier) is applied to the curvature graph. Points with a curvature value greater than the overall threshold and with a local maximum curvature value, are selected and are shown, in Figure 3(c), within small square black areas. Figure 3(d) shows the corner points obtained after non-corner elimination procedure. It can be seen that only three corners are found for the seventh object (counted from left to right) and therefore the object is directly rejected because of its cut corner. To verify the other objects, the size (length and width) of each object is estimated. As a result, all objects have passed the size test. To further inspect the objects, the condition of each object's boundary is checked. By calculating the deviation of boundary points from the straight line formed by two adjacent corners, objects 4 and 6 fail the test and are rejected. The rest of the objects successfully pass the shape-verification test.

3.2 Circular Shapes

The inspection of a circular object again requires the verification of the object's size (radius) and its boundary condition, as well as determining the center point. The

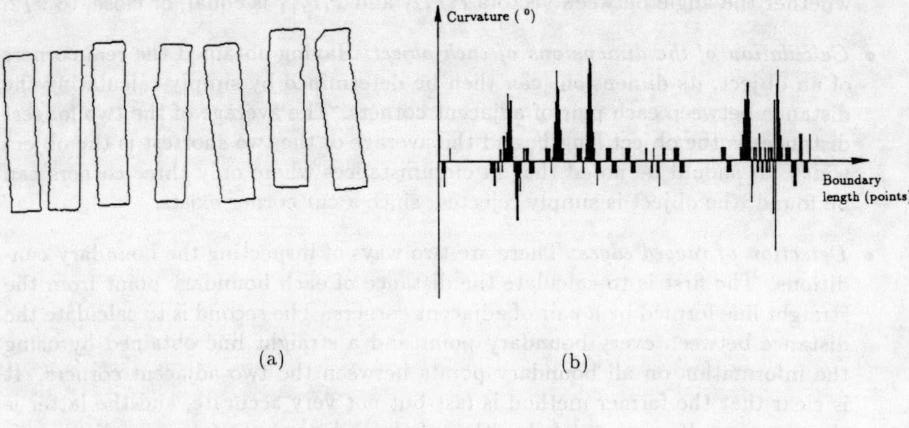

Figure 3: Illustration of corner detection. (a) The object boundaries. (b) The curvature plot for the fourth boundary. (c) Local curvature maxima. (d) Corners detected.

method for inspecting circular shapes is slightly different from that for rectangular shapes. The key issue in inspecting a circular shape is, given a set of discrete points, to find the "real" center of the circular shape. Various methods, e.g., the Hough transform method, for accurately finding the center of a circular shape have been discussed by Cao and Deravi [?, ?]. This subsection presents a method developed with a view to achieving fast processing and reasonably high accuracy. The method developed is based on Landau's algorithm [?], which is briefly reviewed below:

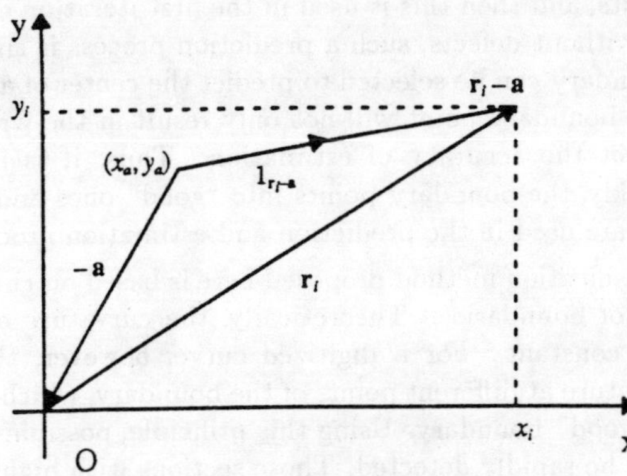

Figure 4: Estimation of a circular arc center and its radius.

As shown in Figure 4, given a set of points $P_1, ..., P_i, ..., P_N$ (N is the number of points on the boundary), with x, y coordinates (x_1, y_1), ..., $(x_i, y_i), ..., (x_N, y_N)$, then $\mathbf{r}_1, ..., \mathbf{r}_i, ..., \mathbf{r}_N$ are defined as the radius vectors to each point, respectively. If define a circular arc with a radius R and a center at (x_a, y_a), then the arc radius and its center can be calculated by randomly choosing the first center point (x_{a0}, y_{a0}) and then iterating through the following two equations,

$$\mathbf{a} = \frac{1}{N}\sum_{i=1}^{N}(\mathbf{r}_i - R\mathbf{1}_{\mathbf{r}_i-\mathbf{a}})$$

$$R = \frac{1}{N}\sum_{i=1}^{N}|\mathbf{r}_i - \mathbf{a}|$$

where $\mathbf{a}$ represents a vector from the coordinate origin to the center point, and $\mathbf{1}_{\mathbf{r}_i-\mathbf{a}}$ is a unit vector in the direction of $\mathbf{r}_i - \mathbf{a}$.

As the number of iterations increases, more-accurate estimations can be obtained, until the required error norm is reached — which is determined according to the accuracy requirements of a specific application.

Unlike many of the current parameter-estimation algorithms [?] which describe the curve as a linear function of a set of parameters and then minimize the least-square error, the above algorithm directly minimizes the radius-estimation error, and the error in the location of the arc center. Therefore, a good estimation can be expected. However, a serious shortcoming of the algorithm is that it is time-consuming and can be inaccurate if it is used to estimate the parameters for a circular arc with a small extent (a small portion of a circle).

To reduce the computational effort and solve the inaccurate estimation problem inherent in Landau's algorithm, it is suggested here carrying out the estimation by incorporating a prediction procedure. First of all the arc's center is predicted to within coarse limits, and then this is used in the first iteration of the estimation. For a circular shape without defects, such a prediction process is simple, since any three points on the boundary can be selected to predict the center of a shape. However, the use of a defective boundary point will not only result in the wrong predicted result, but also will affect the accuracy of estimation. Thus, it is important to *classify* roughly, and quickly, the boundary points into "good" ones and "bad" ones, so that only good points are used in the prediction and estimation processes.

The rapid classification method proposed here is based on curvature measurement and the analysis of boundaries. Theoretically, the curvature of the boundary of a circular shape is constant. For a digitized curve, however, there might be slight variations in curvature at different points of the boundary, which should, of course, be very small on a "good" boundary. Using this principle, possible defective sections on the boundary can be rapidly detected. Those sections with high curvature variations will subsequently be eliminated from the prediction and estimation processes. It should be noted that although this procedure for detecting high curvature-variation segments could be used directly (in some circumstances) to detect defective shapes, generally speaking it is not very reliable since the actual dimensions of a defect cannot be measured using this method.

Having now ensured that a good prediction is used, i.e., assuming that $\widehat{AB}$ is an arc which is a good section, as shown in Figure 5, the procedure for predicting the center and radius for a circular shape can be described as follows:

Let $P_1(x_1, y_1)$ and $P_2(x_2, y_2)$ be, respectively, the middle point of the arc and the middle point of the line linking two end-points, h the distance between the two middle points P_1 and P_2, and R the estimated radius of the arc. The slope k can then be calculated for line P_1P_2 as

$$k = (y_2 - y_1)/(x_2 - x_1)$$

Thus the predicted center x_a, y_a can be determined as

$$\begin{aligned} x_a &= x_2 + (R - h)\cos\theta \\ y_a &= y_2 + (R - h)\sin\theta \end{aligned}$$

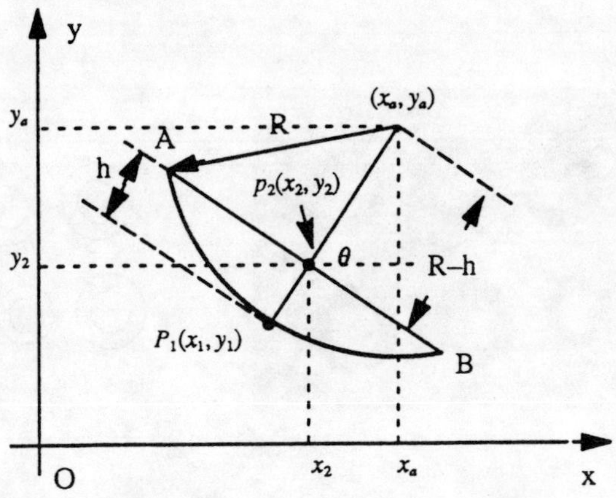

Figure 5: Illustration of circle-center prediction.

where θ is defined differently in terms of different circumstances:

$$\begin{array}{llllll}
1) & \text{if } x_2 - x_1 = 0 & \text{and } y_2 - y_1 > 0, & \theta & = & \pi/2 \\
2) & \text{if } x_2 - x_1 = 0 & \text{and } y_2 - y_1 < 0, & \theta & = & -\pi/2 \\
3) & \text{if } x_2 - x_1 > 0 & & \theta & = & \arctan k \\
4) & \text{if } x_2 - x_1 < 0 & \text{and } y_2 - y_1 \geq 0, & \theta & = & \pi + \arctan k \\
5) & \text{if } x_2 - x_1 < 0 & \text{and } y_2 - y_1 < 0, & \theta & = & -\pi + \arctan k
\end{array}$$

Using this prediction procedure, a point close to the real center of the circle can be found. The estimation can proceed, based on this point, so as eventually to find the real center point. Having found the desired center of a circular shape, the accurate inspection of the circular shape is a relatively easy task. First of all, the distance between each boundary point and the center is calculated, and a histogram of these distances is established, plotted against the sequence of the boundary points. This histogram can be checked to see if the object being inspected is within the given tolerance!

The following describes the application of the above procedure to inspection of mechanical components. Figure 6(a) and Figure 6(b) respectively show a number of washers and their extracted boundaries. Inspection of these washers includes verification of their sizes and the condition of their outer and inner circumferences. Using the procedure developed in subsection 3.2 for circular shape verification, the curvature variation on the object boundaries is first of all measured, so as to classify the boundary points into "good ones" and "bad ones". In Figure 6(c), the small square boxes show the areas where "bad pixels" are found. The good ones are then available for use in the computation of the center of each washer. It should be noted that the

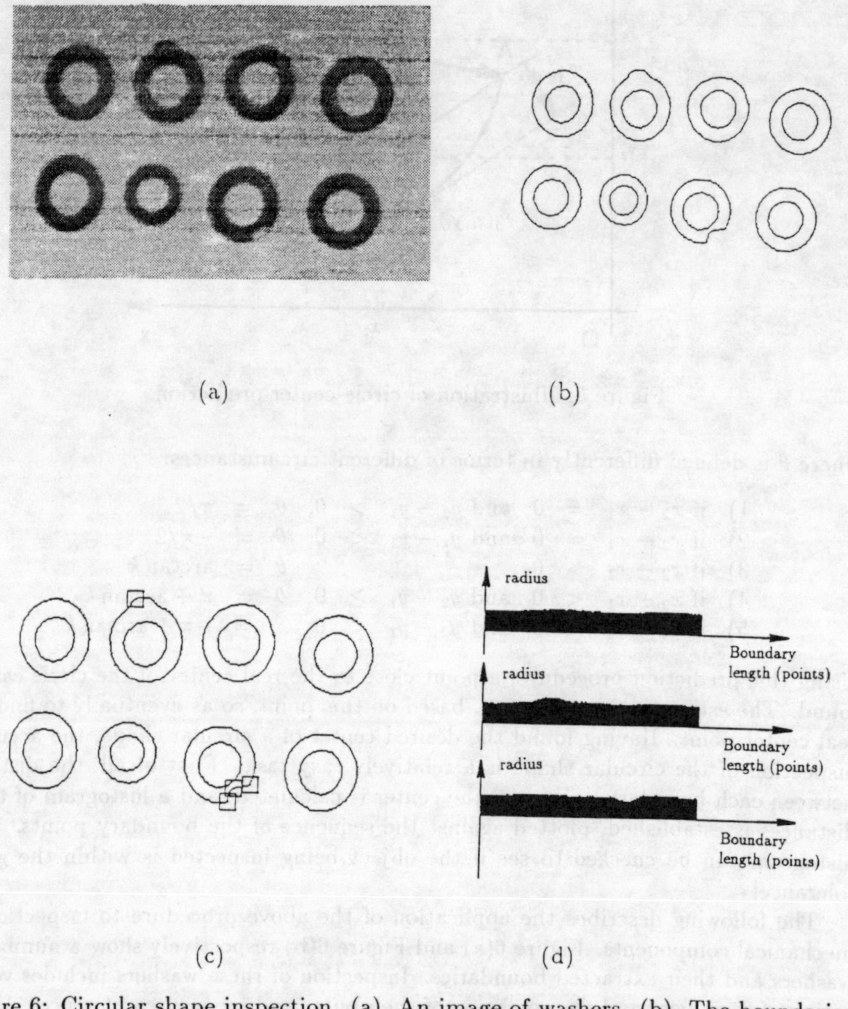

Figure 6: Circular shape inspection. (a) An image of washers. (b) The boundaries of the washers. (c) Segment boundaries, separating "good pieces" and "bad pieces". (d) Centroidal profile histogram.

center of a washer can be estimated using either the pixels on its outer circumference or those on the inner circumference. In this experiment, the one with a higher ratio of "good pixels" to "bad pixels" is used since the center calculated in this way would be more accurate.

The distance between each boundary point and the washer center is calculated and shown in a histogram. In Figure 6(d), the top, middle, and bottom histograms are, respectively, drawn for the outer circumference of the first washer, the outer circumference of the second washer and the outer circumference for the seventh washer. To verify the condition of each washer, a threshold is applied to these histograms. The threshold is determined by the actual tolerance limit of the components. For example, assuming the tolerance limit is 2mm for the washers, a washer is rejected if its distance histogram has values 2mm above or below the standard histogram. Table 2 shows the final measurement and verification results.

Washer 1:	PASS	DIAMETER	31.3699	(outer circumference)
			17.5704	(inner circumference)
Washer 2:	FAIL			(flawed outer circumference)
Washer 3:	PASS	DIAMETER	31.565	(outer circumference)
			18.208	(inner circumference)
Washer 4:	PASS	DIAMETER	31.8242	(outer circumference)
			18.1201	(inner circumference)
Washer 5:	PASS	DIAMETER	30.9209	(outer circumference)
			17.8341	(inner circumference)
Washer 6:	FAIL	(too small)		
Washer 7:	FAIL			(flawed outer circumference)
Washer 8:	PASS	DIAMETER	31.234	(outer circumference)
			18.175	(inner circumference)

Table 2: **Final inspection results (*measured in mm*).**

3.3 Objects Composed of Lines and Circular Arcs

The boundaries of a significant class of industrial parts consist of combinations of straight-line segments and circular arcs, mainly because such primitives are used in many CAD systems. The inspection of this class of objects can be approached in at least two different ways: the first is based on the measurement of the parameters of each primitive segment on the boundary, and the second compares a test contour with a template contour, by normalizing the test contour to the template contour. This latter procedure involves finding the normalization parameter values which minimize a chosen "distance" between the contours. Here, a technique is presented for verification based on the first technique, a measurement of primitive segment parameters.

It is clear that to measure the parameters of each primitive segment, two subtasks are involved. The first is to segment the boundary, and the second is to find an algorithm which accurately estimates the parameters corresponding to each segment. The solutions to these individual problems have been presented respectively in Section 2.4, and Section 3.2. To demonstrate how these solutions are combined to solve the combined verification problem posed here, the method is described below in a practical example.

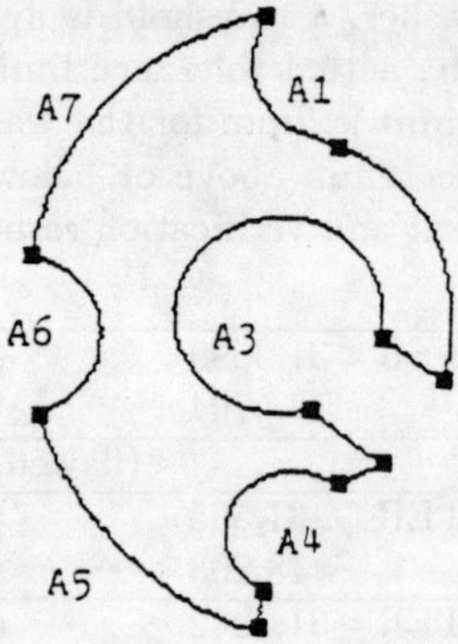

Figure 7: Inspection of objects composed of lines and circular arcs.

As shown in Figure 2, a test object (a mechanical part) is adopted from the literature and recreated it on a CAD system so that the designed parameters can be compared to those estimated using our algorithms. The aim of this experiment is to inspect the boundary by estimating the centers and radii of the circular arcs forming the boundary. The knot-points on the boundary are detected using the proposed knot-point detection algorithm as described in Section 2.4, so that the boundary is broken into primitive line segments and circular arc primitives. The parameters (the line length and the orientation for a line, and the center and radius for a circular arc) are then calculated for each segment. Table 3 shows the actual, the predicted and the estimated centers and radii of all the circular arcs on the boundary. It can be seen that centers and radii of the arcs estimated are equal, or very close to, the actual values. This shows that the parameter estimation algorithm described in Section 3.2 is accurate, and the verification technique proposed here is valuable for applications in which the estimation of centers and radii of circular arcs is required.

4 Conclusions

In this paper, several algorithms are developed for verification of regular-shaped objects by breaking boundaries into fundamental primitives such as lines and circular

Arcs	*Actual*			*Predicted*			*Estimated* (±0.1)		
	radius	*center*		*radius*	*center*		*radius*	*center*	
		x	y		x	y		x	y
arc 1	23	243	66	24	244	68	23	243	66
arc 2	50	225	136	50	224	132	50	225	136
arc 3	28	228	137	28	228	137	28	228	137
arc 4	18	233	197	18	233	197	18	233	197
arc 5	88	246	139	87	243	136	88	246	139
arc 6	23	157	144	23	157	144	23	157	144
arc 7	79	239	130	79	242	126	79	239	130

Table 3: **Comparison of the designed, predicted and estimated parameters.**

arcs. It was first emphasized that critical points derived from object boundaries are important for the recognition and verification of industrial objects. To extract these dominant features, algorithms, for detection of extreme points having high curvatures, were discussed, and an algorithm was then proposed for detecting knot-points formed by basic primitive segments such as lines and circular arcs. The algorithm proposed has following advantages:

- It can locate not only sharp corners but also smooth joins, dents, ends, etc. Our experiments, supported by the literature, have shown that angle-detection schemes can give excellent performance in terms of locating sharp corners — although their use for detecting features such as zero-crossings, which involves complicated analysis of curvature plots and suffers from inaccurate location of the features, is also possible. Smooth joins and ends, dents, cranks, etc., however, can be best located using the scale-space approach. However, the interpretation of an occurrence that consists of two or more closely separated curvature changes by the "Curvature Primal Sketch" [?] is ambiguous, as the Gaussian-filtered responses of these curvature changes interfere with each other. This problem has been solved in the combined scheme since a priori knowledge about the type of occurrence is known before the scale-space approach is applied.

- The computational cost of this scheme is lower than those of scale-space approaches and many other corner-detection schemes [?]. It is well-known that convolution is a very time-consuming process, especially when the filter size (as occurs in many cases of smoothing curves) is large. The knot-point detection scheme described in this paper avoids this computationally-intensive process — since the convolutions of the derivatives of coordinate functions with a Gaussian kernel are required only for certain segments of the boundary (which normally contain far less data than for the complete boundary). It should be noted that the cost of the angle-detection scheme is relatively low because of its computationally simplistic features [?]. Finally, the computational overhead introduced by applying the suggested criteria is not large, as the majority of segments can be identified by making only one or two criterion test(s).

Above all, the results obtained through effective curvature analysis have formed a valuable working base for the development of the verification algorithms. To develop the structural verification algorithm, the existing flat industrial objects were classified into four categories. For each category, corresponding effective verification algorithms were developed, according to the characteristics of the objects in each class.

Acknowledgements

The authors wish to acknowledge the support of their employers, the University of Wales, for the work which has been discussed in this paper. They also wish to acknowledge the ACME Directorate of the Science and Engineering Research Council of the UK which now support the on-going implementation of the work.

References

[1] A. Albano. Representation of Digital Contours in Terms of Conic Arcs and Straight Line Segments. *Computer Vision, Graphics and Image Processing*, vol. 2, pp. 23–33, 1974.

[2] H. Asada and M. Brady. The Curvature Primal Sketch. *IEEE Trans. on PAMI*, vol. PAMI-8, January 1986.

[3] F. Attneave. Some Information Aspects of Visual Perception. *Psychological Review*, vol. 61, pp. 183–193, 1954.

[4] J. Canny. A Computational Approach to Edge Detection. *IEEE Transactions on Pattern Analysis and Machine Intelligence*, vol. 8, pp. 679–698, November 1986.

[5] X. Cao and F. Deravi. Efficient Method for Multiple-Circle Detection. *Internal Report*, 1990.

[6] X. Cao, M.G. Rodd, F. Deravi, and Q.M. Wu. Detection of Multiple Circles Based on Adaptive Hough Transform. *Engineering Application of AI*, vol. 1, pp. 97–101, June 1988.

[7] M.A. Fischler and R.C Bolles. Perceptual Organisation and Curve Partitioning. *IEEE Trans. on PAMI*, vol. 8, pp. 100–104, January 1986.

[8] H. Freeman and L. S. Davies. A Corner-Finding Algorithm for Chain-Coded Curves. *IEEE Trans. on Computers*, vol. C-26, pp. 297–303, March 1977.

[9] M. H. Hueckel. An Operator Which Locates Edges in Pictures. *Journal of the ACM*, vol. 18, pp. 113–125, January 1971.

[10] L. Kitchen and A. Rosenfeld. Graylevel Corner Detection. *Pattern Recognition Letter*, vol. 1, p. 95, December 1982.

[11] U.M. Landau. Estimation of a Circular Arc Center and Its Radius. *Computer Vision, Graphics and Image Processing*, vol. 38, pp. 317–326, 1987.

[12] G. Medioni and Y. Yasumoto. Corner Detection and Curve Representation Using Cubic B-splines. *CVGIP*, vol. 39, pp. 267–278, 1987.

[13] F. Mokhtarian and A. Mackworth. Scale-Based Description and Recognition of Planar Curves and Two-Dimensional Shapes. *IEEE Trans. on PAMI*, vol. PAMI-8, no. 1, pp. 34–43, 1986.

[14] T. Pavlidis and S.L. Horowitz. Segmentation of Plane Curves. *IEEE Trans on Computers*, vol. C-23, August 1974.

[15] J. Powrie. *A Critical Evaluation of Low-Cost Boundary Extraction Schemes for Industrial Inspection Applications.* University of Wales, Department of Electrical and Electronic Engineering, 1990.

[16] C.H. Tek and R.T Chin. On the Detection of Dominant Points on Digital Curves. *IEEE Trans. on PAMI*, vol. 11, pp. 859–872, August 1989.

[17] K. Wilfred. *Advanced Mathematics for Engineers.* Addison-Wesley Publishing Company, 1981.

[18] A. Rosenfeld and A E. Johnston Angle Detection on Digital Curves. *IEEE Trans. on Computers*, pp. 875-878, September 1973.

[19] K.C. You and K.S. Fu A Syntactic Approach to Shape Recognition Using Attributed Grammars. *IEEE Trans on SMC*, Vol. SMC-9, pp. 334-345, June 1979

Appendix. Derivation of Eq.(4) and Eq.(5)

Given the parameters D and h in Figure 8, following equation exists:

$$(D/2)^2 + (R-h)^2 = R^2. \tag{17}$$

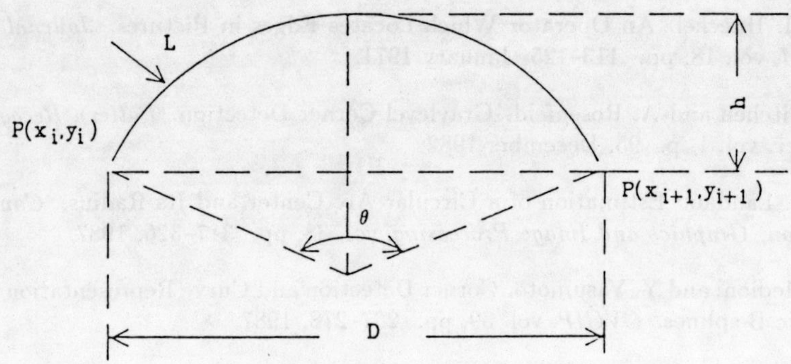

Figure 8: Calculation of radius R and θ

Thus, the radius of the arc passing through two knots and the middle point of the segment can be obtained as

$$R = \frac{(D/2)^2 + h^2}{2h}, \tag{18}$$

and the equation for calculating θ is derived as follows:

$$sin(\theta/2) = \frac{D/2}{R}, \tag{19}$$

$$cos(\theta/2) = \frac{R-h}{R}. \tag{20}$$

Combining Eq. (19) and Eq. (20), it results

$$\frac{D/2}{sin(\theta/2)} = \frac{R-h}{cos(\theta/2)}, \tag{21}$$

i.e.,

$$(D/2)cos(\theta/2) = (R-h)sin(\theta/2). \tag{22}$$

Replacing R in Eq. (22) with $(D/2)/sin(\theta/2)$, then Eq. (22) can be rewritten as

$$(D/2)cos(\theta/2) = \frac{D}{2} - hsin(\theta/2), \tag{23}$$

and $tan(\theta/4) = h/\frac{D}{2}$, i.e., $\theta = 4arctan(h/\frac{D}{2})$ is finally obtained.

On Circles and Circular Arcs Recognition

A. Tamas
Territorial Computing Center
Reublicii 107
3400 Cluj-Napoca, Romania

Abstract

This paper presents a criterion for deciding whether or not a set of lattice points is the digital image of a real circle or a circular arc. For this purpose the digital-inscribable quadrilaterals are used, such that it is not necessary to have a priori information about the center and radius of the circle. The set of lattice points is divided into digital-straight line segments. Then it is verified to determine if all three consecutive digital-straight line segments form a digital-inscribable quadrilateral (the fourth edge of the quadrilateral links the endpoints of the contour). This method also deals with the segmentation of the contour lines into digital-straight line segments and digital-circular arcs.

1 Introduction

Image processing and pattern recognition for technical drawings are mainly concerned with straight line segments, texts and circles. This paper deals with the problem of circles and circular arcs recognition. For this purpose the following problems are discussed:

1. A criterion to establish if a set of lattice points is the digital image of a real circle (recognition problem).

2. An algorithm for computing the center and the radius of a real circle (or circular arc) whose digital image is equivalent to the given set of lattice points (approximation problem).

3. To divide the digital contour into segments such that each segment is either a digital-straight line segment or a digital-circular arc (segmentation problem).

The Hough transform [3], [16], [18] is a well known method to analyse whether or not a set of lattice points is the digital image of a real circle. The disadvantage of this method is that it requires a big computational effort.

Amir [1] proposes a chord method for detection of circles, a version of the Hough transform. In order to reduce the computational effort, the method uses the bisectors of the chords in a circle. Not taking into account the digitization errors, the center of the circle would accumulate all the $N(N-1)/2$ bisectors' intersections.

Kovalevsky [10] first divides the digital curve into digital-straight line segments, based on the notion of the "nearest support" by Anderson and Kim [2]. If the curve is a digital-circular arc then the intersection of all perpendiculars through the middle point of the segments is a convex polygon called *center polygon*. If the curve is not a digital-circular arc then the center polygon is empty.

Nakamura and Aizawa [11] demonstrate a necessary and sufficient condition for a given set of lattice points to be the digital image of a real circle. This criterion, presented in section 3, needs the coordinates of the center of the circle, which are not known a priori.

Chen and Lee [4] use a recognition criterion based on the fact that if a point C is inside a circular object, than each pair of chords P_1P_2 and P_3P_4 passing through C satisfies the equality $P_1CP_2C = P_3CP_4C$. The method computes the center and the radius of each circular object using the least-square error method.

The method presented in this paper first establishes if there exists a circle, preimage of the given set of lattice points. Only if this condition holds are the center and radius of a preimage computed.

The organization of this paper consists of five sections. The following section presents two schemes for digitizing planar curves; the third section briefly describes Nakamura and Aizawa's results. Section 4 presents a necessary and sufficient condition for a set of lattice points to be a digital circle, using digital inscribable quadrilaterals. The last section presents implementation results, followed by conclusions.

2 Schemes for Digitizing Planar Curves

Consider a planar coordinate grid, such that all its nodes are lattice points. Let q be an oriented curve in the plane. For the purpose of computer analysis it is common to represent q by a finite set of lattice points , noted $I(q)$ and called the digital image of q. $I(q)$ can be obtained in different ways. In this section the *grid digitization scheme* and *Kim's grid digitization scheme* for planar oriented arcs are defined.

Definition 2.1 Grid digitization scheme [5]

A set Q of lattice points is the digital image of an arc q, noted $Q = I(q)$, if the following conditions hold:

(i) If a lattice point $p = (h, k)$ is in Q, then q crosses a grid line in a point $z = (x, y)$ such that either $\max\{|h-x|\}, |k-y|\} < 1/2$ or $\max\{|h-x|\}, |k-y|\} = 1/2$ and p lies to the right of q.

(ii) If a lattice point $p = (h, k)$ is not in Q, then there is no crossing point $z = (x, y)$ that satisfies condition *(i)*.

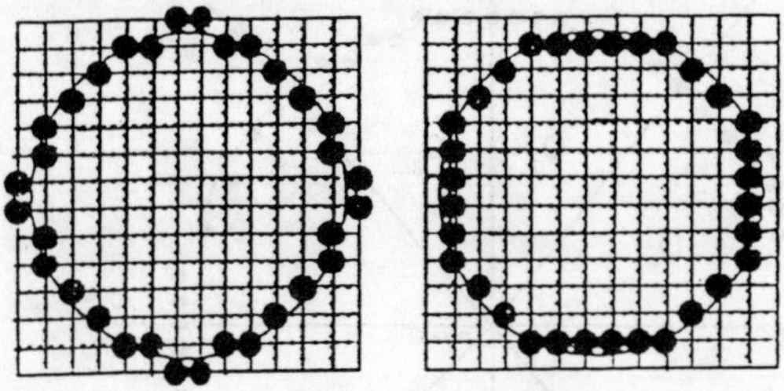

Figure 1: Two different grid digitization schemes.

Definition 2.2 Kim's grid digitization scheme [7]

A set Q of lattice points is the digital image of an arc q, noted $Q = KI(q)$, if the following conditions hold:

(i) If a lattice point $p = (h, k)$ is in Q, then q crosses a grid line in a point $z = (x, y)$ such that either $\max\{|h - x|\}, |k - y|\} < 1$ or $\max\{|h - x|\}, |k - y|\} = 1$ and p lies to the right of q.

(ii) If a lattice point $p = (h, k)$ is not in Q, then there is no crossing point $z = (x, y)$ that satisfies condition *(i)*.

Fig.1 shows the difference between the two digitizing schemes. When Kim's digitizing scheme is used all points of $IK(q)$ lie in the interior of the real circle.

3 Conditions for Digital Images of Circles

Definition 3.1

A set of lattice points Q is called a *digital circle* if there is a real circle q such that $I(q) = Q$.

Definition 3.2

A set of lattice points A is called a *digital-circular arc* if there is a real circular arc a such that $I(a) = A$.

This section briefly presents a very important result of Nakamura and Aizawa [11], which is a necessary and sufficient condition for a set of lattice points to be a digital circle.

Let $p = (a, b)$ be a point in the real plane. As shown in Fig. 2, the octants of the plane defined by p are the 8 angles bounded by the straight lines: $x = a$; $y = b$; $y = x + b - a$; $y = -x + b + a$.

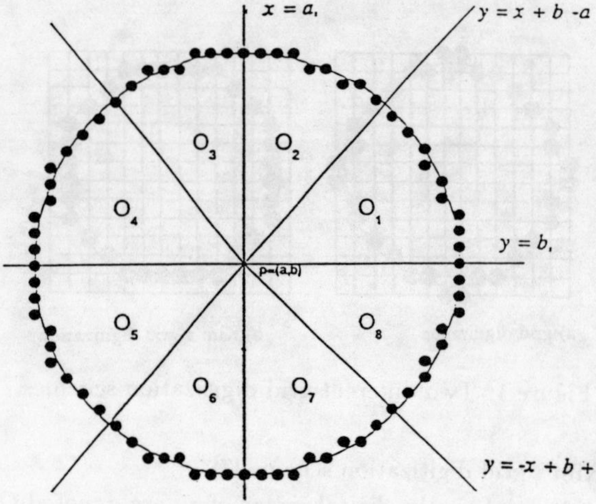

Figure 2: The octants of the plane.

Let Q be a finite set of lattice points, and let c be a given point in the real plane. Let $m = (p_i, q_i)$ and $n = (p_j, q_j)$ the lattice points defined as follows:

$$\begin{aligned} d^2(m,c) &= \max\{d^2(a,c) | a \; from \; Q\} \\ d^2(n,c) &= \min\{d^2(a,c) | a \; from \; Q\} \end{aligned} \qquad (1)$$

where $d(*,*)$ is the Euclidian distance between the two points.

For these two points Nakamura and Aizawa defined two other points m_1 and n_1 which are as follows (see Fig. 3):

$$m_1 = \begin{cases} (p_i - 1/2, q_i) & | \; m \; from \; O_1 \cup O_8 \\ (p_i, q_i - 1/2) & | \; m \; from \; O_2 \cup O_3 \\ (p_i + 1/2, q_i) & | \; m \; from \; O_4 \cup O_5 \\ (p_i, q_i + 1/2) & | \; m \; from \; O_6 \cup O_7 \end{cases} \qquad (2)$$

$$n_1 = \begin{cases} (p_j + 1/2, q_j) & | \; n \; from \; O_1 \cup O_8 \\ (p_j, q_j + 1/2) & | \; n \; from \; O_2 \cup O_3 \\ (p_j - 1/2, q_j) & | \; n \; from \; O_4 \cup O_5 \\ (p_j, q_j - 1/2) & | \; n \; from \; O_6 \cup O_7 \end{cases} \qquad (3)$$

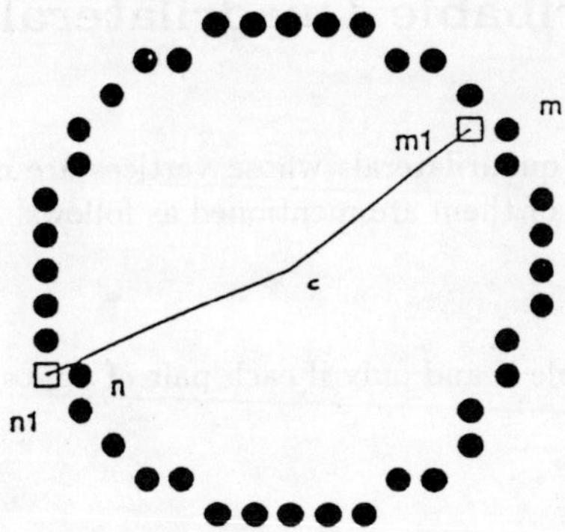

Figure 3: Defining points m_1 and n_1.

Lemma 3.1 [11]

For a set Q of lattice points and a given point c there exists a real circle q, such that $Q = I(q)$ and the center of q is c if and only if the following conditions hold:

(i) Each element of Q has at least two 8-neighbours and Q is in each octant O_i $(i = 1..8)$ a digital arc .

(ii) For the points m_1 and n_1 defined by (2) and (3) the following inequality holds: $d^2(m_1, c) < d^2(n_1, c)$.

For Kim's digitizing scheme point n_2 is defined as follows:

$$n_2 = \begin{cases} (p_j + 1, q_j) & \mid n \; from \; O_1 \cup O_8 \\ (p_j, q_j + 1) & \mid n \; from \; O_2 \cup O_3 \\ (p_j - 1, q_j) & \mid n \; from \; O_4 \cup O_5 \\ (p_j, q_j - 1) & \mid n \; from \; O_6 \cup O_7 \end{cases} \tag{4}$$

Lemma 3.2 [11]

For a set Q of lattice points and a given point c there exists a real circle q, such that $Q = IK(q)$ and the center of q is c if and only if the following conditions hold:

(i) Each element of Q has at least two 8-neighbours and Q is in each octant O_i $(i = 1..8)$ a digital arc.

(ii) For the points m and n_2 defined by (1) and (4) the following inequality holds: $d^2(m, c) < d^2(n_2, c)$.

Section 4 of [11] describes an interesting algorithm for computing the center c of a real circle, which is used in condition *(ii)* of lemma 3.1. But the searches of point c have *exponential complexity*.

4 Digital Inscribable Quadrilaterals and Digital Circles

Inscribable quadrilaterals (quadrilaterals whose vertices are on a circle) have many interesting properties, four of them are mentioned as follows.

Proposition 4.1

A quadrilateral is inscribable if and only if each pair of opposite angles has the sum 180° (see Fig. 4a).

Proposition 4.2

A quadrilateral is inscribable if and only if each angle formed by an edge and a diagonal is equal to the angle formed by the opposite edge and the other diagonal of the quadrilateral (see Fig. 4b).

Proposition 4.3

Let a, b, c, d be the lengths of the edges of an inscribable quadrilateral; let e and f be the lengths of its diagonals. Then the following equality holds:

$$fe = ac + bd$$

Proposition 4.4

Let a, b, c, d be the length of the edges of an inscribable quadrilateral and p its perimeter. Then the radius R of the circle that contains the vertices of the quadrilateral can be computed as follows:

$$R^2 = \frac{(ab + cd)(ca + bd)(ad + bc)}{16(p - a)(p - b)(p - c)(p - d)}$$

The proof of this expression is given in appendix.

Definition 4.1

A set of lattice points Q is called *digital-inscribable quadrilateral* if there exists a real-inscribable quadrilateral q such that $I(q) = Q$.

Another important result obtained by Nakamura and Aizawa [12] is an algorithm that computes, for a digital straight line s and a lattice point d, in s two real-straight lines $L_n(d)$ and $L_m(d)$, which pass through d, such that all real-straight lines "between" them have the same digital image s (see Fig. 5).

Consider a digital-inscribable quadrilateral. According to definition 4.1 there exists a real-inscribable quadrilateral q, preimage of it. For two consecutive edges f_1 and

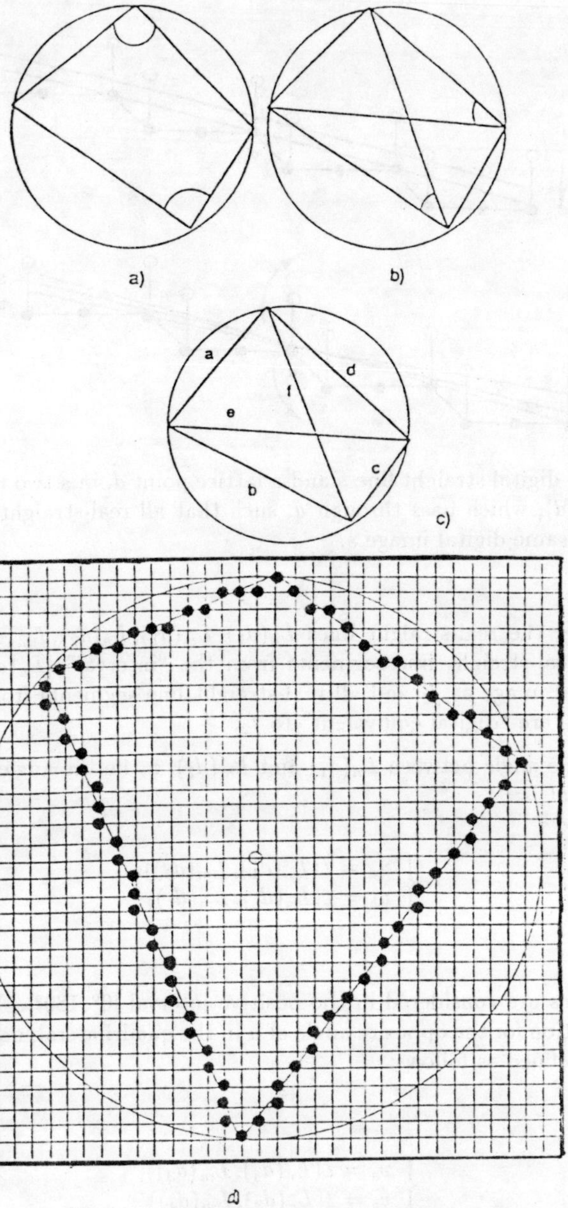

Figure 4: Defining inscribable quadrilaterals. (a),(b), and (c) are real-inscribable qualrilaterals, and (d) is a digital-inscribable quadrilateral.

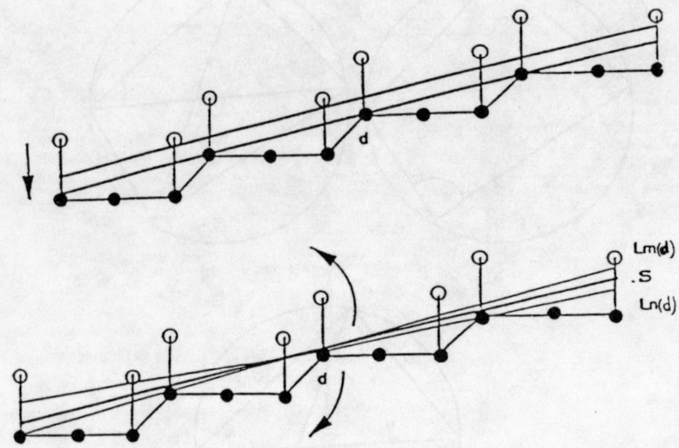

Figure 5: For a digital straight line s and a lattice point d, in s two real-straight lines $L_n(d)$ and $L_m(d)$, which pass through d, such that all real-straight lines "between" them have the same digital image s.

f_2 let's consider the real-straight lines $L_n(d_1), L_m(d_1), L_n(d_2), L_m(d_2)$ (see Fig. 6a) such that all real-straight line segments from the "interval" $[L_n(d_1), L_m(d_1)]$ have the same digital image as f_1 and all real-straight line segments from the "interval" $[L_n(d_2), L_m(d_2)]$ are "digital equivalent" to f_2.

Let u_a be the angle between $L_n(d_1)$ and $L_m(d_2)$, u_b be the angle between $L_n(d_2)$ and $L_m(d_1)$, that is:

$$\begin{cases} u_a = \angle(L_n(d_1), L_m(d_2)) \\ u_b = \angle(L_n(d_2), L_m(d_1)) \end{cases} \tag{5}$$

where the angles are considered in the positive interval $[0°, 180°]$. With almost the same consideration for the opposite angle of u in the quadrilateral, noted v, the angels v_a and v_b are defined as follows:

$$\begin{cases} v_a = \angle(L_n(d_4), L_m(d_3)) \\ v_b = \angle(L_n(d_3), L_m(d_4)) \end{cases} \tag{6}$$

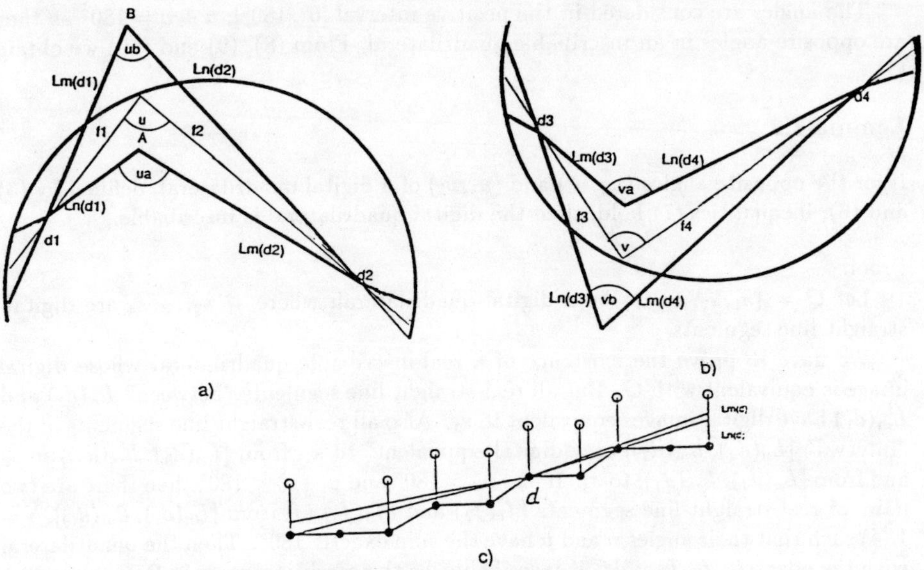

Figure 6: Definition of angles(u_a, u_b) and (v_a, v_b).

Lemma 4.1

In a digital-inscribable quadrilateral, for the opposite pairs of angels (u_a, v_a) and (u_b, v_b) defined by (5) and (6) the following inequalities hold:

$$\begin{cases} u_a + v_a > 180^\circ \\ u_b + v_b < 180^\circ \end{cases} \tag{7}$$

Proof:

The intersection point between f_1 and f_2 lies on the circle, as f_1 and f_2 are edges in a real-inscribable quadrilateral q. Note their angle with u. The intersection point between $L_n(d_1)$ (right most of f_1) and $L_m(d_2)$ (left most of f_2) lies inside the circle (see Fig. 6a) and for their angles u_a we obtain:

$$u_a > u \tag{8}$$

The intersection point B between $L_m(d_1)$ (left most of f_1) and $L_n(d_2)$ (right most of f_2) lies outside the circle (see Fig. 6a) and for their angles u_b we obtain:

$$u_b < u \tag{9}$$

With almost the same considerations for the angles v_a and v_b (see Fig. 6b), the following inequalities hold:

$$\begin{cases} v_a > v \\ v_b < v \end{cases} \tag{10}$$

The angles are considered in the positive interval $[0°, 180°]$; $u + v = 180°$ as they are opposite angles in an inscribable quadrilateral. From (8), (9) and (10) we obtain (7).

Lemma 4.2

If for the opposite angles (u_a, v_a) and (u_b, v_b) of a digital quadrilateral, defined by (5) and (6), inequalities (7) hold, then the digital quadrilateral is inscribable.

Proof:

Let $Q = \{s_1, s_2, s_3, s_4\}$ be a digital quadrilateral, where s_1, s_2, s_3, s_4 are digital-straight line segments.

We have to prove the existence of a real-inscribable quadrilateral whose digital image is equivalent with Q. But all real-straight line segments "between" $L_n(d_1)$ and $L_m(d_1)$ have digital images equivalent to s_1. Also all real-straight line segments in the "interval" $[L_n(d_2), L_m(d_2)]$ are "digital equivalent" to s_2, from $[L_n(d_3), L_m(d_3)]$ to s_3 and from $[L_n(d_4), L_m(d_4)]$ to s_4. If $u_a+v_a > 180°$ and $u_b+v_b < 180°$ then there are two pairs of real-straight line segments (f_1, f_2) and (f_3, f_4) (f_i from $[L_n(d_i), L_m(d_i)]$, $i = 1..4$) such that their angles u and v have the sum exactly 180°. Then the quadrilateral with the edges f_1, f_2, f_3 and f_4 is inscribable, as the condition given in Proposition 4.1 is necessary and sufficient.

From lemmas 4.1 and 4.2 we obtain the following result:

Theorem 4.1

Let $Q = s_1, s_2, s_3, s_4$ be a digital quadrilateral, where s_1, s_2, s_3, s_4 are digital-straight line segments. Let u_a, v_a, u_b, v_b be the angles defined by (5) and (6). Q is a digital-inscribable quadrilateral if and only if the following conditions hold:

$$\begin{cases} u_a + v_a > 180° \\ u_b + v_b < 180° \end{cases} \tag{11}$$

5 Implementation

5.1 Recognition Problem

In order to analyse if a set $Q = \{P_1, P_2, .. P_n\}$ of lattice points is a digital circle (or circular arc) first we have to divide Q into digital-straight line segments. For this purpose was used Sklansky's and Gonzalez's method [17]. This algorithm deals with the computation of the real-straight line segment l_{ij} between two points P_i, P_j and the comparison of the distance of all points between P_i and P_j to l_{ij} with a threshold. If there is a point such that the distance to the real-straight line segment l_{ij} is greater than the threshold, then $\{P_i, P_{i+1}, ..., P_{j-1}\}$ is a digital-straight line segment. If for all points between P_i and P_j the distance to l_{ij} is less than the threshold, then is analysed a "longer" segment, that is $[P_i, P_j + 1]$. Thresholding is a very difficult problem in pattern recognition. We considered in our tests a threshold of 0.1 units for a line segment of unit length.

Then we have to verify for every three consecutive line segments condition (7) (see Fig. 7)

We obtain false circles in the case of very "long", almost colinear segments, as shown in Fig. 8. In order to eliminate these errors an additional tests need to be done: the distance between the middle point of the line segments and the circle must not be greater than a threshold (0.1 for a circle of radius one unit).

5.2 Approximation Problem

In order to compute the center and the radius of a digital circle, the simplest way is to compute first, the radius, using the formula given in proposition 4.4 for all quadrilaterals, and then to consider its average value (see Fig. 9).

5.3 Segmentation Problem

In order to find the endpoint of a circular arc (see Fig. 10) it is analysed the chain of straight line segments obtained with Sklansky's and Gonzalez's method and noted $\{S_1, S_2, ..., S_m\}$.

At least five consecutive line segments which verify condition (7) and distance condition (described in Section 5.1) were considered necessary for a digital-circular arc. We advanced in the chain $\{S_1, S_2, ..., S_m\}$ as much as the two mentioned conditions hold, in order to obtain the "longest" digital-circular arc.

6 Conclusions

Section 4 presents an algorithm that analyses whether or not a given set of lattice points is a digital image of a real circle, using digital-inscribable quadrilaterals. These conditions do not need a priori information about the center and radius of the circle. The set of given lattice points first must be segmented into digital-straight line segments. Then is verified if all three consecutive digital-straight line segments (the fourth edge edge of the quadrilateral links the endpoints of the contour) form a digital-inscribable quadrilateral. In this way is determined if the digital-straight line segments approximate a real circle. This method also permits a simple way to determine the end points of a circular arc (as described in Section 5).

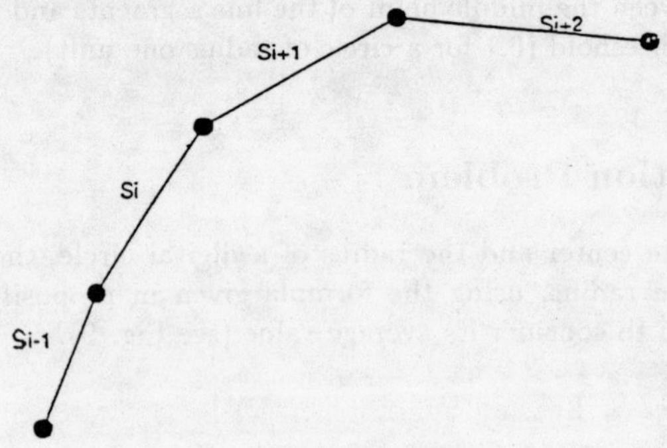

Figure 7: Recognition problem: Construction of straight line segments.

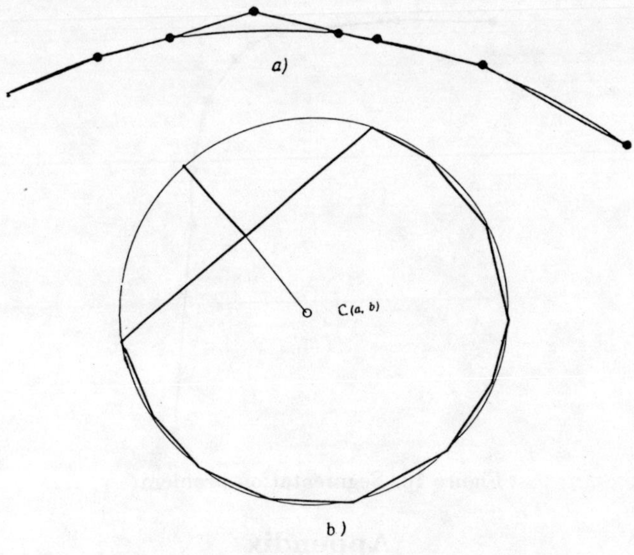

Figure 8: Recognition problem: Distance test.

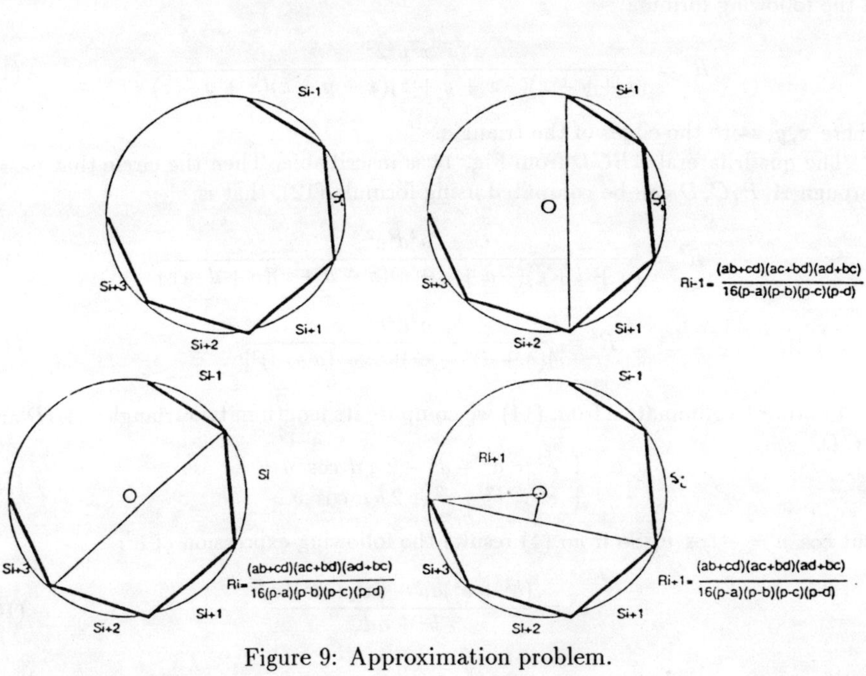

Figure 9: Approximation problem.

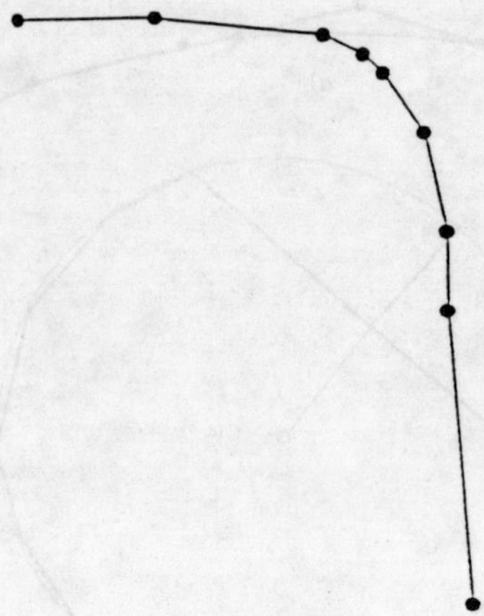

Figure 10: Segmentation problem.

Appendix

The radius R of the circle that passes through the vertices of a triangle is given in the following formula:

$$R^2 = \frac{x^2y^2z^2}{(x+y+z)(-x+y+z)(x-y+z)(x+y-z)} \tag{12}$$

where x, y, z are the edges of the triangle.

The quadrilateral $ABCD$ from Fig. 11 is inscribable. Then the circle that passes through A, B, C, D can be computed using formula (12), that is:

$$R^2 = \frac{a^2d^2e^2}{(a+d+e)(-a+d+e)(a-d+e)(a+d-e)} \tag{13}$$

$$R^2 = \frac{a^2d^2e^2}{[(a+d)^2-e^2)][e^2-(a-d)^2]} \tag{14}$$

In order to eliminate e from (14) we compute its length in two triangles ABD and BCD.

$$\begin{cases} e^2 = a^2 + d^2 - 2\,a\,d\,\cos\,u \\ e^2 = b^2 + c^2 - 2\,b\,c\,\cos\,v \end{cases} \tag{15}$$

But $\cos\,u = -\cos\,v$. So from (4) results the following expression of e^2:

$$e^2 = \frac{(b^2+c^2)ad + (a^2+d^2)bc}{bc+ad} \tag{16}$$

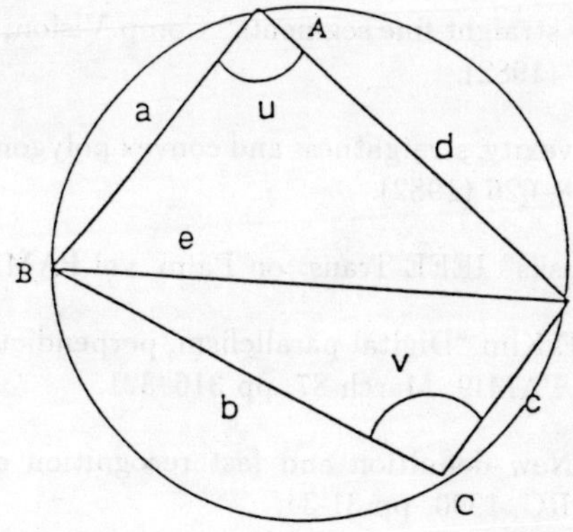

Figure 11: Inscribable quadrilateral $ABCD$.

Using (14) and (16) results the following expression of R^2:

$$R^2 = \frac{(ab+cd)(ca+bd)(ad+bc)}{(-a+b+c+d)(a-b+c+d)(a+b-c+d)(a+b+c-d)} \tag{17}$$

Let p be the semiperimeter of the quadrilateral, that is $a+b+c+d=2p$, then:

$$R^2 = \frac{(ab+cd)(ca+bd)(ad+bc)}{16(p-a)(p-b)(p-c)(p-d)} \tag{18}$$

References

[1] Amir, I. "Algorithm for finding the center of circular fiducials" Comp.Vision, Graphics, Imag.Proc., vol 49, 1990, pp 398–406.

[2] T.A.Anderson, C.E.Kim "Representation of digital line segments and their preimages" Comp.Vision, Graphics, Imag.Proc., vol 30, no 3, pp 279-288.

[3] D.H.Ballard "Generalizing the Hough transform to detect arbitrary shapes" Pattern Recognition 13, no 2, 1981, 111–122.

[4] L.H.Chen, K.L.Lee "A new method for circular object detection and location" Pattern Recognition Letters no II, 1990, 691–697.

[5] H.Freeman "On encoding an arbitrary geometric configuration" IRE Trans. Elec. Comp., vol EC10 (1961) pp 260–268.

[6] C.E.Kim "On cellular straight line segments" Comp.Vision, Graphics, Imag.Proc., vol 18, no 4, pp 369–381 (1982).

[7] C.E.Kim "Digital convexity, straightness and convex polygons" IEEE Trans. PAMI, vol PAMI 4, no 6, pp 618–626 (1982).

[8] C. E. Kim "Digital disks" IEEE Trans. on Pami, vol PAMI6, no 3, 1984.

[9] K.Krishnaswamy, C.E.Kim "Digital parallelism, perpendicularity and rectangles" IEEE Trans. PAMI, vol PAMI9, March 87, pp 316–321.

[10] V.A. Kovalevsky "New definition and fast recognition of digital straight line segments and arcs" ICPRC, 1990, pp 31-34.

[11] A.Nakamura,K.Aizawa "Digital circles" Computer Vision, Graphics and Imag. Proc., vol 26, pp 242–255 (1984).

[12] A.Nakamura,K.Aizawa "Digital images of geometrical pictures" Computer Vision, Graphics and Imag.Proc., vol 30, pp 107–120 (1985).

[13] A.Rosenfeld "Digital straight line segments" IEEE Trans. , vol C23, no 12, pp 1264–1269.

[14] A.Rosenfeld "Arcs and curves in digital pictures" J.Assoc. Com. Mach., vol 20, 81–87.

[15] A.Rosenfeld, C.E.Kim "How a digital computer can tell us whether a line is straight" Amer. Math Monthly vol 89, no 4, pp 230–235 (1982).

[16] J. Sklansky "On Hough technique for curve detection" IEEE Trans. on Comp., vol C27, 1978, 923–926

[17] J.Sklansky, V.Gonzalez "Fast approximation of digitzed curves" Pattern Recognition, vol12, no 5, 1980, pp 327–332.

[18] X. Wu, J.G.Rokne "Double-step incremental generation of lines and circles" Comp. Vision, Graphics and Image Proc. vol 37, pp 331–334, Mar 87.

Model-based Synthesis of Vision Routines

Tilo Messer
Bayerisches Forschungszentrum für Wissensbasierte Systeme
Orleansstr. 34, W-8000 Müenchen 80, Germany
messer@forwiss.tu-muenchen.de

Abstract

In this paper we present a knowledge-based approach for recognizing objects by generating sequences of vision routines. The system *FIGURE* generates a sequence of vision operations with respect to a query about a particular object. The resulting computer vision program detects the object, which the query was about, in a grayscale image. We propose a three-stage procedure. First, we use a 2-D representation of real objects modelling them in a model object base. A query to the model object base about an object procures evidence features of the goal object, one or more features. The second stage involves a set of the deduced evidence features and their values forming a generic model of the real object. The generic model serves as a specification for a vision program. Finally, a configuration module comprised of low-level and medium-level vision operations, segmentation algorithms and morphological operations is used to generate a vision program. The configuration process is supported by various rules. These rules are used to set appropriate parameter values for the vision operations and to control how vision operations are combined to form a program.

1 Introduction

Combining computer vision and expert system techniques leads to improvements in both automatic image interpretation and automatic vision program configuration. Automatic configuration implies the arranging of sequences of vision operations by means of rules which are activated by knowledge about various domains. Above all, two domains have to be emphasized: the computer vision domain and the application domain. Research results concerning automatic vision system configuration contain, on the one hand, interactive user consultation systems for developing vision programs and, on the other hand, automatic adaption of vision operations to new scene contents ([Ender & Liedtke 86, Liedtke & Ender 86]).

2 Related Work

During the last 10 years, efforts have been made trying to create an automatic process for arranging vision programs. Sueda ([Sueda 85], [Matsuyama 86]) developed a consultation system for image processing. In his system, a step in the automatic vision program configuration is taken once appropriate commands have been automatically selected from a large number of available commands and appropriate arguments for these commands have been specified. A user of an interactive image processing system is also confronted with this problem. The consultation system uses help facilities and restricted configuration knowledge for guiding the user to the appropriate choice. A further step towards automatic configuration has been taken by Sakane, Nazif and Matsuyama: Sakane and Tamura ([Sakane & Tamura 85]) propose an automatic vision program generation system. They define an intermediate language to express a command sequence which is translated into SPIDER code ([Tamura et al. 83]). Nazif ([Nazif & Levine 84]) and Levine describe a rule-based segmentation system to create effective segmentation algorithms. Production rules, regarding the splitting and merging of regions and lines, serve as a basis for the inference mechanism. Matsuyama ([Matsuyama 86]) uses a network arrangement to represent image features and transfer-processes as parts of the knowledge needed for a segmentation process. The specification for the automatic image segmentation expert is given by the user. Tomita ([Tomita 88]) combines vision program modules with machine learning techniques. The user can teach the system the way of recognizing objects. The system uses the built-up model to analyse images automatically and to recognize the expected objects. Vogt ([Vogt 88]) also classifies objects by training sequences of morphological algorithms. Joo and Haralick ([Joo & Haralick 89]) illustrate the ability of mathematical morphology to extract shape information from grayscale and binary images. They point out that the problem they want to solve is to reduce any machine vision task to a sequence of morphological operations. They propose four knowledge sources which influence the reduction: description of the input image, goal, morphological operations, and knowledge and reasoning. Syska and Neumann ([Syska 86], [Neumann 87]) construct an image recognition system by combining the design, scope, and organization of the images' component parts. The work of Risk and Börner ([Risk & Börner 89]) also deals with the configuration of a complete vision system (X-ray inspection of cast metal parts). They argue that the result of the configuration process does not have to be revised. In case of a wrong result, computer vision experts are asked to bring in their knowledge and experience to modify the knowledge base of the vision operations.

3 *FIGURE* – Architecture and Control

We have implemented a system called *FIGURE* for the automatic configuration of vision programs. *FIGURE* is divided into four modules which consist of: a model object

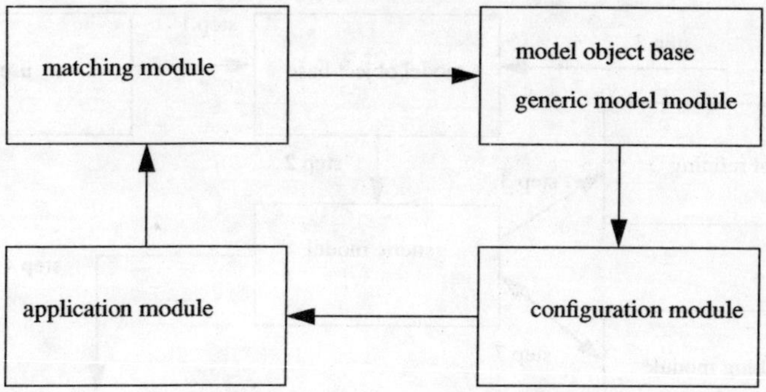

Figure 1: Architecture of *FIGURE*. The direction of the arrows means the flow of control.

base together with a generic model module, a configuration module, an application module and a matching module (Fig. 1).

Let us have a closer look at the architecture of *FIGURE* (Fig. 2).

Extending the concepts of existing systems our approach generates vision programs using a specification which results from a query to a knowledge base about a goal object. Examples of queries (step 1) which may be put to the object knowledge base are: "Is an object x in the scene?", "Where is the object x?". The query can only refer to objects which are modelled in the model object base. The query results in generic model features of the object. A set of pairs of generic model features with their values is a specification of the goal object (step 2). This specification is then used as the generic model of the object. Some of the specified features and values may be similar to those of other objects not asked for. In order to cope with this problem, a mechanism for refining the specification is provided (step 3). Finally, the specification is used in controlling the configuration of the vision program (step 4). Running this vision program on the input image (step 5) results in instances of the goal object. Instances are regions which are computed by the sequence of morphological and mathematical set operations (step 6). Identification faults can be detected by matching the resulting instance with the previous generic model features (step 7). The model refinement mechanism mentioned above can also be used to get an instance which fits the adapted generic model better (step 3).

One important aspect of our approach is to design a model object base with appropriate facilities to make the relationship clear and predictable which exists between the specification of the vision program and the result of its application to the input image.

This paper focuses on steps 1, 2, 4 and 7 according to Fig. 2 and is structured as follows. First of all, the structure of the model object base will be explained.

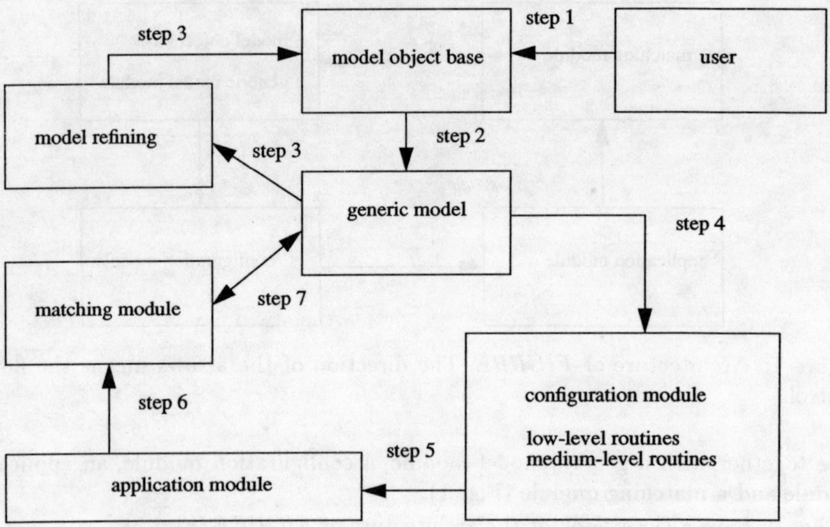

Figure 2: A closer look at *FIGURE*.

Thereafter, a brief survey of the computed generic model features is given. The way these features are used to control the configuration process of the vision program is then described. Finally, a number of photographs illustrate some intermediate results.

4 Model Object Base

In order to deduce generic model features, a *model object base* is built up. 3-D objects of the real world are modelled as 2-D objects in the model object base. The transformation from 3-D to 2-D is done manually. Due to our special application domain we have to generate only very few 2-D objects for each real object. [1] The model object base is built up by hierarchical taxonomy and a 'part-of'-hierarchy. The objects in the model object base are called *boa-objects*. [2] Boa-objects can have boa-objects as their parts forming the part-of hierarchy (Fig. 3a). A boa-object can also be a specialization of another boa-object (Fig. 3b). A boa-object is essentially described as a set of *evidence features*.

Beginning with the boa-object which is the subject of the query, the control al-

[1] We use natural traffic scenes for our experiments. The images of the scene are taken by a camera which is fastened in the front of the car and which is directed straight forward.

[2] Boa is an acronym of a German term.

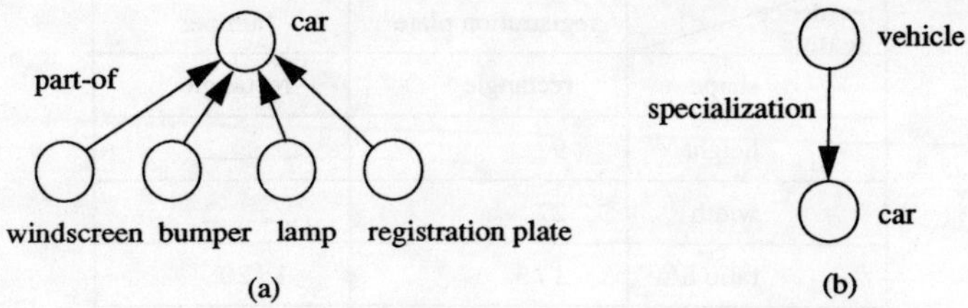

Figure 3: (a) parts of the boa-object 'car'; (b) specialization of 'vehicle'.

gorithm implemented in the model object base runs through all boa-objects which are connected with the starting boa-object by specialization and 'part-of' relations. The process terminates when all reachable boa-objects have been examined. These boa-objects are the result of the query. The *evidence features* of these objects are combined to form a *generic model.*

5 Evidence Features

Evidence features have already been used by Hoffman and Jain ([Hoffman & Jain 88]) in order to manage the difficulties of representing objects and object parts in a fully automatic vision system. [3] Given two areas P and Q, examples of evidence features used by Hoffman and Jain would be:

- the diameter of P
- the surface area of P, and
- the maximum distance between P and Q.

[3]Hoffman and Jain classify 3-D objects by a learning approach using evidence features.

boa-object / evidence feature	registration plate	bumper
shape	rectangle	rectangle
height	9	———
width	27	———
ratio h/w	1 / 3	1 / 20
grayscale	210	185
view	frontal	frontal

Figure 4: Boa-objects and their evidence features.

Examples of evidence features and possible values used in *FIGURE* are given below:

evidence feature	possible values or value type
shape:	line, rectangle, circle, triangle
width:	number of pixels
height:	number of pixels
ratio height/width:	scalar
radius:	number of pixels
angle of the longer principal axis to baseline:	degrees
grayscale:	value between 0 and 255
absolute reflectance:	high, medium, low
relative reflectance:	lighter, darker, opposite
absolute contrast:	high, medium, low
relative position:	above, below, right-of, left-of, inside-of, single
view:	frontal, side

A table of boa-objects with their evidence features is built up. The entries of the table are, on one side, boa-objects resulting from a query to the object knowledge base about a real object, and, on the other side, the values of the evidence features of these boa-objects. An example of such a table is given in Fig.4.

6 Relations between Objects

Each evidence feature of a boa-object has maximum one value. If a query results in more than one boa-object it is obvious that there could be more than one entry for an evidence feature in the resulting table. In this case, the representation of boa-objects and their evidence features enables us to build up relationships between various boa-objects. The relationships are built by examining the values of the evidence features of the boa-objects. Various boa-objects can be connected if the values of some evidence features meet certain conditions. These connections lead to two advantages, whereas the first is concerning the configuration process and the second one has to do with the interpretation of computed instances of the generic model.

The former advantage lies in the ability to use the condition *boa-similar* in the configuration process. Such a relationship allows a much more detailed specification of the configuration process. The latter advantage lies in making it possible for a verifying mechanism to recognize possible identification faults. The configured vision program could produce a resulting instance which cannot be identified as an instance of the goal object. Exploiting the relationship between various boa-objects can make it possible to avoid identification faults. Thus we now want to give a proper definition of the terms *boa-similar* and *non-boa-similar*.

Two boa-objects are boa-similar if they have at least one evidence feature in common with equal values. For example, the boa-objects 'registration plate' and 'bumper' are boa-similar because they have the same shape (evidence feature 'shape' has the value 'rectangle') (Fig. 4). [4]

Formal definition:

Let O_1 and O_2 be two boa-objects, $\{e_1^1, \ldots, e_n^1\}$ and $\{e_1^2, \ldots, e_m^2\}$ the evidence features of O_1 and O_2 respectively, $v(e)$ the value of an evidence feature e.

O_1 and O_2 are similar if:

$$\exists e_i \in \{e_1^1, \ldots, e_n^1\} \cap \{e_1^2, \ldots, e_m^2\}: v(e_i^1) = v(e_i^2)$$

Two boa-objects are non-boa-similar if they have no evidence feature in common or, if each common evidence feature has a different value for the two boa-objects.

Formal definition:

O_1 and O_2 are *non-similar* if

$$\{e_1^1, \ldots, e_n^1\} \cap \{e_1^2, \ldots, e_m^2\} = \emptyset$$

or if

$$\forall e_i \in \{e_1^1, \ldots, e_n^1\} \cap \{e_1^2, \ldots, e_m^2\}: v(e_i^1) \neq v(e_i^2).$$

---

[4] If the table of boa-objects and evidence features is treated as a table of a relational database, the procedure of getting similar objects could be interpreted as a join operation.

7 Generic Model

The *generic model* of a query consists of all evidence feature values of the deduced boa-objects. We call them *generic model features.* The generic model features describe a prototype of the goal object. They are composed of *generic feature vectors* referring to each boa-object. This way of structuring fits well with the composition rules of the configuration module. The rules are activated by the values of the generic model features. The order of composition of the vision operations is also influenced by the number of the generic feature vectors.

Two categories of generic feature vectors exist: *positive generic feature* vectors and *negative generic feature vectors.* Positive generic feature vectors contain the values of the generic model features which contribute information about objects to be detected and identified to the configuration process. Negative generic feature vectors contain generic model feature values describing boa-objects which are *not intended* to be detected and identified. Negative generic feature vectors are used to control the configuration process in order to prevent detecting and identifying objects, esp. those which are similar boa-objects. Similar boa-objects have evidence features with different values or different evidence features from the searched-for object.

8 Configuration Module

The configuration module contains a collection of vision operations. For our purpose morphological operations and mathematical set operations are predominantly used. Examples of morphological operations are:

- dilation, erosion
- opening, closing
- morphological sieve operation
- operations using letters of the Golay alphabet
- operations producing a skeleton of a region

Examples of mathematical set operations are:

- intersection, union, complement

Low-level vision operations and segmentation operations are applied to the grayscale input image in order to get a binary image. We do not concentrate on this part of the configuration module because it is not the topic of this paper.

Morphological operations for binary images are composed. *Composition rules* and *composition meta rules* control the configuration. An example of a composition rule is:

If	*the value of the generic model feature 'width' exists*
then	*this value is a lower bound of the height of a mask for an opening operation*
and	*connect the opening operation with a complement operation.*

Knowledge about the appropriate application of the morphological vision operations is expressed by *composition rules.* Due to the fact that the parameters concerning masks are almost the only ones which can be set [5] to control the configuration process, a lot of rules exist concerning mask creation. The values of the generic model features are the principal values used to control the specification.

Composition meta rules represent knowledge about how to arrange operations or the order of sequences of operations. If it is necessary to consider values of a generic model feature which compete with each other during the configuration process, operation sequences are run concurrently and the instances are combined resulting by their application to the image with an union operation. Negative generic feature vectors activate composition meta rules concerning the complement operation. These types of rules use the generic feature values as knowledge sources. An example of a composition meta rule is:

If	*two or more positive generic feature vectors exist*
then	*select one positive generic feature vector*
and	*create a new branch in the composed sequence.*

After the configuration process is finished an instance of the generic model is computed. Fig. 5 shows the described whole process in a simplified, schematic form.

[5] Another variable parameter is the nember of iterations done in a morphological operation. BOA does not yet take into consideration morphological operations which use letters of the golay alphabet as structuring elements. Input and output images are not treated as parameters which can be used for the knkowledge-based configuration process.

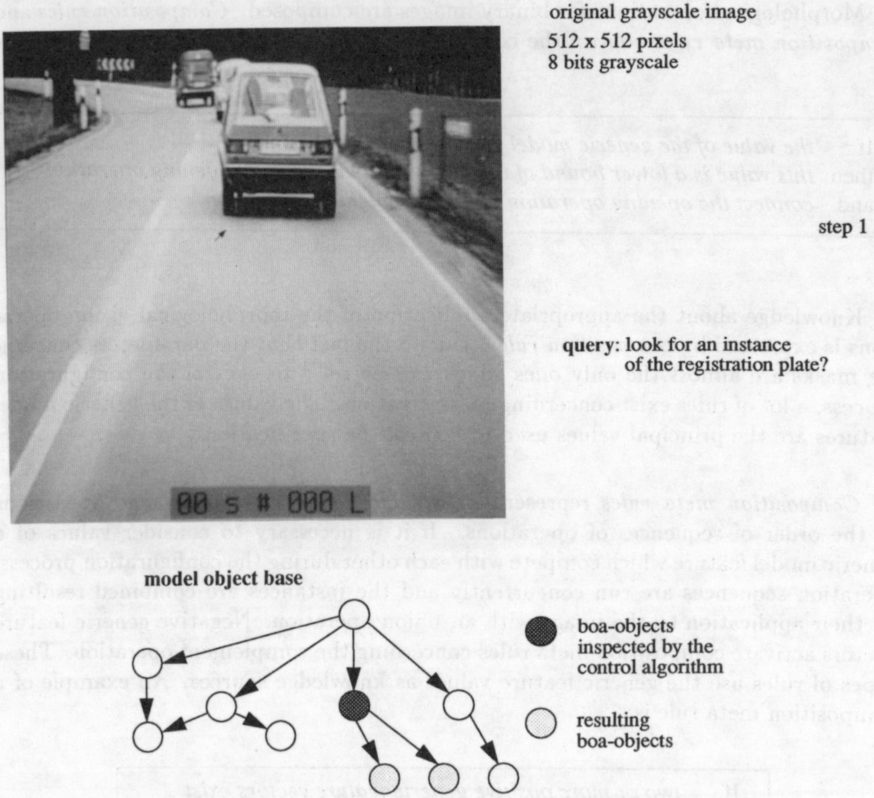

	registration plate
shape	rectangle
height	9
width	27
ratio h/w	1 / 3
grayscale	210
view	frontal

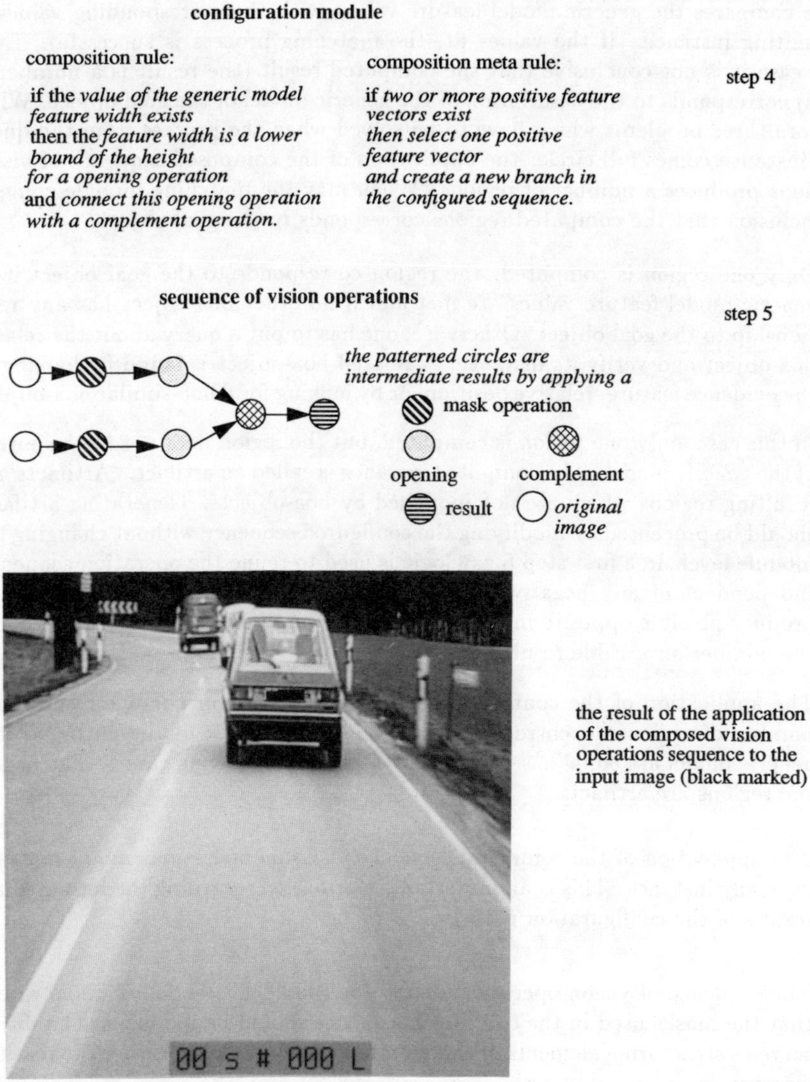

Figure 5: An exemplary process activated by a query to *FIGURE*.

9 Instances

To make sure that the computed result is an instance of the generic model, a matching process compares the generic model feature values with the corresponding values of the resulting instance. If the values fit, the matching process is successful. Even in this case it is not conclusive that the computed result (the result is a number of regions) corresponds to the description of the generic model of the goal object. Why? There are three problems which have to be solved when the process from the query to the instance comes full circle: the application of the composed sequence of vision operations produces a number of regions. When may the matching module come to the conclusion that the computed regions corresponds to the goal object?

1. Only one region is computed: the region corresponds to the goal object if all generic model feature values are met and if no other boa-object has any relationship to the goal object. Otherwise, one has to put a query about the related boa-object and verify its instance. A related boa-object is found by looking at the evidence feature 'relative position' or by looking for a boa-similar boa-object.

2. In this case, only one region is computed, but the region does not fit the entries of the generic model. The computed instance is called an artifact. 'Artifacts' are resulting regions which are not modelled by boa-objects. Generating artifacts should be prevented by modifying the configured sequence without changing the module level. In a first step fuzzy logic is used to refine the operation sequence. Independent of any negative generic feature values, masks are generated to produce possible opposite instances. Using a complement operation can reduce the number of possible resulting artifacts.

3. The application of the configured sequence to the original image results in a number of regions: each region which values and those of the entries of the generic model are equal has to be treated like that following case 1. The rest of the regions are artifacts. [6]

   The application of the composed sequence of vision operations might not produce any instance. This is an important information regarding the intermediate results of the configuration process.

4. The sequence of vision operations has to be modified. We follow the strategy that the masks used in the composed sequence should be modified. The masks serve as structuring elements of the morphologic operations. Either the mask of

[6] If it is assumed that a value of a generic model feature has the value 10 taken from a range of possible widths from 0 to 100, then the complementary value for this range of possible values is defined as 90. Assuming the possible range remains constant, it is obvious that, e.g., the value 55 cannot have a significant complementary value.

the closing operations are too restricted or the masks of the opening operations connected with an complement operation are too large.

A detailed description of the production rules regarding the configuration sequence can be found in [Messer & Nause 92].

10 Experimental Results

One part of *FIGURE* is implemented using NEXPERT Object for the representation of the object knowledge base and the processing of the generic feature models. The configuration module is implemented in IF/Prolog. The vision operations are taken from the image processing tool HORUS ([Eckstein 88]). The images of the traffic scenes which are the basis of the experiments were put at our disposal by BMW. [7] The image size is 512 × 512 pixels and the gray level is 8 bits.

As an example, intermediate results during the configuration process for detecting and identifying a registration plate are demonstrated. Fig. 6(a) shows the original input image which is preprocessed and segmented, as shown in Fig. 6(b).

The first step of the program synthesis is a binary preprocessing. A closing operation fills up small holes in the binary image (a mask of 3 × 9 pixels is used) (Fig. 7(a)). The size of the mask is orientated by the evidence feature 'ratio of width and length' of the boa-object 'registration plate' (see also Fig. 5).

The following step of the program synthesis removes long vertical regions by running an opening operation using a mask measuring 27 × 1 pixels. The width of the registration plate is a lower bound for the height of the mask (Fig. 7(b)).

As a result, long vertical regions are removed from the image in Fig. 7(a) by a complement operation (Fig. 8(a)).

The last step of the program synthesis selects those regions which are larger than the appropriate mask. An opening operation with 9 × 27 mask is configured (Fig. 8(b)). The result is a region, which is an instance of the generic model of the given task.

The following example shows regions which are instances of the searched-for object, artifacts and instances of boa-objects, which are not searched-for. If the goal object is a 'bumper' the result given below demonstrates the effect of instances produced by other boa-objects (e.g. 'registration plate') or non boa-objects (artifacts). The structure of the operation sequence corresponds to that for detecting the registration plate, but uses, of course, different masks (Fig. 9).

---

[7]We thank BMW AG, Munich, for making the images available to us. The images are parts of some image sequences.

Figure 6: (a) original grayscale image (upper); (b) the result of the segmentation operation with absolute values for the threshold (lower).

Figure 7: (a) Closing operation with mask 3×9 in order to fill up holes(upper); (b) opening operation with mask 20×1 (lower).

Figure 8: (a) complement of the closing and the opening operations (upper); (b) the resulting region (lower).

Figure 9: Resulting regions including artifacts and other boa-objects than the goal object (here: bumper).

11 Conclusion

In this paper, we have discussed a knowledge-based approach for detecting and identifying objects in a grayscale image by generating a sequence of vision operations and applying them to the image. Our approach is restricted to scenes including objects which have simple shape and topological relations. This is due to the simple segmentation operation and the use of morphological algorithms. This restriction, however, could be prevented by including other vision operations which are considered in the configuration process.

If we applied the system to image sequences identification faults could be reduced. Assuming, a resulting instance is not unique to be identified as the goal object, a further instance, which is computed in a following image of the image sequence, could be used to clarify the identification of the previous instance. Furthermore, it is possible that a goal object can only be identified when considering it in its context. As a consequence, contextual knowledge which goes beyond the evidence feature 'relative position' should be modelled in the object knowledge base.

In conclusion, the system is well applicable to scenes with objects of a restricted complexity, but could be adapted to more complex scenes by modifying parts of the system.

Acknowledgements

We gratefully acknowledge the constructive and critical comments of Bernd Radig. Many thanks also go to Michael Schubert, who is the coauthor of an earlier version of the paper published at the Vision Interface 91.

References

[Eckstein 88] Wolfgang Eckstein: Das ganzheitliche Bildverarbeitungssystem HORUS, in: H. Bunke, O. Kubler, P. Stucki (eds.): Proc. of the 10th DAGM-Symposium, Sep. 27 - 29, 1988, Zurich, pp. 53 - 59.

[Ender & Liedtke 86] M. Ender, C.-E. Liedtke: Reprasentation der relevanten Wissensinhalte in einem selbstadaptierenden regelbasierten Bilddeutungssystem, in: G. Hartmann (ed.): Proc. of the 8th DAGM-Symposium, Sep. 30 - Oct. 2, 1986, Paderborn, pp. 219 - 223.

[Hoffman & Jain 88] Richard Hoffman, Anil K. Jain: Learning Rules for 3D Object Recognition, in: Proc. of the Conference on Vision and Pattern Recognition 88, June 5 - 9, 1988, Ann Arbor, pp. 885 - 892.

[Joo & Haralick 89] Hyonam Joo, Robert M. Haralick: Applications of Mathematical Morphology to Machine Vision, in: H. Burkhardt, K.H. Hoehne, B. Neumann (eds.): Proc. of the 11th DAGM-Symposium, Oct. 2 - 4. 1989, Hamburg, pp. 1 - 27.

[Liedtke & Ender 86] C.-E. Liedtke, M. Ender: A Knowledge-based Vision System for the Automated Adaption to New Scene Contents, in: Proc. of the 8th International Conference on Pattern Recognition 86, Oct. 27 - 31. 1986, Paris, pp. 795 - 797.

[Matsuyama 86] T. Matsuyama: Expert Systems for Image Processing: An Overview, in: I.T. Young et al. (eds.), Signal Processing III, New York, 1986, pp. 869 - 872.

[Messer & Nause 92] T. Messer, U. Nause: MORSY - Morphological Object Recognition System, submitted to Vision Interface 92.

[Nazif & Levine 84] A.M. Nazif, M.D. Levine: Low Level Image Segmentation: An Expert System, in: IEEE Transactions on Pattern Analysis and Machine Intelligence PAMI-6 (5), 1984, pp. 555 - 575.

[Neumann 87] Bernd Neumann: Towards Computer Aided Vision System Configuration, in: Ph. Jorrand, V. Sgurev (eds.): Artificial Intelligence II: Methodology, Systems, Applications, Sep. 16 - 19, 1986, Varna, 1987, pp. 385 - 393.

[Risk & Börner 89] I.C. Risk, H. Börner: VIDIMUS: A Vision System Development Environment for Industrial Applications, in: W. Brauer, C. Freksa (eds.): Proc. of the Kongress fur Wissensbasierte Systeme 1989, Oct. 16 - 17, 1989, Munchen, pp. 477 - 486.

[Sakane & Tamura 85] K. Sakane, H. Tamura: Automatic Generation of Image Processing Programs by Knowledge-Based Verification, Proc. of the Conference on Computer Vision and Pattern Recognition, June 19 - 23, 1985, San Fransisco, pp. 189 - 192.

[Sueda 85] N. Sueda: An Expert System for Image Processing, in: Image Technology and Information Display 17 (9), 1985, pp. 19 - 22 (Japanese).

[Syska 86] Ingo Syska: Ein Expertensystemansatz fur die automatische Konfigurierung von industriellen Bildverarbeitungsanlagen, Diplomarbeit, Fachbereich Informatik, Universitat Hamburg, 1986.

[Tamura et al. 83] Hideyuki Tamura, Shigeyuki Sakane, Fumiaki Tomita, Naokazu Yokoya: Design and Implementation of SPIDER - A Transportable Image Processing Software Package, in: Computer Vision, Graphics, and Image Processing 23 (1), 1983, S. 273 - 294.

[Tomita 88] Fumiaki Tomita: Interactive and Automatic Image Recognition System, in: Machine Vision and Applications 1, 1988, pp. 59 - 69.

[Vogt 88] R. Vogt: Automatic Configuration of Simple Morphological Algorithms, in: Proc. of the Conference on Vision and Pattern Recognition 88, June 5 - 9, 1988, Ann Arbor, pp. 760 - 765.

Wavelets and "Chirplets": Time–Frequency "Perspectives" With Applications

Steve Mann
Massachusetts Institute of Technology
Cambridge, Massachusetts, 02139
e–mail: steve@media.mit.edu

Abstract

There has been a tendency among individuals, in a variety of technical fields, to think in either the time domain (the world as seen through an oscilloscope) or the frequency domain (the world as seen through a spectrum analyzer[1]). Many labs are equipped with modern oscilloscopes and spectrum analyzers, each telling its story about what is actually happening. To the contrary, however, physical phenomena quite often fit neither the particle ("things") nor the wave ("stuff") descriptions. Is light "stuff" or a collection of "things"? A framework for this duality has existed since the eleventh century, where musicians have been using time–frequency representations. Engineers and scientists are now realizing the full utility of time–frequency space.

In this chapter, various Time–Frequency (TF) methods are introduced and extended, within a unified general framework: the chirplet. A brief historical perspective, touching on some of the early ideas (such as the work of Fourier and Gabor) is presented in a new way, leading to a smooth extension into the chirplet domain.

Applications of this new time–frequency paradigm are also discussed.

1 Introduction: Time–Frequency Space

1.1 Brief Overview

There has recently been much debate in the literature regarding the relative merits of the wavelet transform and its classical counterpart, the Short Time Fourier Transform

[1]A spectrum analyzer does actually display time evolution of the spectrum over a short time interval, but the time dependency is generally not exploited fully in the display.

(STFT). The STFT, better known to some as the *spectrogram*, may also be thought of as a bank of constant bandwidth filters, or a sliding data window, of fixed size.

The wavelet family, which is derived through affine transformations of the physical coordinate space of one *mother wavelet*[1][2], represents a constant–Q filter bank. It is also known as a *pyramidal decomposition* or *subband coding*, so named because its lower frequency bases are greater in physical extent (time or space). Thus the wavelet transform is like a STFT where the window size automatically grows in inverse proportion to the frequency at which the analysis is being made.

A new interpretation of this relationship between constant–Q and constant bandwidth is provided, where an analogy to projection is given. First, functions of one coordinate (1–D functions) are considered in detail.

A function of a single physical coordinate gives rise to a Time–Frequency (TF) space which occupies two coordinates and may thus be conveniently represented on the printed page. Within the new framework, the wavelet family may be derived through the use of both a 1–D orthographic projection and an inverse 1–D orthographic projection, which, together, simply define an affine change in the one physical coordinate. In this chapter, the utility of wavelets is extended by also applying orthographic projection to the Fourier domain, in addition to the physical domain, to arrive at a set of "wavelets" which are constrained neither by constant–Q nor by constant bandwidth. **Throughout the chapter, the term "wavelets" is used in quotes, to denote this more general class of functions, which, interestingly enough, include the bases of the STFT as a special case.**

Because dilation in time leads to contraction in frequency, one does not immediately attain the two affine parameters from each domain. Thus instead of having four parameters to describe a "wavelet", one most often only has three.

The STFT may be regarded as a tiling of the TF plane by identically–shaped tiles[3], while the wavelet transform embodies tiles of varying aspect ratio, with the additional constraint that the aspect ratio depends on the distance from the $f = 0$ axis. "Wavelets", on the other hand, represent a tiling with arbitrarily chosen, unconstrained aspect ratios.

Allowing the TF tiles to undergo an affine coordinate transformation themselves leads to the next generalization. While wavelets are affine in the physical coordinate, and "wavelets" are affine in both the physical and Fourier coordinates, chirplets are affine in the time–frequency plane. In fact the reason for coining the term "chirplet"[4] is based on the fact that the additional TF–affine degrees of freedom are attained through the use of chirping data windows.

The TF–affine chirplet has been found to have practical applications in the field of radar. There is an intimate relationship between the chirping nature of the TF–affine chirplets and the Doppler return from accelerating objects.

A second means of extension of the wavelet, through the use of projective geometry, is presented. The additional extension is combined with the TF–affine extension, leading to a TF–projective family (also known as a family of TF–*perspectives*[5]).

TF–projective chirplets have eight parameters. The discrete TF–projective chirplet transform of a time series (vector) is then an array which has eight indices. While the array may seem impossible, or at best, difficult, to compute or store with a reasonable grid density, adaptive methods have been developed for pruning the search space[6]. Alternatively, lower–dimensional manifolds[7] may be selected from this analysis space. Furthermore, chirplets represent a general framework in which to consider many other TF methods as special cases.

Lastly, the chirplet is extended to functions of two coordinates. Applications in machine vision and image processing are suggested, in particular, with respect to a novel imaging method, "dusting". Dusting adds a new dimension to the concept of taking a picture. It has its roots in the "painting with light" paradigm, where a photographer walks around in a darkened space illuminating objects with long/multiple exposure(s), while the camera is mounted on a tripod. Digital "dusting" is based on the Computer Enhanced Multiple Exposure Numerical Technique (CEMENT), an algorithm for merging multiple image files from different parts of the same scene.

1.2 The Fourier Operator

The familiar Fourier transform will be presented in a slightly different manner, both to illustrate some of the mathematical conventions introduced in this chapter, and because the Fourier operator is a useful link between classical theory and the chirplet transform.

The following are notational deviations from the usual conventions:

- In an integration, the *measure* of the integration will appear immediately after the integral symbol, as in $\int dt\ s(t)$.
- In a Dirac notation, the convention will be to conjugate the first of the two entries.
- In a Dirac notation the two entries will be separated by a vertical bar, rather than a comma, so that the first and second halves can be treated separately, without confusion.

The Discrete Fourier Transform (DFT) of a signal, $s(t)$, may be expressed as follows[2]:

$$S[k] = \frac{1}{\sqrt{N}} \sum_{n=0}^{N-1} s[n] exp(-j2\pi k \frac{n}{N}), k = 0, 1, 2, ..., N-1 \qquad (1)$$

[2]The convention, common in the electrical engineering field, of denoting the Fourier transform of a variable by an upper case character, and the corresponding variable itself (e.g. the time series) by a lower case character, is adopted throughout this chapter. Furthermore, TF–distributions, such as wavelet transforms, whether they be discrete matrices with two indices, or continuous functions of two variables, will be denoted by bold–face upper case characters.

where $j = \sqrt{-1}$.

In order to preserve the energy equality between the physical and transform domains (Plancherel's theorem) the usual factor of $1/N$ has been replaced with $1/\sqrt{N}$ (and the factor of $1/\sqrt{N}$ also appears in the definition of the inverse DFT).

In matrix notation, equation 1 becomes:

$$\mathbf{S} = \mathcal{F}\mathbf{s} \tag{2}$$

where the matrix $\mathcal{F}$ contains the discrete Fourier bases, of length N samples, in each of its M rows, and $\mathbf{s}$ is N samples of the signal (time series), arranged as a column vector. The matrix $\mathcal{F}$ may be regarded as a "Fourier operator", and the column vector which follows it is the "operand". $\mathcal{F}$ is unitary (its conjugate transpose is its inverse).

In general, any expansion onto a set of bases may be represented in this way, but the inverse operation may not always be so simple. For example, the system may be *over–determined*, or *under–determined*, $(M \neq N)$ where a non–orthogonal set of bases is deliberately selected. *Over–determined* means that there are more bases than are required for reconstruction of the original function.

These over–determined *frames* are well–known in wavelet theory; quite often the number of *analysis functions* exceeds the number necessary for specifying the signal under analysis. While a pseudo inverse would be computed to obtain the original function back, quite often, the original function need never be recovered, and these frames are often more useful for signal analysis, than, for example, data compression, where an orthonormal basis is desired. Orthonormality (or at least bi–orthonormality) is frequently desired when fast reconstruction of the original signal is desired. Within the scope of this chapter, however, signal analysis is the focus, and, thus non–orthogonal over–determined frames will generally be used.

The continuous Fourier transform of an arbitrary signal, $s(t)$, may be expressed:

$$S(f) = \int_{-\infty}^{\infty} dt \ exp(-j2\pi ft)s(t) \tag{3}$$

Note that the signal, $s(t)$, appears last, and the rest of the equation may be regarded as an operator acting on $s(t)$.

Expressed in the Dirac notation:

$$S(f) = < e^{+j2\pi ft}|s(t) > = \mathcal{F}|g > \tag{4}$$

The first part of the inner product $< exp(+j2\pi ft)|$ is the "Fourier operator", which may be re–written:

$$\mathbf{F} = e^{-j2\pi|f><t|} \tag{5}$$

where $|f >< t|$ is the outer product of the time and frequency parameters themselves.

The Fourier operator may be visualized as an image on a video display, or density plot on a printed page, and appears on the book cover, where the density corresponds to its magnitude, and the colour is an encoding of the phase in a perceptually uniform (*Munsell*) space. The crosshairs denote the time and frequency axes. Later, it will become evident that transformations, such as rotations, acting on the Fourier operator, lead to a set of chirping basis functions, which are intimately related to the chirplet transform.

The Fourier operator is a very special Time–Frequency (TF) distribution which may extend over all time and over all frequency, or may be localized by various window functions, which act like a viewport, selecting specified regions of Time–Frequency (TF) space.

1.3 Gabor's "logon" paradigm

Gabor proposed an expansion of a signal onto a set of bases which he illustrated by way of "logons", which are portions of a time frequency (TF) space[3]. He showed that an arbitrary signal can be decomposed onto a set of functions which are all just modulated versions of a single Gaussian envelope. The STFT, quite popular in electrical engineering, may also be thought of as a set of bases which are all just modulated and time–shifted versions of one parent window. But modulation is nothing more than translation in frequency, and, thus, the STFT may be viewed as a tiling of the TF–plane where the "tiles" may be moved along either of the two domains. This was the approach taken by Gabor[3].

Gabor emphasized the use of a Gaussian window, since it minimizes the uncertainty product $\Delta t \Delta f$ (provided that these support measures are quantified in terms of the root mean square deviations from their mean epochs[3]). The time frequency logon diagrams, depicted in Figure 1, show how the Gaussian bases, or any other appropriate window function, cover the time frequency space, and how good frequency resolution may be traded for improved temporal resolution, and vice–versa. Gabor originally used rectangles to designate each of these *elementary signals.* If each of these rectangles were a pixel, and its brightness was set in accordance with the appropriate coefficient in the signal expansion (for example by replacing the rectangle with a dither pattern), the logon diagram would be a density plot ("image") of the TF distribution. Figure 2 illustrates examples of the logon bases of Figure 1 (a), while Figure 3 illustrates examples of the logon bases of Figure 1 (b).

1.4 Wavelets

It is assumed that the reader has some understanding of the wavelet transform. An excellent introduction, with an emphasis on signal processing applications, is provided by Mallat[1].

[3]Although the logon representation is not a "frame"[8], and is therefore not numerically stable, we proceed in the development of the line of thought, and will later address the issue of numerical stability, by *over–representing* the expansion.

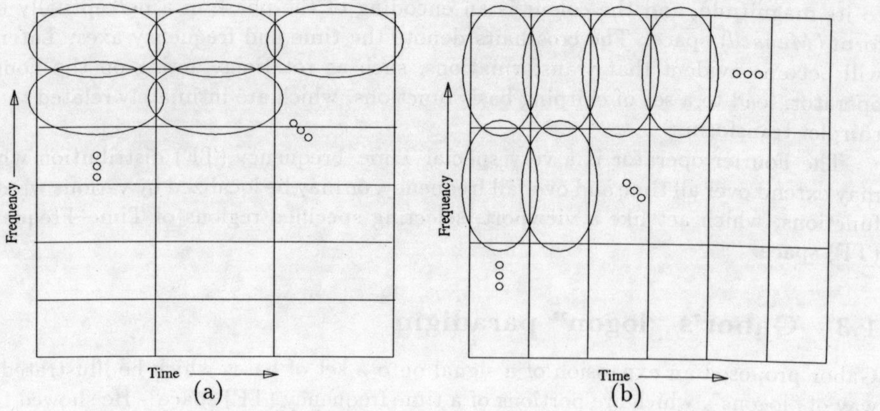

Figure 1: Single scale Gabor expansions, onto two fixed sets of *elementary signals.* In both cases, the frequency axis is bi–polar (zero in center, positive frequencies in top half, negative frequencies in bottom half). (a) "Complete" set of Gabor bases, each of which is a tone packet of relatively long duration and narrow bandwidth. (b) Another "complete" set of bases where better temporal resolution is attained at the expense of frequency resolution.

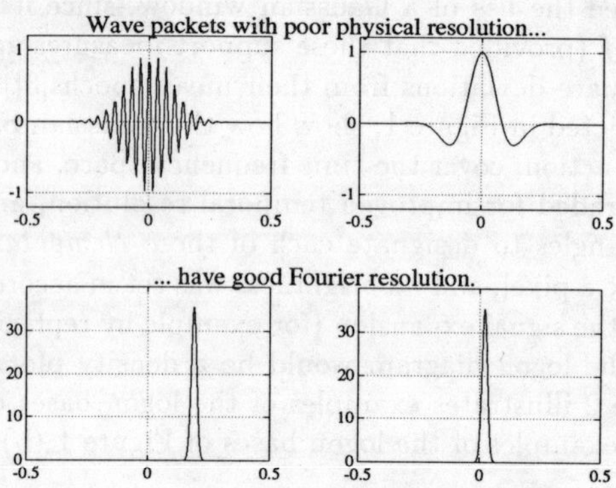

Figure 2: Two examples of *elementary signals* of relatively long duration and narrow bandwidth. The top two plots are the time–domain waveforms (as would appear on an oscilloscope). The bottom two plots are the Fourier magnitude spectra. The leftmost plot pair corresponds to a single logon ("tile") from the tiling in Figure 1(a), located at the top of the center column of logons. The rightmost pair corresponds to a different logon also from the tiling of Figure 1(a), but located three down from the top, in the center column of logons.

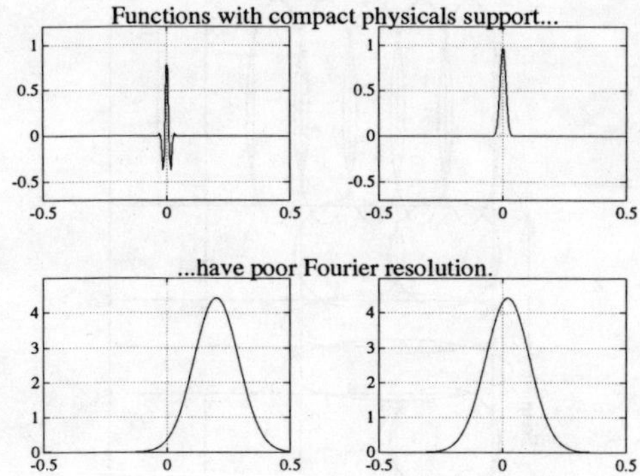

Figure 3: Two examples of short tone packets. They have more compact physical support and thus have greater Fourier support. Again, the top two plots are the time–domain functions, and the bottom two plots are the magnitude spectra. The leftmost plot pair corresponds to a single logon from the topmost row of the tiling in Figure 1(b), while the rightmost plot pair corresponds to a single logon from the middle row of the same tiling (also Figure 1(b)). This more ephemeral basis, gives improved temporal resolution at the expense of degraded frequency resolution.

Briefly, the wavelet transform is an expansion onto a set of bases given by translates and dilates of some *mother wavelet*, $\psi(t)$. The entire family of wavelets, $\psi(\mathbf{A}\mathbf{t}+\mathbf{b})$, may thus be specified by a linear parameter, $\mathbf{A} \in \mathbb{R}^{m \times n}$, and a translation parameter, $\mathbf{b} \in \mathbb{R}^m$. For the time being, the discussion will be limited to wavelets which are functions of a single *physical* variable, "time". Thus the parameters $\mathbf{A}$ and $\mathbf{b}$ will be scalars, say, a and b.

Frequently it is desired to maintain a constant L^2 norm (energy) among all the members of the wavelet family, so a normalization constant is used, based on the dilation parameter. The family of wavelets is thus given by:

$$\psi_{ab}(t) = \sqrt{a}\psi(at+b) \tag{6}$$

For a given signal (function), $s(t)$, the 1–D *wavelet transform* is the expansion onto the family $\psi_{ab}(t)$, which is given by:

$$\mathbf{W}(a,b) = <\psi^*_{ab}(t)|s(t)> \tag{7}$$

The 1–D wavelet transform may also be described as a "Time–Scale (TS) space". More generally, such an expansion is often referred to as the *analysis space*, where the dilation and translation operations are referred to as the *analysis mapping.*

In a manner similar to the Fourier operator, the *continuous wavelet operator* may be defined as $\mathcal{W} = <\psi^*_{ab}|$, so that the 1–D wavelet transform becomes $W(a,b) = \mathcal{W}|s>$.

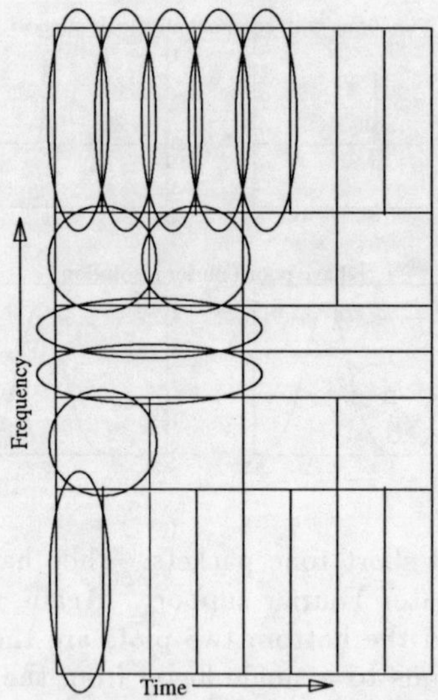

Figure 4: Logon representation of the set of basis functions corresponding to a wavelet transform. The wavelet transform corresponds to a "tiling" of the TF–plane in logons which have a constant–Q. Note the two–sided spectrum (zero in center), indicating the use of wavelets which lie in the Hardy space (*analytic wavelets*). These complex wavelets do not suffer from mirroring about the $f = 0$ axis.

$\mathcal{W}$ is analogous to a 3–D array, and the result, W, is analogous to a 2–D array (matrix), in the same way that $\mathcal{F}$ of Equation 5 was analogous to a 2–D array, and the result of $\mathcal{F}$ acting on a signal was analogous to a 1–D array (vector).

An interesting way to think of the STFT is that the second dimension (to attain the 2–D TF space from the Fourier operator) is attained by windowing $\mathcal{F}$, while the wavelet operator, $\mathcal{W}$, immediately provides the desired 2–D TF space, by its two–parameter definition.

Within Gabor's logon framework, the wavelet transform may be thought of as a constant–Q tiling of the TF–plane, as illustrated in Figure 4. In the TF plot of Figure 4 the frequency is shown on a linear scale. Thus one can readily see that the aspect ratios of the bases are proportional to the distances from the $f = 0$ axis. The aspect ratios are still constrained, but the constraint is different: instead of the constant size constraint of the Gabor type expansion (see Figure 1) one has a constant "Q" (ratio of bandwidth to center frequency) constraint. Thus instead of having identical "tiles", one has "tiles" which become more slender as they move away from the time axis.

2 "Wavelets" versus Wavelets

In the development of a Generalized Logon Transform (GLT)[7], the constraints of constant aspect ratio or constant Q were first removed, resulting in an arbitrary expansion onto a basis of "wavelets" of varying shape, with the only constraint being that each "wavelet" be a Gabor function, or some other similar function of time, frequency, and scale. In Mann and Haykin[7], the full set of *TF–affine parameters* were also added to the GLT, and will be discussed later.

In the past, the term "wavelet" often denoted any arbitrary basis function which acted as a bandpass filter (when used as a convolution kernel). More recently, however, the term *wavelet* has been used to denote a basis function of constant shape (from an affine group of translates and dilates of one *mother wavelet*), as defined in Equation 6. Indeed, in the earlier papers, the term "wavelets of constant shape" was often used to denote the affine basis explicitly. In this chapter, whenever the word "wavelet" appears in quotes, it denotes "wavelet" in the wide sense: any ephemeral "burst of energy", or "piece of a wave". "Wavelets" include the bases of the STFT and the true wavelets.

Actually one may be a bit more restrictive in the use of the term "wavelets", and still fulfill the desire to span both the STFT space and the wavelet space. The classical wavelet is affine in the one–dimensional space (affine transformations amount to translation and dilation: two degrees of freedom). "Wavelets" are members of a family of bases which may be derived using 1–D affine coordinate transformations in both the time domain and the frequency domain, applying these transformations to a particular *mother wavelet.* Later, a unifying affine relationship, in the TF domain, will be introdued. "Wavelets" are defined by:

$$\psi_{abc}(t) = \psi(at + b)exp(jct) \tag{8}$$

The new parameter, c, is the frequency center. Thus, in terms of temporal center, t_c and frequency center f_c, the "wavelet" bases are given by:

$$\psi_{a,t_c,f_c} = \psi_{a,t_c,f_c}(t) = \psi(a(t - t_c))exp(j2\pi f_c t) \tag{9}$$

Now the "wavelet" transform of a signal, $s(t)$, is given by:

$$W(a, t_c, f_c) =< \psi^*_{a,t_c,f_c}(t)|s(t) > \tag{10}$$

The "wavelet" transform thus treats the two domains, time and frequency, in a symmetrical manner. The logons are free to move about in the TF–plane, and assume an arbitrary aspect ratio. There are no artificial constant bandwidth or constant–Q constraints. On another philosophical level, one may view "wavelets" as wavelets–in–time and wavelets–in–frequency, since application of Equation 6 in both domains leads to a translation and a dilation in both domains. The dilations are redundant (dilation in one domain by a results in dilation in the other domain by $1/a$), but the translations afforded in the frequency domain gives "wavelets" one additional degree of freedom that wavelets do not have. Thus "wavelets" are related to each other by affine transformations of the marginals of the TF-distribution. In the next section, "wavelets" are further extended by allowing the complete set of affine transformations in the TF-plane itself.

3 The TF-affine Chirplet

An expansion of an arbitrary function onto a basis of multi-scale chirps (swept frequency wave packets) is presented.

Chirplets have been applied to a variety of practical problems such as the detection of floating objects, in marine radar, by their "acceleration signatures". Doppler acceleration is related to the "chirpiness" of radar backscatter. To simplify the presentation, the 1-D chirplets (functions of one variable, say, "time") are first introduced. Extensions to higher dimensions, along with applications in imaging processing are presented later.

It has already been noted that the one-dimensional wavelet has two free parameters, the two affine transformations of the line: translation and dilation. A particular *mother wavelet* is chosen, and all other members of the family of basis functions are obtained by stretching or shifting it in time. Thus the corresponding wavelet transform is a space with two degrees of freedom. It may, for example, in the discrete case, be stored in an array with two indices.

The TF-affine chirplet transform is derived through affine transformations in the TF plane. Because there are up to six[4] such transformations, the resulting *chirplet space* has indexdimension[5] up to six (depending on the particular "mother chirplet" chosen).

The TF-affine chirplet paradigm allows for a unified framework because it embodies many other methods as special cases. Both the wavelet transform, and the Short Time Fourier Transform (STFT), which is based on Gabor's expansion onto a basis of elementary "logons", are special cases of the chirplet transform.

Fourier methods are limited to use with stationary signals, while the STFT applies to signals which exhibit so-called short term stationarity. The chirplet, however, enables us to deal with signals which are not even locally stationary.

[4]The uncertainty product, of course, limits the minimum area, but the area is free to increase beyond this minimum value. When the area is kept constant, the chirplets have an elegant mathematical similarity. In order to vary the area occupied in the TF plane, it is necessary to break away from the pure notion of a single *mother chirplet* and use multiple primitives.

[5]The terms "indexdimension", D and "vectordimension", $\mathcal{D}$ are used to distinguish between the number of indices an array has (or the number of parameters a function has) and the dimensionality of the vector space. A 3-by-1 array (*vector*), which is a 1-D array, lies in a 3-$\mathcal{D}$ vectorspace. Likewise, a continuous function of one variable has an infinite *vectordimension* but has *indexdimension* one.

Various slices[6] through the chirplet space are considered. In particular, the TF plane is one such slice, the time–scale plane (wavelet transform) another. Less trivial slices include the Dispersion Transform[9] (DT) and the Bowtie Chirplet Transform[7] (BCT).

3.1 Linearly Varying Instantaneous Frequency

A pure tone is given by:

$$\psi(t) = e^{+j2\pi f_c t} \tag{11}$$

where $j = \sqrt{-1}$ and f_c is a constant (the frequency of the tone in cycles per second).

A chirp is a swept frequency where the *instantaneous frequency* varies with time. Chirps are most familiar as the sound made by a bird or a bat. Two ways of specifying the linear chirp will be used in the chapter. The first parameterizes the chirp in terms of its f–intercept b and half–slope a, as follows:

$$\psi(t) = e^{j2\pi(at+b)t} \tag{12}$$

where the phase is given by $\varphi = 2\pi(a^2 t + bt + \varphi_0)$, so the frequency, $d\varphi/dt$ is $\dot{\varphi} = 2\pi(2at + b)$ which indicates that the intercept is $2\pi b$. The time rate of change of frequency is a constant: $\ddot{\varphi} = 2\pi(2a)$.

The second representation, particularly useful when the chirp is windowed (to form a "chirplet"), characterizes the chirp in terms of the instantaneuos frequencies at the boundaries of the window. The parameters f_{beg} and f_{end} are the instantaneous frequencies at the beginning and end of the windowed chirp.

3.2 "Le Pépielette"; "The Chirplet"

Yves Meyer coined the term *ondelette* (from the French word for *wave*, <<onde>>, and the diminutive <<ette>>). The closest English translation, *wavelet*, became the accepted word within the English language.

Similarly, the term <<pépielette>> (combining the diminutive with the French word <<pépier>> or <<pépiement>>) was coined to denote a "piece of a chirp", the English version being "chirplet". Figure 5 shows that the relationship of a chirplet to a chirp is analogous to that of a wavelet to a wave. The first published reference is in Mann and Haykin[7]; the term also appears elsewhere in the literature[10].

The essence of chirplets is an extension of the wavelet from its *affinity* in the physical ("time") domain to an affinity in the TF domain. In other words, chirplets consist of all the members of a particular time domain signal, which are affine transformations of each other when viewed in TF space.

Philosophically there are three ways to think of the chirplet:

---

[6] The term "slice" is used to denote a subset of the *analysis space*, which may be derived using a smaller number of degrees of freedom than the full available number of degrees of freedom.

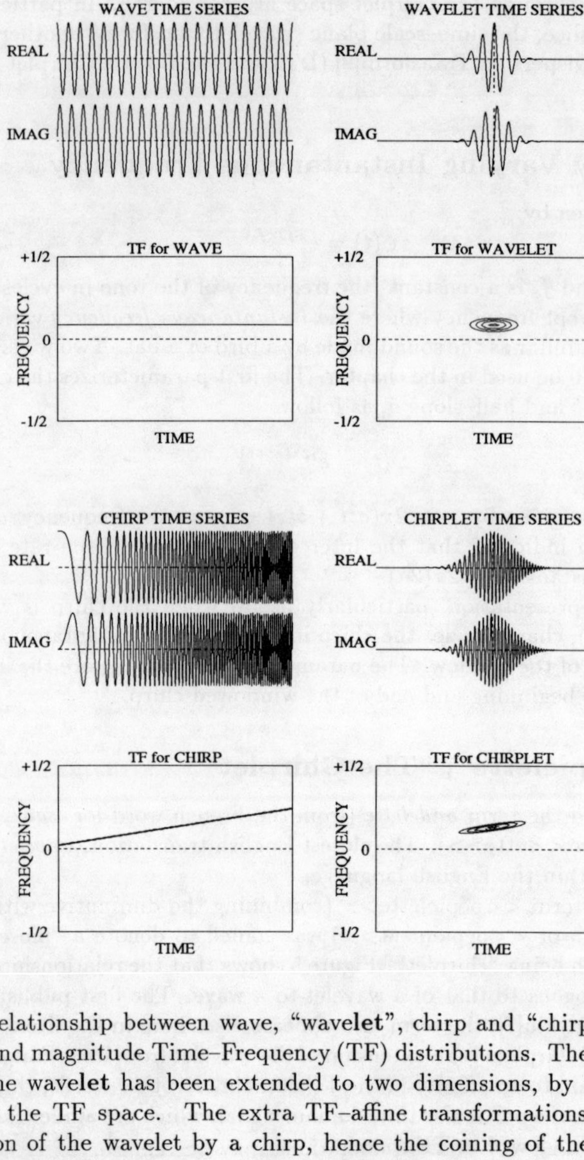

Figure 5: Relationship between wave, "wave**let**", chirp and "chirp**let**", in terms of time series and magnitude Time–Frequency (TF) distributions. The one–dimensional affinity of the wave**let** has been extended to two dimensions, by incorporating the affinity into the TF space. The extra TF–affine transformations are achieved by multiplication of the wavelet by a chirp, hence the coining of the term "chirp**let**" (e.g. a chirp–in–time performs a *shear* along the frequency axis). The nature of the Gaussian function eliminates two of the six TF–affine degrees of freedom which a more general chirplet may have. Thus the Gaussian chirplet only has four degrees of freedom: (t,f) translation, rotation, and aspect ratio.

1. Affine transformations of the Fourier operator, already discussed.
2. Using the "piece–of–a–chirp" framework.
3. Thinking in terms of affine coordinate transformations acting on the TF distribution of a particular *mother chirplet.*

3.3 TF–Affine Geometry

The affine transformation on the TF space is defined:

$$[t_2, f_2]^T = \mathbf{A}[t_1, f_1]^T + [t_c, f_c]^t, \mathbf{A} \in \mathbb{R}^{2\times 2} \tag{13}$$

where $[t_2, f_2]$ represents the original coordinate system (from the TF distribution of the *mother chirplet*), and $[t_2, f_2]$ represents the transformed *chirplet space.* The first term is a linear part, with the second term being a translation to a new center, designated by the letter "c".

It is commonly known (for example, Barnsley[11]) that any matrix A may be expressed as:

$$\mathbf{A} = \begin{bmatrix} r_1 cos(\theta_1) & -r_2 sin(\theta_2) \\ r_1 sin(\theta_1) & r_2 cos(\theta_2) \end{bmatrix} \tag{14}$$

Equation 14 may be re–written:

$$\Box = \begin{bmatrix} \Delta_t cos(tan^{-1}(f_{\text{dif}})) & -\Delta_f sin(tan^{-1}(t_{\text{dif}})) \\ \Delta_t sin(tan^{-1}(f_{\text{dif}})) & \Delta_f cos(tan^{-1}(t_{\text{dif}})) \end{bmatrix} \tag{15}$$

where the angle θ_1 is equal to the arctangent of the frequency "swing" of the chirplet, f_{dif}, the angle θ_2 is equal to the arctangent of the temporal "swing" (shear in time, t_{dif}) of the chirplet, and $\Box$ denotes an arbitrary TF operator.

The parameter Δ_t stands for amount of dilation–in–time (the parameter of $\boxed{\leftrightarrows}(\Delta_t)$), and the parameter Δ_f is the parameter of the $\boxed{\uparrow\downarrow}(\Delta_f)$ operator. The quantity t_{dif} is given by:

$$t_{\text{dif}} = t_{\text{end}} - t_{\text{beg}} \tag{16}$$

and the quantity f_{dif} is given by

$$f_{\text{dif}} = f_{\text{end}} - f_{\text{beg}} \tag{17}$$

which is the span of the instantaneous frequency of the chirp over its time window.

When the dilations are related reciprocally, so that the area of the logon remains constant, the TF–affine operator is unitary. (The matrix, $\Box$, is unitary.) A unitary TF–affine operator will not alter the time–bandwidth product of the *mother chirplet* upon which it is acting, and is said to be *area–preserving.*

The TF–shear and TF–dilation operations may be represented either by matrix multiplications in the TF domain, or, by operators acting entirely in the time domain. It is the power of the latter representation which enables the chirplet to be fully exploited in a very simple manner.

3.4 The TF–affine transformations:

The TF–affine chirplet space is characterized by up to six degrees of freedom. Two of these degrees of freedom are achieved in the time domain, through multiplication by a chirp. The STFT, for example, is achieved by fixing the scale parameter constant, allowing the time parameter to vary, and multiplying by a special case of a chirp, namely a pure tone (a complex exponential).

A set of six symbols, which are pictorial representations of the action[7] in the TF plane, are frequently used to designate each of the six time–domain operators. In other words, while the symbols depict graphically, what is happening in the 2–D TF–plane, they act in the physical 1–D domain.

3.4.1 Translation in Time

An operator, which performs both translation in the time domain, and translation–in–time in the TF domain, is defined, in two ways, as follows:

$$\boxed{\rightrightarrows}(t_c)s(t) = s(t - t_c) = \mathcal{F}^{-1}M(-t_c)\mathcal{F}s(t) \tag{18}$$

where $\mathcal{F}$ is the Fourier operator $\int_{-\infty}^{\infty} dte^{-j2\pi ft}$, $\mathcal{F}^{-1}$ the inverse Fourier operator, and $M(f_c) = e^{j2\pi f_c}$, the constant frequency modulation operator. The second definition, while seemingly more obscure, is presented as a framework for later generalizing to a frequency–dependent time shift (shear in time).

In its most common everyday use, Frequency Modulation (FM) is usually applied to signals in the time domain (causing a time–dependent shift in frequency), but here "frequency" modulation has been applied in the frequency domain to give rise to a (possibly frequency–dependent) shift in time. Philosophically, one may feel more comfortable if this operation were called "Time Modulation (TM)", but the mathematical definition of the operator is the same as FM, so it will still be called "FM". The reasoning will become evident shortly. Also note the sign change of t_c.

The translation–in–time operator, when viewed in TF space, manifests itself as a shift right. In terms of Equation 13, the matrix **A** is just the identity matrix, and the vector, b is given by $\mathbf{b} = [t_c, 0]$.

The arrows point right to indicate the convention of using positive quantities to denote right shifts, and negative quantities to denote left shifts. Of course, the arrows do not necessarily indicate the direction of the shift, but, rather they indicate which sign convention is used.

[7] Formally, the relations are exact for the Wigner–Ville distributions, but the same approximate behaviour is observed for the TF distribution calculated by almost any method. The sliding window implementation of Thomson's method is favoured for illustrative purposes, as it is free of the cross terms associated with the "raw" Wigner distribution, and is non–parametric, thus eliminating the need to select a particular window, or a particular Wigner smoothing kernel.

3.4.2 Translation in Frequency

The frequency translation operator performs a shift–in–frequency within the TF plane, and also a shift in frequency in the frequency domain. The operator is defined in time domain, acting on a particular function of time, $s(t)$, as follows:

$$\boxed{\uparrow\uparrow}(f_c)s(t) = s(t)e^{j2\pi f_c} \tag{19}$$

Again, the arrows point in the direction of positive shift. In terms of Equation 13, the matrix $\mathbf{A}$ is just the identity matrix, and the vector, $\mathbf{b} = [0, f_c]$.

3.4.3 Shear in Frequency

The frequency shear operator performs a shearing in frequency in the TF domain through a multiplication of the operand by a *linear FM* chirp. Thus the operator is defined:

$$\boxed{\downarrow\uparrow}(f_{\mathrm{beg}}, f_{\mathrm{end}}) = e^{j2\pi(f_{\mathrm{dif}}t+f_c)t}s(t) = e^{j2\pi(\frac{f_{\mathrm{end}}-f_{\mathrm{beg}}}{2}t+\frac{f_{\mathrm{end}}+f_{\mathrm{beg}}}{2})t}s(t) \tag{20}$$

The operator may either embody only shear, or it may embody both shear and translation in frequency. When the operator is to be used in its purest form (shear only), it must be written as a function of of two equal and opposite quantities, so that the second argument is the same as the first, except for a sign change: $f_{\mathrm{beg}} = -f_{\mathrm{end}}$. When it is to be defined as a generalization of Equation 19, it may be written with two equal parameters or with only one:

$$\boxed{\uparrow\uparrow}(f_c)s(t) = \boxed{\downarrow\uparrow}(f_c, f_c)s(t) = \boxed{\downarrow\uparrow}(f_c)s(t) \tag{21}$$

This relationship is nothing more than the fact that a steady tone is a special case of a chirp.

3.4.4 Shear in Time

The shear in time is performed by a *linear FM* operation in the Fourier domain, where it might better be referred to as a linear Time Modulation (TM). The shear in time operator, acting on $s(t)$, is defined:

$$\boxed{\rightleftarrows}(t_{\mathrm{beg}}, t_{\mathrm{end}})s(t) = \mathcal{F}^{-1}\boxed{\downarrow\uparrow}(-t_{\mathrm{beg}}, -t_{\mathrm{end}})\mathcal{F}s(t) \tag{22}$$

Again, the arrows indicate the direction of positive shear: to the right (later in time) at higher frequencies, and to the left (earlier in time) at lower frequencies.

If desired, the operator defined in Equation 22 may actually embody both shears and translations in time. In other words, when Equation 18 is to be written as a special case of Equation 22, it will be written:

$$\boxed{\rightrightarrows}(t_c)s(t) = \boxed{\rightleftarrows}(t_c, t_c)s(t) = \boxed{\rightleftarrows}(t_c)s(t) \tag{23}$$

3.4.5 Dilation

If a time domain function is dilated in time (stretched out) its frequency domain representation is contracted in frequency ("squashed in"). Thus the two dilation operators counteract each other in the following manner:

$$\boxed{\leftrightarrows}(\Delta t) = \boxed{\updownarrows}(1/\Delta t) \tag{24}$$

so that one operation "undoes" the other.

In many applications, this limitation may not be a problem, and the resulting 5–D constant–area (*symplectic*) chiplet space is adequate. In other applications, however, it may be desired to isolate the two operations (subject, of course, to the minimum *uncertainty product*). Two methods have been discovered for circumventing the counteraction of the two operators. One of these methods is now presented.

3.5 The Prolate Chirplet

Families of Discrete Prolate Spheroidal Sequences (DPSSs) are of special interest in the signal processing community because of their collective behaviour when used as a set of multiple data windows. The family leads to optimal energy concentration within a specified band of frequencies, when a uniform penalty is given for energy falling outside the band (regardless of where outside the band) and an equal reward is given for energy content inside the band, regardless of where. Details are provided in Landau, Pollack, and Slepian[12][13][14][15][16]).

Thomson's multiple window method[17] has been extended[18] to the TF–affine chirplet domain.

Averaging is done over multiple chirplet windows, computed with not a single *mother chirplet* but, rather, with a set of prolates which have each undergone the range of affine transformations in the TF–plane, using the six TF–affine operators (or a subset of these). Thus the family of chirplets is, in essence a family of families of prolates, where each family represents a particular set of affine transformations. The six indexdimensional chirplet space (or a chosen slice through this space) is computed using each of the $2NW$ lowest order prolate chirplets. Then the magnitude of the $2NW$ spaces is taken, and an average[8] is performed over these $2NW$ chirplet spaces to arrive at the result.

The term "prolate chirplet" has been coined to designate an entire family of multiple window functions which have collectively undergone a single TF–affine coordinate transformation.

The prolate chirplet embodies all six TF–affine degrees of freedom, although the TF–area must vary in discrete steps. (See Figure 6.)

[8] More precisely, an eigenvalue–weighted average is performed, but the $2NW$ approximation is generally adequate.

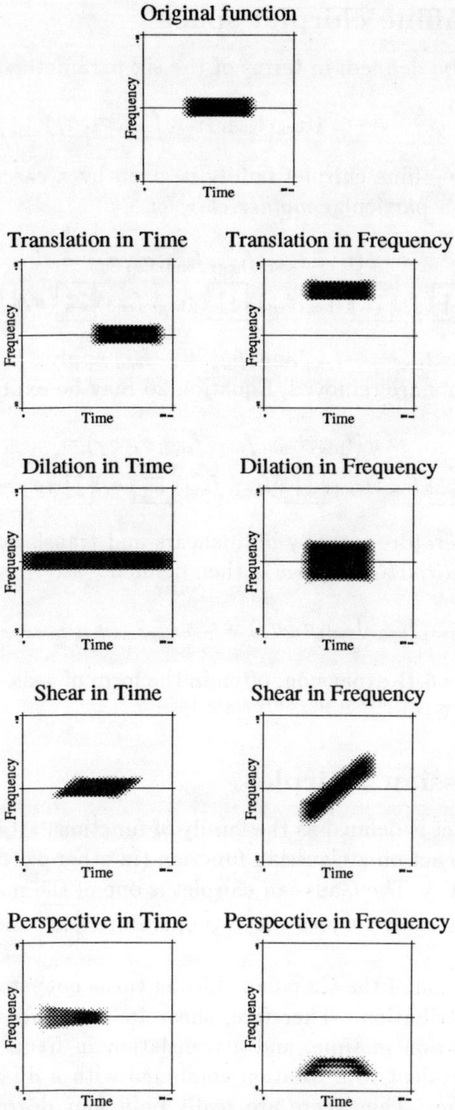

Figure 6: Chirplet transformations in TF space. The Discrete Prolate Spheroidal Sequences (DPSSs) are used to arrive at the "prolate chirplet". In this example, each chirplet actually consists of 30 separate functions which must each be acted on by identical chirplet operators. The last two examples are associated with the "projective chirplet" which will be introduced in the next section.

3.6 The TF–affine chirplet space

The chirplet is now be defined, in terms of the six parameters, as follows:

$$\psi = \psi(t_{\text{beg}}, t_{\text{end}}, f_{\text{beg}}, f_{\text{end}}, \sigma_t, \sigma_f) \tag{25}$$

Thus the general TF–affine chirplet family is given by a cascading of the six linear operators, acting on a particular *mother chirplet*, ψ:

$$\psi(t_{\text{beg}}, t_{\text{end}}, f_{\text{beg}}, f_{\text{end}}, \sigma_t, \sigma_f) = \\ \boxed{\rightrightarrows}(t_c)\boxed{\upuparrows}(f_c)\boxed{\rightleftarrows}(t_{\text{beg}}, t_{\text{end}})\boxed{\downarrow\uparrow}(f_{\text{beg}}, f_{\text{end}})\boxed{\leftrightarrows}(\sigma_t)\boxed{\updownarrow\updownarrow}(\sigma_f)\psi(t) \tag{26}$$

where the constraints $t_{\text{beg}} = -t_{\text{end}}$ and $f_{\text{beg}} = -f_{\text{end}}$ apply.

If those constraints are removed, Equation 26 may be expressed as:

$$\psi(t_{\text{beg}}, t_{\text{end}}, f_{\text{beg}}, f_{\text{end}}, \sigma_t, \sigma_f) = \\ \boxed{\rightleftarrows}(t_{\text{beg}}, t_{\text{end}})\boxed{\downarrow\uparrow}(f_{\text{beg}}, f_{\text{end}})\boxed{\leftrightarrows}(\sigma_t)\boxed{\updownarrow\updownarrow}(\sigma_f)\psi(t) \tag{27}$$

so that the shear operators embody both shears and translations.

The *TF–affine chirplet transform* is then defined [9] as:

$$\underline{\mathbf{C}}(t_{\text{beg}}, t_{\text{end}}, f_{\text{beg}}, f_{\text{end}}, \sigma_t, \sigma_f) = < \psi_{t_{\text{beg}}, t_{\text{end}}, f_{\text{beg}}, f_{\text{end}}, \sigma_t, \sigma_f}(t) | s(t) > \tag{28}$$

where $\underline{\mathbf{C}}$ is a complex 6-D expansion, often in the form of a six–index complex–valued array of the form $\underline{\mathbf{C}} \in \mathbb{C}^{\,m_t \times n_t \times m_f \times n_f \times m_d \times n_d}$.

3.7 The Gaussian Chirplet

The Gaussian chirplet is defined as the family of functions attainable by allowing the chirplet operators to act on a Gaussian function (in other words, using a Gaussian as the "mother chirplet"). The Gaussian chirplet is one of the many chirplets for which the uncertainty constraint couples the two operators; $\boxed{\leftrightarrows}$ and $\boxed{\updownarrow\updownarrow}$ counteract each other.

The TF distribution of the Gaussian chirplet turns out[10] to be a bivariate normal (2–D Gaussian) distribution. Therefore, shear–in–time turns out to be equivalent to a rotation, a dilation in time, and a translation in frequency. Similarly shear–in–frequency is equivalent to a rotation combined with a dilation in frequency, and a translation in time. Thus there are really only four degrees of freedom for the bivariate Gaussian logon in TF space; the two operators $\boxed{\rightleftarrows}$ and $\boxed{\downarrow\uparrow}$ are redundant:

---

[9] Functions of more than two variables and arrays of more than two indices are designated by underlined boldface upper case characters.

[10] Again, the shape is exactly 2–D Gaussian for the Wigner–Ville TF distribution, and almost exact for most other distributions.

all shearing of the Gaussian chirplet may be performed using the $\boxed{\downarrow\uparrow}$ operator, which is the easier of the two to perform, assuming the time series, $s(t)$, is given, rather than $S(f)$.

A Gabor function may be regarded as a Hadamard[11] product of a complex exponential (wave) and a Gaussian window (envelope). For the time being, the relative phase between the Gaussian window and the wave is neglected, as the magnitude chirplet transform is the primary interest. This simplification is possible because these functions are *analytic signals*; the contribution due to one of the bases is very insensitive to the relative phase between the wave and its envelope.

Thus the Gaussian chirplet, for all practical purposes, has the following four degrees of freedom:

1. $\boxed{\rightrightarrows}(t_c)$ controls the temporal location: location of the chirplet's mean epoch within the time series being analyzed.

2. $\boxed{\uparrow\uparrow}(f_c)$ controls the center frequency: the chirplet's mean epoch in Fourier space.

3. $\boxed{\leftrightarrows}(\Delta t)$ controls the chirplet's physical support, Δt, otherwise known as the effective pulse width or *scale.* Δt (the parameter c in Equations 30 and 31) is usually measured in terms of the RMS deviation from the mean epoch (the "second–order energy–moment").

4. $\boxed{\downarrow\uparrow}$ controls the "chirpiness"; the chirplet's rate of change of the instantaneous frequency with respect to time. If the chirpiness is zero, one has an ordinary Gabor function (a pure tone modulating a Gaussian envelope).

Generally the Guassian is formulated to have unit L^1 norm, so that it will represent a valid Probability Distribution Function (PDF), but it will be re–formulated here, so that it has unit L^2 norm. Thus the *energy* of the Gaussian chirplet, rather than its magnitude, will be a valid PDF in the new formulation.

$$\text{env} = \sqrt{\frac{1}{\sqrt{2\pi}\sigma} e^{-\frac{1}{2}\left(\frac{t}{\sigma}\right)^2}} = \frac{1}{\sqrt{\sqrt{\pi}c}} e^{-\frac{1}{2}\left(\frac{t}{c}\right)^2} \tag{29}$$

where $c = \sqrt{2}\sigma$.

The Gaussian chirplet is now defined in terms of the FM half–slope, a and intercept, b as:

$$\psi_{ab} = (\text{env})e^{j2\pi(at+b)t} \tag{30}$$

[11]Neglecting the following subtle difference: if the functions are continuous, consider their product, if they are discrete time series (vectors), consider their *element by element multiplication* or *Hadamard* product.

If a dilation parameter is desired, it may be defined as:

$$\psi_{abc} = \boxed{\substack{\leftrightarrow \\ \leftrightarrow}}(c)(\text{env})e^{j2\pi(at+b)t} = (\text{env})e^{j2\pi(a(t/c)+b)(t/c)} \tag{31}$$

where c is the dilation parameter. ψ_{abc} is a kernel which may be moved to the left and right (translated in time). In other words, there is an implicit fourth degree of freedom, the temporal center of the chirplet, given mathematically by:

$$\psi_{abcd} = \boxed{\rightrightarrows}(d)(\text{env})e^{j2\pi(a(t/c)+b)(t/c)} \tag{32}$$

Note also that shear in time and shear in frequency both result in a simultaneous rotation and elongation, so there are only four degrees of freedom (neglecting, of course, the complex amplitude as a "nuisance parameter"). Setting $a = 0$ in Equation 30 reduces the chirplet transform to the sliding window Fourier transform (with Gaussian window) with the frequency given by b and the time, τ, given by parameter d, the location of the window, $env(t-\tau)$, or the bases $\psi_{abc}(t-\tau)$. There is a subtle difference between moving a window past a pure tone (wave) and moving both the window and the wave together. For the time being, this phase shift will be overlooked. Thus there are three philosophical views of the STFT, which are equivalent:

- Fourier transformation of segments of the data after overlapping windows are applied to it

- an expansion onto a set of bases defined as an appropriate product of the wave (complex exponential) and the envelope, "env" (the data window)

- a bank of constant bandwidth filters

Allowing c of Equation 31 to represent the scale parameter, one may obtain a wavelet decomposition.

Allowing each of the bases to be positioned at time τ, gives:

$$\psi = e^{-\frac{1}{2}(\frac{t-\tau}{c})^2} e^{j2\pi(a\frac{t-\tau}{c}+b)\frac{t-\tau}{c}} \tag{33}$$

(where the first term is the "envelope" and the second term is the "wave".)

Fixing the value of the "chirpiness" (Doppler acceleration), a, to zero, and the center frequency b to a constant (say 1), we have:

$$\psi_{c\tau} = e^{-\frac{1}{2}(\frac{t-\tau}{c})^2} e^{j2\pi\frac{t-\tau}{c}} \tag{34}$$

which is a wavelet transform (with Gabor wavelet) with scale parameter c and location parameter τ.

4 The Projective Chirplet

The TF–affine chirplet is one generalization of the wavelet. A second generalization of the wavelet is the projective chirplet. Let us, for the moment, consider the expression:

$$\psi_{z_1,z_2,\theta}(t) = \psi(z_2 \tan(\tan^{-1}(t/z_1)) - \theta) \tag{35}$$

which can be shown to be another generalization of the wavelet (Equation 6), simply by observing that, for some arbitrarily large quantity, q:

$$\lim_{q\to\infty} \psi_{qz_1,qz_2,\theta/q}(t) = \psi(qz_2(\frac{t}{qz_1} - \theta/q)) = \psi(at+b) \tag{36}$$

where $a = z_2/z_1$ and $b = -z_2\theta$.

Equation 35 is nothing more than an equation of projective geometry in the line, and the relationship between Equations 6 and 35 is nothing more than a statement of the fact that line–affine geometry is a special case of line–projective geometry, or that orthographic projection of the line is a special case of perspective projection of the line, where *focal length* becomes arbitrarily large, and the perspective angle small.

Combining the extension afforded by Equation 35 with the six TF–affine parameters provides only one new degree of freedom, characterized by θ; it turns out that z_1 and z_2 are redundant in the face of the TF–affine parameters.

The same operation may be performed in the frequency domain. Projection in the frequency domain leads to one more degree of freedom, the Fourier perspective, designated by the symbol ϕ. Altogether, the 1–D chirplet space has eight free parameters. The two new parameters are designated by the symbols: $\boxed{\updownarrow \cdot}(\theta)$ and $\boxed{\dot{\leftrightarrow}}(\phi)$.

It is interesting to note that a photograph of a newspaper (rectangular shape) embodies seven degrees of freedom, but a photograph of that photograph embodies the full eight (co–linear geometry). No additional free parameters are attained by further re–projections. Thus twice–applied projective geometry, with its eight degrees of freedom (say the coordinate pairs of the four corners of the rectangle), is as general as projective geometry applied many times over. Likewise, in the TF plane, the eight parameter chirplet space is sufficiently general for any application in TF perspectives.

5 Extensions to Higher Dimensions

The 2–D chirplets may be represented in a 4–D phase space; there are two physical (e.g. spatial) coordinates, say, x and y, and two Fourier (spatial frequency) coordinates, say, u and v. For convenience, this space will be referred to as a (x, y, u, v) space in a manner analogous to the TF space already introduced.

The 2–D (x, y, u, v)–affine chirplet consists of 20 parameters which are itemized below:

- 6 affine parameters in the physical space (designated by the term *physical–affinities*).
- 6 affine parameters in the Fourier space (*Fourier–affinities*).
- 4 parameters to describe the linear spatial FM: $x_{\mathrm{beg}}, x_{\mathrm{end}}, y_{\mathrm{beg}}, y_{\mathrm{end}}$
- 4 parameters to describe the spatial–frequency domain chirping: $u_{\mathrm{beg}}, u_{\mathrm{end}}, v_{\mathrm{beg}}, v_{\mathrm{end}}$

The last eight parameters consist of shears, and will be collectively referred to as the *shear space.*

The (x, y, u, v)–projective chirplet has 24 parameters; the additional four parameters, p_x, p_y, p_u and, p_v are due to the two *physical perspectives* and the two *Fourier perspectives.* These 24 parameters will likewise be divided into the eight *physical perspectives*, the eight *Fourier perspectives*, and the same eight shears resulting from the 2–D linear FM.

In image processing, both Fourier and wavelet (pyramidal) methods are used. Again, as with the 1–D chirplet, there is a desire to circumvent the usual debate between those in favour of the windowed Fourier transform and those who prefer the wavelet. Rather than pursue this first desire, the current focus of attention is on a more important argument. In order to fully develop this argument, only the second of the two wavelet generalizations will be extended to higher dimensions in this chapter. In other words, the discussion is limited to the *physical perspectives*, an eight parameter chirplet subspace within the 24 parameter space, analogous to the three parameter physical perspectives of the 1–D chirplet (the second of the two wavelet–generalizations presented earlier). Therefore, **the same eight symbols used to denote the entire 1–D chirplet space are used to denote the 2–D physical perspective subspace.**

Consider a brick wall. A picture of that wall, taken with the camera's optical axis exactly perpendicular to the wall, and the film exactly parallel to the wall, will produce a periodic pattern on the film. On the other hand, for any arbitrary camera angle, there will be some perspective. Some bricks will appear larger in the picture than others. In fact, the spatial frequency of the pattern on the film will rise or fall as a function of location, in a manner analogous to the chirping functions of the previous section.

Figure 7 is a good example of the two scenarios. The first scenario (a) is known as *one point perspective.* Many early works in perspective (such as "The Last Supper" by Leonardo da Vinci, and "The School of Athens" by Raphael) limit themselves to one point perspective. But even these paintings depict scenes with *depth* and thus the Fourier basis is still not appropriate. The image of the patterns on the ceiling in "The Last Supper" rises in spatial frequency toward the center of the picture (the *vanishing point*). Thus only in cases of one point perspective, where scenes have very little depth, does the 2–D Fourier basis approximate periodicity in the scene itself.

(a) (b)

Figure 7: The need for the chirplet, in image processing, is quite evident. (a) Example where the camera is lined up parallel to the wall, and there is very little depth. In this case, which is rare, the sliding window Fourier transform, or a pyramidal decomposition performs optimally. (b) Generally, however, camera orientation is arbitrary, but the chirplet transform is still able to exactly characterize periodicity in perspective. In this example, most of the perspective is embodied in the $\boxed{\updownarrow \cdot}$ operator.

It turns out that the 2–D projective chirplets, precisely match the range of all possible camera orientations.

In Figure 8 a number of planechirps are illustrated. The chirplet is formed by windowing these chirps. If the Gaussian window is chosen, it will remain a Gaussian under perspective projection. Imagine elliptic contours as conic sections, subtending the center of projection of a camera. Any ellipse may be associated with an arbitrary circle on the cone; the two ellipses in question may be regarded as projections of the same hypothetical circle. Therefore the contours corresponding to the Gaussian (bivariate normal distribution) are simply projected to another set of elliptical contours which also share a common center of projection.

5.1 Projective Geometry

Following Horn and Weldon [19] a camera–based coordinate system is defined. The optical center of the camera is at the origin (see Figure 9 (a)) and the Z–axis is along the optical axis of the camera. Consider an ideal pinhole camera model (no diffraction) where rays of light from objects in the scene form an image on the image plane, which is perpendicular to the Z–axis. The distance from the optical center to the image plane is denoted by z. For simplicity in the mathematical derivations, imagine that both the scene, and the image plane (film or CCD array) are out in front of the camera, so points in the scene and points on the image plane both have

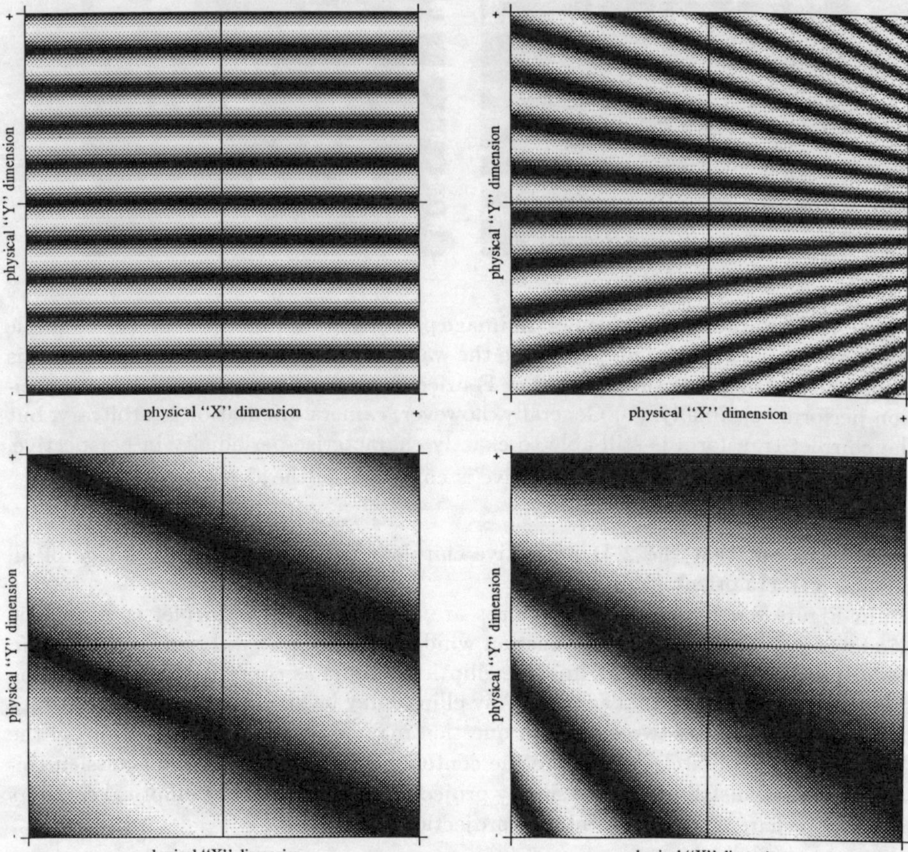

Figure 8: Physical–perspectives of planewaves. These "planechirps" form the basis of the 2–D chirplet transform in the same manner that planewaves form the basis of the 2–D Fourier transform. Appropriate window functions are applied to planechirps to obtain the 2–D chirplet bases in the same way that window functions are used to physically localize the 2–D Fourier bases. Planewaves are special cases of planechirps, as can be seen from the two planechirps at the left, which are also planewaves.

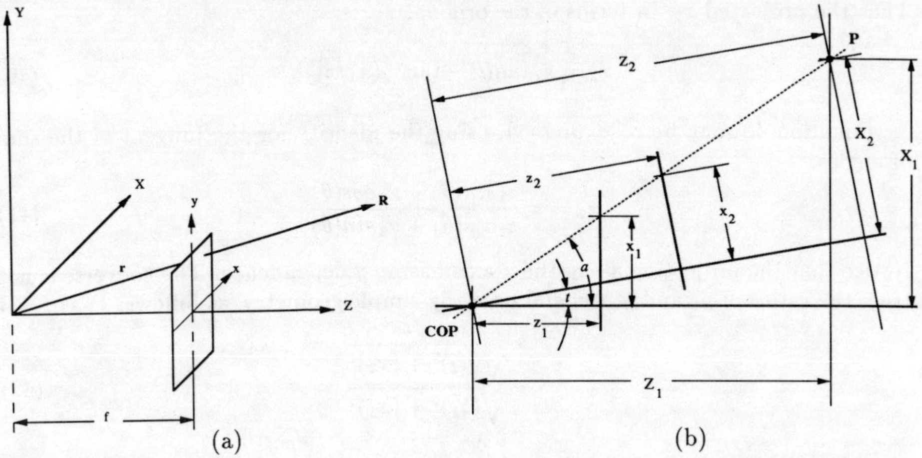

Figure 9: (a) Camera–based coordinate system. (b) Geometry of plane–perspective projection. The camera is rotated about its Center Of Projection (COP). The dotted line denotes a ray of light traveling from object point **P** to **COP**. Heavy lines denote camera optical axis and film plane. The Y axis points out of the page.

positive Z coordinates.

The following is a derivation of the relationship between an image taken by a camera at one azimuth and elevation, and the same camera at a different azimuth and elevation. The derivation is simplified if the two image planes are extended to intersect; a new set of coordinates is constructed with an X–Z plane perpendicular to this line of intersection. (See Figure 9 (b).)

The magnification a lens gives is equal to the ratio of the effective focal length, to the distance from the object:

$$x_1 = \frac{z_1}{Z_1}X_1 \text{ and } x_2 = \frac{z_2}{Z_2}X_2$$

First consider the X–Z plane, as follows:

$$\tan(\alpha) = \frac{X_1}{Z_1} = \frac{x_1}{z_1} \tag{37}$$

$$\tan(\theta - \alpha) = \frac{X_2}{Z_2} = \frac{x_2}{z_2} \tag{38}$$

Re-arranging Equations 37 and 38 gives:

$$tan^{-1}(x_1/z_1) = \theta - tan^{-1}(x_2/z_2) \tag{39}$$

Thus the projected x_2, in terms of the original x_1, is:

$$x_2 = z_2 \tan(\theta - \tan^{-1}(x_1/z_1)) \tag{40}$$

Equation 40 may be re-expressed, using the identity for the tangent of the sum of angles:

$$x_2 = z_2 \frac{z_1 sin(\theta) - x_1 cos(\theta)}{z_1 cos(\theta) + x_1 sin(\theta)} \tag{41}$$

Notice that the projection along the x axis has no y dependence. The converse is not true; the ratios of y_1 and y_2 depend on x, by simple geometry, as follows:

$$\frac{y_2}{y_1} = \frac{\sqrt{(f_2)^2 + (x_2)^2}}{\sqrt{(z_1)^2 + (x_1)^2}} \tag{42}$$

Combining this equation, with Equation 40 gives:

$$y_2 = \frac{-f_2 y_1}{z_1 cos(\theta) + x_1 sin(\theta)} \tag{43}$$

Thus Equations 40 and 43 define the plane projective geometry with which the projective chirplet is defined.

Selecting a particular *mother chirplet* $\psi(x,y)$, the chirplet family becomes:

$$\psi_{t_x,t_y,d_x,d_y,s_x,s_y,p_x,p_y} = \boxed{\substack{\rightarrow\\ \rightarrow}}(t_x)\boxed{\uparrow\uparrow}(t_y)\boxed{\substack{\leftrightarrow\\ \leftrightarrow}}(d_x)\boxed{\updownarrow\updownarrow}(d_y)\boxed{\rightleftarrows}(s_x)\boxed{\downarrow\uparrow}(s_y)\boxed{\updownarrow\cdot}(p_x)\boxed{\leftrightarrow}(p_y)\psi(x,y) \tag{44}$$

where t stands for translation, d stands for dilation, s stands for shear, and p stands for perspective projection. The angle θ, together with the angle of rotation applied to the coordinate system of Figure 9, represent the two degrees of freedom from which p_x and p_y are derived, while again, the parameters z_1 and z_2 are redundant in the presence of the six affine parameters. As before, the chirplet transform is also defined as the inner product of $\psi(x,y)$ with the signal, $s(x,y)$ (generally an image).

Possible choices of *mother chirplet* include the two–dimensional Gaussian–windowed planewave, the 2–D slepian function, or the signal $s(x,y)$ itself. Using the signal itself as the *mother chirplet* casts the chirplet transform in a new light, by making it an extension of the well–known radar *ambiguity function*[20]. (The *ambiguity function* is the Fourier transform of the Wigner distribution.) The term *auto–chirplet transform* is used to denote the choosing of $\psi(x,y) = s(x,y)$. Furthermore, the chirplet may also act as a generalization of the cross–Wigner distribution, with respect to its use to piece together pictures, when two images are interchangeably used as $\psi(x,y)$ and $s(x,y)$.

6 Applications in Higher Dimensions: Dusting and CEMENT

An interesting application of the 2–D chirplet, pertaining to a novel image acquisition method, "dusting", is very briefly described. Dusting is an extension of the well–known photographic concept of "painting with light", where multiple or long exposures are accumulated from a common vantage point.

The dusting apparatus may be broken down into three basic components.

- Communications and control: network consisting of radio channels to carry information to and from a camera, at a fixed location. User free to wander about with a small portable TV for video–preview (current or past buffers).
- CAmera Direction And Ranging (CADAR)
  - based on the principles of radar: camera operates in *scanning* or *staring* mode
  - CADAR positions camera in arbitrary direction (pan and tilt), and zooms in or out as desired
  - CADAR electro–magnetically positions filters and vignettes as desired
- The Computer Enhanced Multiple Exposure Numerical Technique (CEMENT)
  - allows multiple images to be *cemented* together in a common perspective
  - images must be taken from the same vantage point (usually via *CADAR*)
  - one image is generally taken at a wide–angle setting with bright lights (or in the daytime), and the rest of the images are acquired in the dark (using "registration stakes" if necessary, to mark the location)
  - cement computes 4–$\mathcal{D}$ vector for each image; elements are: [azimuth; elevation; rotation; zoom]; computation takes long time
  - once vector appended to each image, semi–interactive mixing may proceed at a reasonably quick pace

A wide range of related innovations have been developed, in conjunction with dusting. For example, the Sequential Wave Imprinting Machine (SWIM) is an array of lights which is connected to a digitally addressed communications port. As the array is moved about, a small radar unit monitors its speed, and controls the rate of readout of the "swim buffer". Depending on the contents of the buffer, various waves and other shapes may be imprinted into the current image. Swim may even create text, in a manner similar to the operation of a giant dot–matrix printer.

6.1 Limitations

The first dusting system, developed in the 1980's, suffered from the obvious limitations of insufficient memory, and the great computational requirement. The system was far from real–time operation. Images were typically collected on video tape, or by conventional photographic means, and *cemented* together at a later time. Thus the user was not able to view the finished result until much later. It would take many hours to assemble an image, and only very crude "experimental" images were possible. While some commercial work was possible, work of this calibre was only made feasible by use of photographic films as initial acquisition media. The full utility of video was limited by the low resolution of sub–NTSC quality cameras. Currently cameras with much higher resolutions are available.

Furthermore, with the advent of increased disk storage capability and better computing facilities, it is hoped that "dusting" will soon be possible in real–time. The original cement interface was based on a command–line system which would only somewhat interactively allow the user to experiment with different weightings. These weightings could be made more interactive, and perhaps automated to a certain degree.

7 Conclusions

Existing Time–Frequency (TF) methods have been presented, in a different light, to facilitate both a unified perspective, as well as an extension into new territory. The wavelet, with its affinity in the physical domain, has been extended to a class of functions with an affinity in the time–frequency domain. Furthermore, that generalization has been further extended to a complete eight parameter transform space in the 1–D case, and a 24 parameter transform space in the 2–D case.

The full utility of the chirplet transform lies, not in its high dimensionality per–se, but rather in its lower–dimensional subspaces. Many other techniques may be expressed as lower–dimensional manifolds in chirplet space. The obvious examples are the wavelet and STFT. Further examples, such as the Generalized Logon Transform[7] (Gaussian chirplet) exist as chirplet subspaces simply by choice of *mother chirplet*. Other examples, such as the Dispersion Transform[9] (DT) and the Scale–Frequency[7] (SF) space have been useful in the analysis of radar signals. The Bowtie Chirplet Transform[7] (BCT), which has been shown to be optimal for the detection and classification of accelerating radar targets, is a 2–D manifold in the chirplet space (i.e. on a discrete grid, it is simply a matrix of numbers which may be easily displayed or printed).

The 24 parameter 2–D chirplet space is briefly discussed before turning the focus to an eight parameter manifold in that space, namely the *physical perspectives.* Other manifolds in that space, such as the *Fourier perspectives* and the 8–D *shear space* are simply mentioned but not discussed at length.

An application to the Computer Enhanced Multiple Exposure Numerical Technique (CEMENT) is described, where the physical perspectives are related to camera *egomotion*. The camera, like a radar, scans out the scene from a fixed point in space, thus allowing CEMENT to produce a composite image, without the problems associated with translation of the camera's optical center. Another subspace, the *physical–affinities*, is effective only where a telephoto lens is used (orthographic case).

Dusting, a new artistic medium, as well as a new imaging paradigm, is also described.

Acknowledgements

The author would like to thank Simon Haykin of McMaster University for having been an excellent M.Eng thesis advisor, and for his encouragement, enthusiasm, and advice in the field of radar. Thanks also to Kent Nickerson, of McMaster University, who helped with the design and building of components within some of the early dusting systems. Over the last thirteen years, Kent has always been a good "sounding board", with his great wealth of knowledge. Thanks to those at McMaster for keeping the computers "alive and well" and tolerating large disk files for seemingly frivolous endeavours outside the scope of the immediate research goals. The author would also like to thank Richard Baraniuk of University of Illinois, and, more recently, Berthold Horn, Rosalind Picard, Walter Bender, and Shawn Becker, of Massachusetts Institute of Technology, for many stimulating discussions. These people will no doubt keep the author's interests alive in years to come.

References

[1] Stephane G. Mallat. A theory for multiresolution signal decomposition: The wavelet representation. *IEEE Transactions on Pattern Analysis and Machine Intelligence*, 11, July 1989.

[2] Ingrid Daubechies. *The Wavelet Transform: a method for time-frequency localization, Simon Haykin, Editor*. Prentice–Hall, 1991.

[3] D. Gabor. Theory of communication. *J. Inst. Elec. Eng.*, Vol.93(Part III):429–457, 1946.

[4] Steve Mann and Simon Haykin. Chirplets and Warblets: Novel Time–Frequency Representations. *Electronics Letters*, 28(2), January 1992.

[5] Steve Mann and Simon Haykin. Time–frequency perspectives: The chirplet transform. *IEEE ICASSP–92*, March 23–26 1992.

[6] Steve Mann and Simon Haykin. An Adaptive Wavelet Like Transform. *SPIE, 36th Annual International Symposium on Optical and Optoelectronic Applied Science and Engineering*, 21-26 July 1991.

[7] Steve Mann and Simon Haykin. The Generalized Logon Transform (GLT) . *Vision Interface '91*, June 3-7 1991.

[8] Ingrid Daubechies. Orthonormal bases of compactly supported wavelets. *Communications on Pure and Applied Mathematics*, XLI:909–996, 1988.

[9] Richard Baraniuk and Douglas Jones. New dimensions in wavelet analysis. *IEEE ICASSP–92*, March 23–26 1992.

[10] D Mihovilovic and R.N. Bracewell. Adaptive chirplet representation of signals on time–frequency plane. *Electronics Letters*, 27(13):1159–1161, 1991.

[11] Michael Barnsley. *Fractals Everywhere.* Academic Press, Inc., 1988.

[12] D. Slepian and H.O. Pollack. Prolate spheroidal wave functions,Fourier analysis and uncertainty- I. *Bell System Technical Journal*, 40:43–64, January 1961.

[13] H.J. Landau and H.O. Pollack. Prolate spheroidal wave functions,Fourier analysis and uncertainty, II. *Bell System Technical Journal*, 40:65–84, January 1961.

[14] D. Slepian and H.O. Pollack. Prolate spheroidal wave functions,Fourier analysis and uncertainty, III: The dimension of essentially time-and band-limited signals. *Bell System Technical Journal*, 41:1295–1336, July 1962.

[15] D. Slepian. Prolate spheroidal wave functions,Fourier analysis and uncertainty, IV:Extensions to many dimensions;generalized prolate spheroidal functions. *Bell System Technical Journal*, 43:3009–3058, November 1964.

[16] D. Slepian. Prolate spheroidal wave functions,Fourier analysis and uncertainty, V:The discrete case. *Bell System Technical Journal*, 57:1371–1430, May-June 1978.

[17] D.J. Thomson. Spectrum estimation and harmonic analysis. *Proc. IEEE*, 70(9):1055–1096, September 1982.

[18] Steve Mann; Supervisor, Simon Haykin. Chirplets and warblets, novel time–frequency methods with applications in Doppler radar. Master's thesis, McMaster University, Communications Research Laboratory, August 1991.

[19] Berthold K.P. Horn and E.J. Weldon JR. Direct Methods for Recovering Motion. *International Journal of Computer Vision*, 1988.

[20] Merrill Skolnik. *Introduction to Radar Systems* . McGraw Hill, second edition, 1980.

Electromagnetic Models for Perceptual Grouping

Thierry Pun
AI and Vision Group
Computing Science Center
University of Geneva
12, rue du Lac, CH-1207 Geneva Switzerland
Tel: +41(22) 787 65 82, e-mail: pun@cui.unige.ch

Abstract

The use of electromagnetic analogies for perceptual grouping of image primitives is presented. This approach assumes that grouping is a low-level, data-driven and global process, with all image tokens interacting in some way.

Two computational models are introduced, which allow determination of proximity and directionality of image primitives. With the second model, prominent image features are considered as being electrical charges. According to Poisson's equation, they generate a scalar potential and an associated "electrical" vector field. The potential and field, determined by the combined global influence of all image features, are well defined over the entire image. The scalar potential can be used for proximity grouping, while the local direction of the field allows grouping of primitives according to their common directional tendency.

Implementation problems and solutions are presented. Various results are shown and discussed.

1 Introduction

Low-level vision aims at extracting primitives as significant as possible out of an observed scene, using mostly or only bottom-up mechanisms. Following [Mar82] [Zuc83], "grouping" is used as a generic term for a set of processes that construct more abstract entities out of more primitive ones. Such processes are crucial in analyzing visual inputs; they start at an early stage and occur recursively at different levels.

Concepts from electromagnetic theory, namely electrical potential and field, allow proximity and directional grouping respectively. The former type of grouping links tokens that are close to each other, while the latter links tokens according to a measure of their common directional tendency.

Section 1 introduces low-level perceptual grouping, proximity and directionality, as well as discusses the interest of determining directionality. Section 2 first states some basic motivations for the proposed electromagnetic approach. They lead to two simple computational models; implementation problems and solutions are detailed for one of them. Section 3 presents and discusses results using abstract images.

1.1 Low-level perceptual grouping

Even when looking at images composed only of very simple 0D or 1D tokens (dots, segments), there is clear evidence that the human visual system accomplishes more than simply contrast detection. Global groupings that are perceived in such circumstances have been extensively reported. Psychologists from the Gestalt school are famous in this respect [Wer23] [Pom77]; they theorize on the spontaneous appearance of a "good shape" from seemingly unconnected tokens. This is often reported as "emergent features effect", "configural superiority", or "the whole is more than the sum of its parts" (holism). This emergent percept hypothetically relies on the well known Gestalt "laws", such as proximity, continuity, similarity, common direction or common fate, closure, etc.

Although the Gestaltists show convincing evidence of the phenomenon, little is said about its nature. Attempts at modeling the reciprocal influence of patterns by a diffusion process [Koh20], or brain potentials and DC currents [Koh44] did not go very far. Neurophysiological research has made the hypothesis of direct current flows highly unlikely [Bec82], although magnetic fields, oscillatory mechanisms, latencies and phase coherence mechanisms are known to exist in the cortex [Kau84] [Sch90] [Bur91a] [Bur91b]. Nevertheless, such Gestalt physics and the quest for physical properties yielding order and regularity have influenced many computer vision researchers.

Grouping is a form of perceptual organization. Studies aiming at explaining perceptual organization have been conducted from various points of view. The information theoretic approach [Att54] [Lee71] [Res82] considers good figures as those having a rather regular, predictable shape, hence minimizing some form of information measure. The transformational invariance approach [Pal83] defines these good figures as being the most stable, i.e., the most invariant to geometrical transformations. Grouping occurs through parallel evaluation of these invariances over many stimuli: grouped primitives are those related by little or no transformation. Another approach relying on geometric groups properties puts emphasis on how local processes in the visual system work cooperatively to produce global phenomena of perception [Hof85]; the Lie transformation group is used as a model [Dod83]. All these approaches offer interesting and relevant notions; not much however is given in computational terms.

Different researchers have dealt with the extraction of regional or global characteristics out of images composed of distributed, local tokens; noted examples are detailed in [Gla69] [Ste78] [Kas85]. Often however these characteristics are global properties, such as orientation, rather than explicit groupings. There, grouping is not seen as the fundamental problem, but rather as a consequence, deriving from the results.

Other approaches are more directly related to perceptual grouping. In [Low82], statistical distributions of features are computed, such as local orientation; non-parametric tests yield significant image relations. Texture segmentation, where emergent features play an important role, has also been related to grouping problems [Jul83] [Bec83]. Grossberg et al (e.g. [Gro88]) propose an approach based on a cooperative/competitive hierarchy of receptors; all their models have a strong physiological flavor. Sha'ashua and Ullman [Sha90] use a relaxation network to extract global salient features from very noisy figures; their iterative scheme allocates more resources to structurally salient tokens, finally producing figure-ground separation.

In [Zuc83], Type I and Type II groupings are defined. In essence, Type I processes are one-dimensional and have well-defined spatial supports [Zuc85]. Type II groupings operate on Type II processes, that is dense families of implicit contours. The present approach combines characteristics from both Type I and II groupings. In particular, although the percept is essentially two-dimensional, grouping is characterized by high positional and orientational change specificity.

In summary, physiology offers no general explanation to the mechanism of perceptual grouping. Neither does psychology, where there still are two competing schools, namely perception (bottom-up) and cognition (top-down). Computationally speaking, there is no general model available; most of the current approaches are bottom-up and often of a very ad-hoc nature. The present model, which does not particularly try to model the human visual system, puts the emphasis on how perceptual grouping can be computationally modeled in a simple, coherent and effective manner.

1.2 Proximity and directionality

In the process of grouping tokens, spatial *proximity* plays an important role. The first tokens to be grouped, with or without directional information, are the closest ones. Amongst the Gestalt principles [Wer23], proximity is certainly the first to operate. It is obvious in any grouping, at any level of complexity. It is an underlying paradigm for both cooperative and competitive mechanisms, since both deal with tokens that are close together.

Directionality refers to the global feeling of an underlying direction that some patterns give (Fig. 1): circular Moiré patterns [Gla69], hairs [Zuc83], dot patterns, etc. Although fairly global, the sensation is rather resistant to a reduction of the field of view (Fig. 2). In addition, it is robust to perturbations.

Patterns such as those shown in Fig.1 somehow give the impression of a "flow"

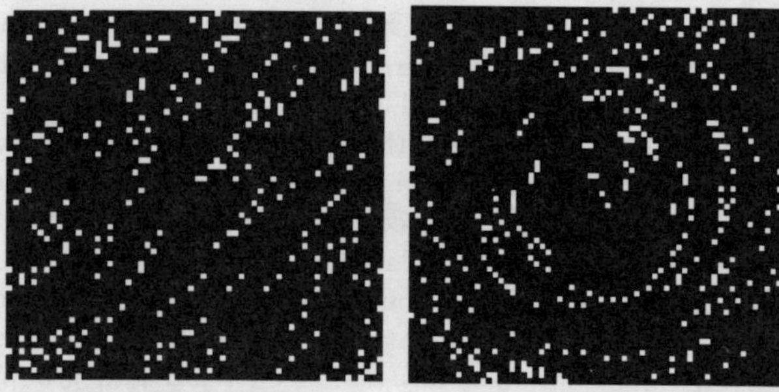

Figure 1: Dot patterns with strong directionality sensation; a) direction at approximately 45°, 60x60 pixels; b) Moiré pattern, 64x64 pixels, composed by superimposing 150 randomly distributed dots and a 4° rotated copy of them.

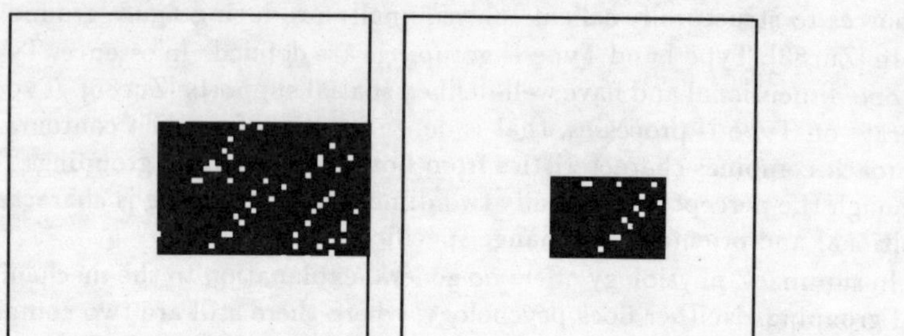

Figure 2: Empirical illustration of the resistance to the reduction of field of view.

of particles, linear for Fig.1.a and circular for Fig.1.b. This analogy with a flow is however misleading, since in the present case everything is static. The matter of interest here is therefore not the extraction of the velocity vector associated with each token, but the determination of the underlying direction(s) that emerges upon examination of the image.

Directionality grouping is meant as the process of joining tokens in order to compose groups that lie along the *perceived* emergent direction. The *supports of the local direction vectors* are defined as the line segments linking pairs of grouped tokens. Despite the globality of the overall percept, this grouping process has a well defined local behavior: groups of tokens presenting small orientational differences with the overall direction are easily perceived.

Proximity as well as directionality can be simply and efficiently computed using the proposed electromagnetic framework: whereas the electrical potential provides means of inferring proximity relationships, the electrical field allows determination of directional information.

The assumptions underlying the computational model presented here are that:

- grouping is a low-level, data-driven mechanism;
- grouping is a global process, where all image tokens interact in some way; the closer they are, the more they do so. Globality yields a relative insensitivity to most small perturbations; this is in accordance with the very qualitative nature of the human visual system;
- the global measures on which the grouping process is based are well defined locally over the entire discrete image. Possibility of a local definition is necessary, since certain minute pattern variations have very significant effects.

1.3 Why determining directionality?

Directionality provides means of defining a *functional grouping*, which links together tokens having similar "behavior"; this is reminiscent of the Gestalt "Law of common destiny" [Wer23]. After linkage, patterns thus constituted can be fed to a higher-level process of a computer vision system. Furthermore, tokens which show a different behavior are located. These are of importance in a vision system, for example by offering a means of directing a focus of attention mechanism [Mil90] [Pun90] [Mil91].

Inferring directional information is useful in itself, since it provides a valuable characterization of image properties. In particular, the present approach yields the supports of the local direction vectors.

The directional grouping approach cannot deal with effects such as similarity, or symmetry. It seems plausible that *hierarchical grouping* will have to be used in order to solve these issues (if it is postulated that they are the result of bottom-up processes). The first two levels could be based upon proximity and directionality. Similarity, then symmetry, could be computed by more involved grouping stages, on the basis of techniques such as transformational measures [Pal83] [Ead88] [Yue90].

2 Electromagnetic models

2.1 Motivations

The primary motivation for the electromagnetic approach is related with the globality of perceptual grouping. In order to capture such property, global modeling of interactions between all elements in a given image region is necessary. This is precisely what is accomplished by a potential and its associated vector field: each token is subject to the combined influence of all others. There is no need to define size or orientation of some operators; the effect of each element is implicitly taken into account.

Another motivation is that although potential and field have global influences, they are well defined locally. This allows integration of local, proximity based processes, to coherent global operations. As argued in Section 1.2 both of these aspects, global and local, are necessary for directional grouping. They are inherent to the present approach.

The type of solutions for potential and field that the present model yields (e.g. potential surface and field lines) verify a minimum energy principle. This is in accordance with the soap bubble paradigm [Att82] [Wit83], which calls for perception of as much regularity as data allows. Minimum properties are of interest because grouping certainly relies on regularity.

The electromagnetic approach is not a visual process theory, in the sense that there is no implication that the human visual system follows the same principles (see however [Kau84] [Sch90] on the role of electromagnetic fields in the cortex). It provides nevertheless a coherent and elegant formalism to model how local processes work cooperatively to produce global phenomena of perception.

Section 2.2 presents two possible electromagnetic analogies for modeling image and tokens: the Laplace model and the Poisson model (some basic notions of electromagnetism are recalled in Appendix).The Poisson model is then used in Section 2.3 for proximity grouping, and in Section 2.4 for directionality determination and grouping.

2.2 Two possible models

Each important image feature, called below a *generating feature*, generates an electrical field and consequently a scalar potential. Two electromagnetic analogues for these generating features can be considered: either electrical conductors at a given potential (Laplace model, [Pun89]), or space charges (Poisson model, [Pun91]). These two analogies are described below, with more emphasis on the second model which is the one used in Sections 2.3 and 2.4

2.2.1 Laplace model: image features as electrical conductors

This section describes the first analogy, namely the Laplace model. Some key features are first selected in the original image. In the case of an abstract dots image (Fig. 1), these "strong tokens" are the dots themselves. More generally, they are image elements at locations with high gradient (contours). Each of these token has an initial potential value V, function of its luminance l :

$$V = l, \ \ or \ \ V = \log(l) \tag{1}$$

The image boundary is set at potential $V = 0$. All these initial conditions being established, it is possible to determine the potential V at all points. In electrical terms, the image can be seen as a plane intersecting electrical conductors being at various potentials; each intersection is one token (Fig. 3). In this model, no space charge is assumed.

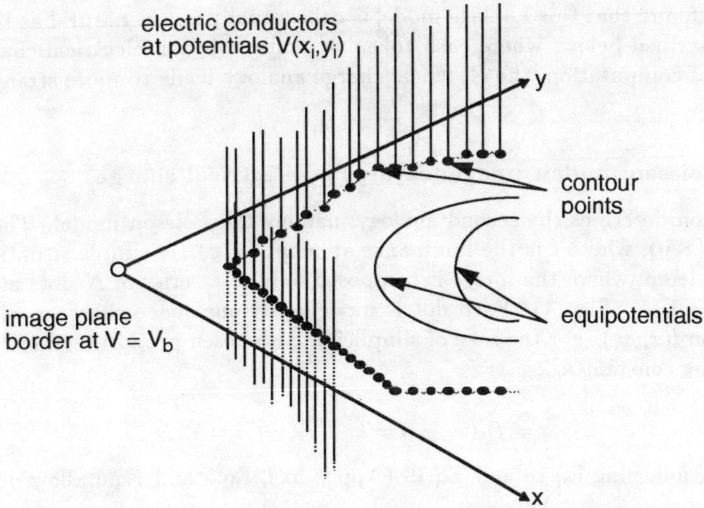

Figure 3: Image features modeled as electrical conductors. The scalar potential they produce defines equipotential curves.

The potential $V(x, y)$ at each point (x, y) must satisfy Laplace's equation:

$$\Delta V = 0 \tag{2}$$

Laplace's equation yields the smoothest interpolation surface for a set of initial conditions. V values are computed using discrete relaxation, which can be seen as propagation of known image values to an increasingly global neighborhood. A simple implementation of the relaxation using a top-down, left to right scanning of the image, is given by:

$$V^{i+1}(x,y) = \frac{V^{i+1}(x-1,y) + V^{i+1}(x,y-1) + V^{i}(x+1,y) + V^{i}(x,y+1)}{4} \tag{3}$$

with the iterations stopping when $|V^{i+1} - V^{i}| < \varepsilon$ for all (x, y) . To speed-up relaxation, V values at the unknown points may initially be set to the actual image luminances, rather than to 0.

Equipotential curves, which are necessarily closed, can then be determined as those curves that link locations with a given potential $V = V_{eqp}$.

The potential values at the strong tokens, initially equal to image luminances, are *not modified* by the procedure: they are fixed values (all other values are modified by the relaxation). This produces a result that is different from a simple low-pass filtering where all values vary at each step.

It was found that this Laplace model is conceptually not as natural as the Poisson model described below, where each token is modeled by an electrical charge. Also, in terms of computation, the electrical charge analogy leads to more straightforward algorithms.

2.2.2 Poisson model: image features as electrical charges

This section describes the second analogy, namely the Poisson model. The image is denoted $l(x, y)$, where l is the luminance at point (x, y). A simple situation can be first considered, where the image is composed only of a series of N dots at positions $(x_i, y_i), i = 1..N$ (Fig. 4). Each dot is modeled by one space charge q whose value depends on $l(x_i, y_i)$. For the sake of simplicity, q is chosen proportional to l (with the normalizing constant k):

$$q = f[l(x_i, y_i)] = k \cdot l(x_i, y_i) \quad i = 1..N \tag{4}$$

Hence, following Eq.18 and Eq.19 (Appendix), Eq.5 and Eq.6 allow direct computation of the potential $V(x, y)$ and the electrical field $\vec{E}\ (x, y)$ at each point:

$$V(\vec{r}) = k \cdot \sum_i \frac{l(x_i, y_i)}{\left|\vec{r} - \vec{r}_i\right|} = k \cdot \sum_i \frac{l(x_i, y_i)}{\sqrt{(x - x_i)^2 + (y - y_i)^2}} \tag{5}$$

and

$$\vec{E}\ (\vec{r}) = k \cdot \sum_i l(x_i, y_i) \frac{\vec{r} - \vec{r}_i}{|\vec{r} - \vec{r}_i|^3} \tag{6}$$

where $\vec{r}$, $\vec{r}_i$ and $\left|\vec{r} - \vec{r}_i\right|$ respectively correspond to the position (x, y) of the point where $V(x, y)$ and $\vec{E}\ (x, y)$ are to be determined, the position of charge i , and the distance between the point and the charge i, all in a common absolute coordinate system.

It appears from Eq.6 that the electrical field is not defined on a charge. As described below however (§2.4), it is necessary to be able to define a pseudo-field $\vec{E}$ at the token locations. This pseudo-field is determined by adding the contributions of every charge but the one positioned at the token location:

$$\vec{E}\ (\vec{r}_j) = k \cdot \sum_{i \neq j} l(x_i, y_i) \frac{\vec{r} - \vec{r}_i}{\left|\vec{r} - \vec{r}_i\right|^3} \tag{7}$$

Using Eq.7 rather than Eq.6 combines both global (sum $\sum$) and localized ($i \neq j$) behaviors. In this way, $\vec{E}$ is well defined over the entire discrete image (Fig. 5).

By generalizing the somehow artificial case of dot patterns, more complex grey-level images could possibly be treated. The following problem however has to be dealt

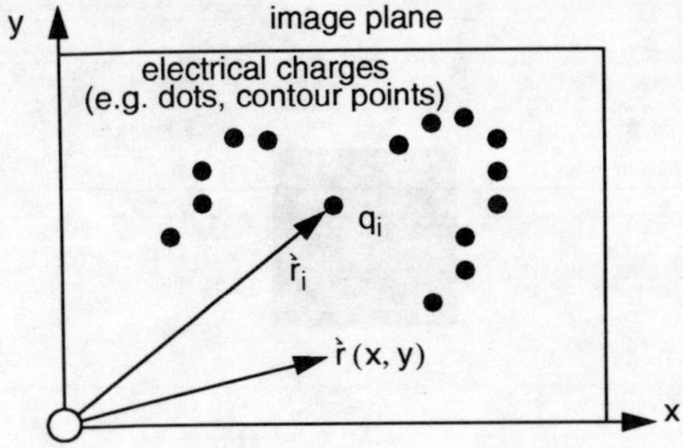

Figure 4: Image features modeled as electrical charges. They produce a scalar potential as well as an electrical vector field in the image plane.

with: with the present synthetic images, dots (tokens) are assimilated to charges and generate a Laplacian field; this field is assumed to carry the underlying structural information contained in the image. The trouble with real images is that there is no guarantee that the grey-levels carrying the pictorial information will follow a smoothly varying function, more precisely a function with zero Laplacian (Laplacian field). How then could the present model be used, since it makes the assumption that underlying features satisfy Laplace's equation?

What needs to be done is to assume that Laplace's equation is locally satisfied, i.e. that the image is locally smooth. The generating features should then be positioned at locations where the image changes from one smooth patch to another; they would therefore correspond to points on object contours. Their associated charge could be the luminance of the *original* image, at the contour location. Some experiments along this line have been presented in [Pun89], although making use only of the potential V (proximity grouping).

In summary, generating features such as points or contours are represented by dots. Each such dot generates a field; the global solution is obtained by superimposition of all individual effects.

2.3 Poisson model: proximity grouping

Proximity grouping is the process which links tokens which are close together. The simplest way of achieving this is by using as a grouping criterion the Euclidian distance separating the tokens. An alternate approach, presented in this section, is to make use of the electrical potential determined using either the Laplace model or the Poisson model. General principles are recalled here; more details can be found in [Pun89].

The potential $V(x, y)$ needs first to be determined over the entire image (Eq.5). Then, for each constant potential value V_{eqp} in the dynamic range $V_{min} \leq V_{eqp} \leq V_{max}$, a)

b)

c)

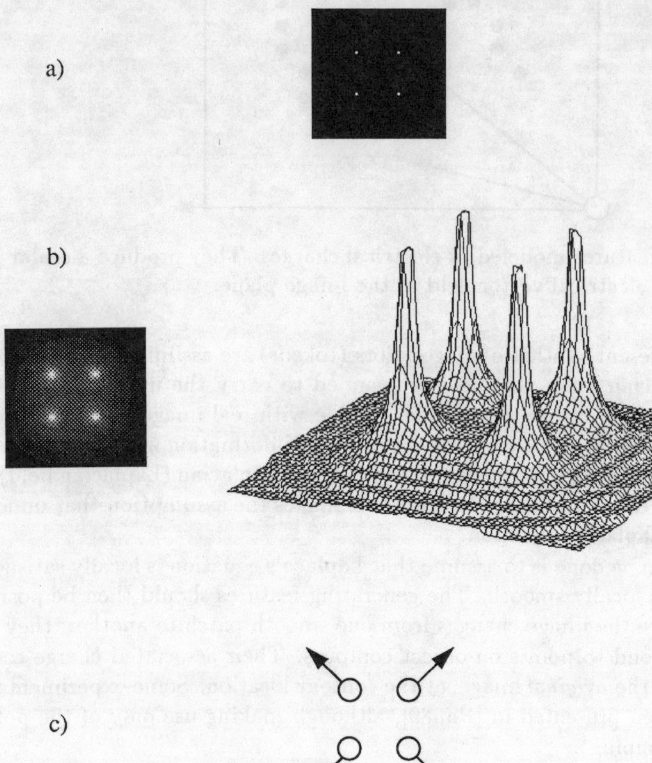

Figure 5: Example of determination of potential and field; a) 4 dots, i.e. 4 charges; b) potential V, in grey-levels and in perspective; c) field $\vec{E}$.

there exists at least one equipotential curve defined by:

$$V(x, y) = V_{eqp} \tag{8}$$

The equipotentials are closed; depending on V_{eqp} , they surround more or less image elements. These surrounded image tokens are those that are grouped together.

In addition, links between the grouped regions could be defined using field lines. Field lines are curves that are in each point tangent to $\vec{E}$: they are therefore orthogonal to equipotentials. In the cartesian referential (x, y) , each element (dx, dy) of a field line verifies:

$$\frac{dx}{E_x} = \frac{dy}{E_y} \tag{9}$$

Since field lines are normal to equipotentials, they can be seen as links between the tokens (generating features), as well as between the equipotentials. They could therefore be used to indicate which group of primitives is to be related to which other.

2.4 Poisson model: directional grouping

2.4.1 Supports of the direction vectors and directionality grouping

Directionality grouping is the process of joining tokens in order to compose groups lying along the emergent direction. The line segments linking pairs of grouped tokens will define the supports of the local direction vectors.

In computational terms, directional grouping links together those segments having the *most similar* electrical fields $\vec{E}$ in direction and orientation. As an example, Fig.6 shows 4 charges with $\vec{E}$ computed using Eq.7. Such 4-point patterns appear at various locations in the test images and are representative of a frequently occurring situation. $\vec{E_1}$ and $\vec{E_2}$ are the most similar, therefore tokens q_1 and q_2, rather than q_1 and q_3 (nearest neighbors) will be grouped. In the same manner, $\vec{E_3}$ and $\vec{E_4}$ are similar; q_3 and q_4 are therefore grouped. Finally, the supports of the local direction vectors are the lines q_1q_2 and q_3q_4.
Fig.6 illustrates two important points:

- directionality grouping is different from proximity grouping: the latter would join $q1$ with $q3$, $q2$ with $q4$;
- the supports of the local direction vectors are not colinear with the direction of the local field $\vec{E}$. The $\vec{E}$ field direction corresponds to the gradient of the local potential function; this is not the case with these local direction vectors.

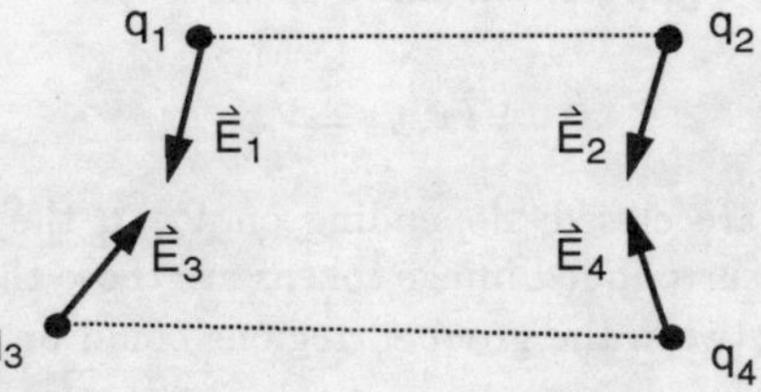

Figure 6: 4 image tokens q_1, q_2, q_3, q_4; electrical fields $\vec{E_1}$, $\vec{E_2}$, $\vec{E_3}$, $\vec{E_4}$ (Eq.7). The dotted lines are the supports of the local direction vectors.

2.4.2 Scale and similarity

Pairs of tokens that will be grouped are those for which the $\vec{E}$ field directions are the most similar. The measure of similarity s between two charges is defined as the normalized scalar product of the field directions:

$$s(q_i, q_j) = \frac{\vec{E_i} \cdot \vec{E_j}}{\left\| \vec{E_i} \cdot \vec{E_j} \right\|} = \cos(\vec{E_i}, \vec{E_j}) \tag{10}$$

The similarity varies between -1 and +1. A perfect match is indicated by +1; opposite and orthogonal directions are respectively -1 and 0.

The *scale of observation* also needs to be selected; this is actually the only parameter that is required. Tokens q_i and q_j are grouped only if they are closer than a given distance d_{max}:

$$\left\| \vec{r}\ (q_i) - \vec{r}\ (q_j) \right\| \le d_{max} \tag{11}$$

3 Implementation, results and discussion

Dot patterns have been chosen as test images in order to circumvent other issues, such as contours finding. The methodological choice of using abstract images has already been extensively addressed (e.g. [Zuc83] [Ull84]). Section 3.1 below presents implementation and details results. Section 3.2 discusses important features of the approach.

3.1 Implementation and results

Two of the images used for the experiments are shown in Fig.1. In Fig.1.a, which is similar to Fig.8 in [Zuc83], there is an overall direction at approximately 45 degrees. Fig.1.b, inspired by [Gla69] or Fig.3 in [Zuc83], consists of the superimposition of a

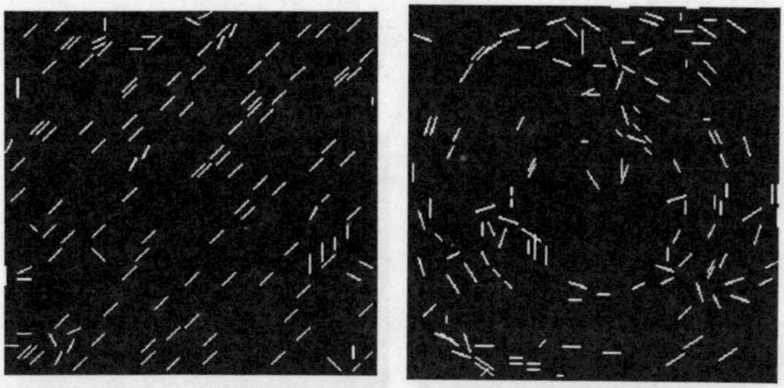

Figure 7: Directional grouping: a) on Fig.1.a (d_{max} = 3): average and mode of the distribution are 45 degree (84% have an orientation of 45° ± 18°); b) on Fig.1.b (d_{max} = 4). Line segments indicate token pairs as well as supports of local direction vectors. These vectors correspond well with the underlying global direction.

random field comprising approximately 150 dots with a rotated version of itself (angle 4 degrees). The Moiré figure thus constituted gives a strong circular feeling, that is to say that the local direction vectors lie along the tangents to the circles.

Values of tokens are arbitrarily set at 255, the background being 0. $\vec{E}$ at each location is computed from Eq.7. Tokens groups, that define the supports of the direction vectors as well, are built by pairing dots that have the most similar $\vec{E}$ directions (Eq.10) and are closer than a fixed d_{max} (Eq.11).

Fig.7 presents the result of directional grouping on Fig.1 patterns. Situations similar to the one schematically depicted in Fig.6 appear at various places: dots that are linked together are not necessarily the closest ones. For the sake of comparison, Fig.8 shows the result of grouping dots on the basis of proximity, i.e. by linking nearest neighbors as long as their distance does not exceed a limit d_{max}. Results are clearly different. In Fig.8, as opposed to Fig.7, the line segments do not match with the perceived directional sensation.

Fig.7.a and Fig.8.a have been quantitatively compared by analyzing the distribution of line segments orientations (Fig.9). For Fig.7.a, the average and mode of this distribution are at 45 degrees; 73% of the segments have the "correct" 45° angle, 84% have an orientation of 45° ± 18° . This corresponds very well with the overall direction sensation. By contrast, for Fig.8.a, there are 3 almost equal modes at 0°, 45° and 90°; only 30% of the angles are 45°. The latter result is much less in accordance with the overall direction.

The difference between Fig.7.b and Fig.8.b is more difficult to quantify. Qualitative differences are however easily perceived, as in the lower-left part of the image, or more generally on the outside of the circular Moiré. Directional grouping yields segments consistently tangential to the "circles", therefore well in accordance with the perceived circular pattern.

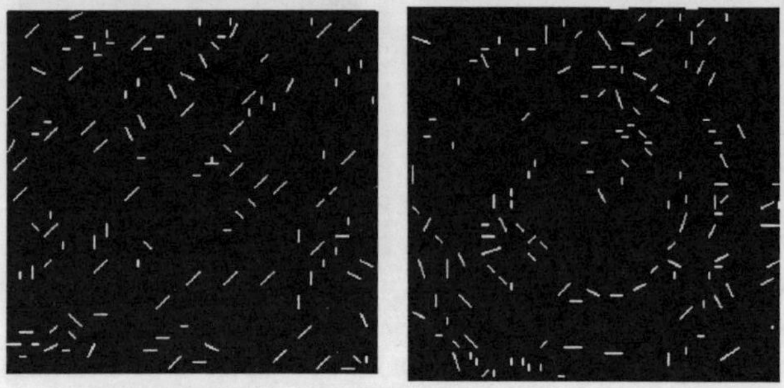

Figure 8: Proximity grouping: a) on Fig.1.a ($d_{max} = 3$): 30% of the segments have the 45° angle; b) on Fig.1.b ($d_{max} = 4$). Values of d_{max} are the same as those used in Fig.7. Here, the local direction vectors do not correspond with the underlying global direction.

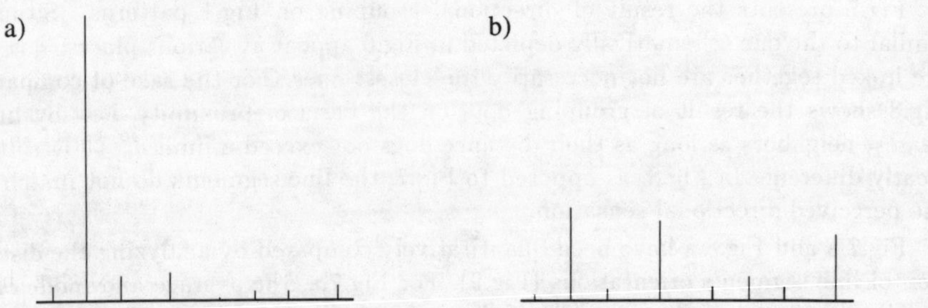

Figure 9: Distributions of the line segment orientations; a) for Fig.7a; b) for Fig. 8a.

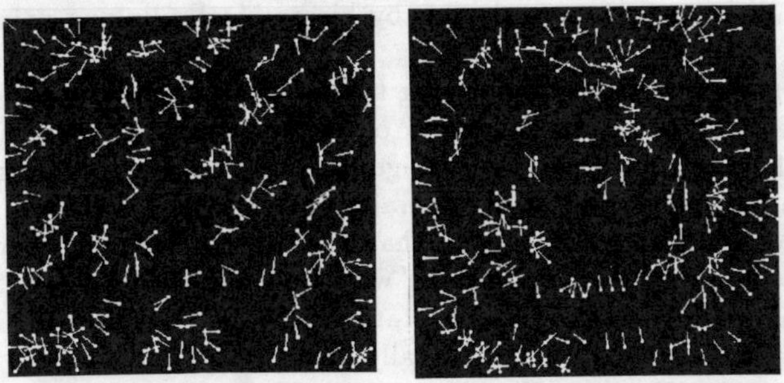

Figure 10: Local $\vec{E}$ directions for Fig.7.a and Fig.7.b (normalized vectors).

Fig.10 is a drawing of the normalized vectors $\vec{E}$ at each dot location. This figure clearly illustrates the fact that the electrical field vectors are not parallel to the local direction vectors. They are not orthogonal either: there is no obvious systematic relationship between them. What is achieved is functional grouping, that is pairing tokens that share similar "behaviors".

3.2 Discussion

Results obtained with directionality grouping (Fig.7) markedly differ from those obtained with proximity grouping (Fig.8). Directional grouping provides results qualitatively similar to the global perception; this is not the case with proximity grouping. The schematic four tokens pattern of Fig.6 exemplifies the basic reason for the difference; such patterns can be found at various locations in Fig.7 and Fig.8. This difference between proximity and directionality grouping makes them complementary and calls for a two-stage grouping model where advantages of each mechanism could be jointly used. This aspect is currently being investigated.

The approach needs no tuning or parametrization, apart from the choice of d_{max}. Various means for automatically determining an appropriate d_{max} value are being experimented. One possibility is to select the first possible grouping, as being the one which groups more than $x\%$ of all tokens. From a more general perspective though, the problem of selecting d_{max} is reminiscent of the pervasive scale determination problem. This question arises at every level in a vision system, iconic as well as symbolic. It is therefore possible that an appropriate observation scale could be determined by an entirely different approach, for example based on measures in scale space [Koe84] [Bru90].

Grouping is here limited to linking tokens by pairs, although the similarity measure (Eq.10) could be used as well to group n-tuples of dots. Groups of $n > 2$ dots would however not allow definition of the local direction vectors, whose supports are pairs of tokens. Grouping such pairs into more complex patterns is certainly of interest, but it seems more appropriate to do this by hierarchically linking pairs two by two rather than by joining them all at once.

The line segments obtained by pairing tokens do not define direction vectors in the strict sense, but rather the supports of these vectors: there is a 180 degrees ambiguity in their orientation. This ambiguity simply reflects the ambiguity of the pattern. When looking at Fig.1.a, the overall direction can equally well be perceived as going towards the upper-right corner than towards the lower-left one. The Moiré of Fig.1.b can be perceived clockwise as well as counter-clockwise. Such type of ambiguity is inherent to images that are static. Patterns varying with time would be needed to raise it; they would unequivocally define an orientation, such as towards the upper-left with a 45° angle. The time varying Maxwell's equations could possibly be used to help determine the optical flow vectors.

It can be noticed that Poisson's equation is a particular case of the heat equation [Koe84]. There is however an important difference. In the present electromagnetic model, initial conditions remain constant: they are fixed (given) space charges. In terms of heat transfer, the present approach corresponds to a situation with permanent, constant heat sources in the medium [Léo85].

Finally, another characteristic of the present approach that deserves notice is that it is invariant under monotonous transformations of interpoint distances. If images are scaled by a factor α, and d_{max} also is scaled to $\alpha \cdot d_{max}$, results will be the same.

4 Conclusion

A robust, deterministic computational model for directional grouping has been presented. This model integrates local processes into coherent, global operations. Although extremely simple in concept as well as in implementation, it appears as an elegant and powerful means for:

- determining the supports of the local direction vectors;
- hierarchically linking tokens that lie along the perceived global direction.

This functional grouping relies on directionality of patterns, and is markedly different from what a nearest-neighbor based grouping would provide. The differing characteristics of proximity and directionality grouping therefore pave the way towards a two-stage model combining their respective properties.

Acknowledgements

The support of the Swiss National Research Foundation (FNRS) through grants 20-26475.89 and 4023-027036 is gratefully acknowledged. The author thanks Pierre-Yves Burgi, Juan Mosig and Philippe Narbel for many stimulating discussions and comments, as well as Michel Regamey for having written part of the software.

Appendix: Fundamentals of Maxwell's Equations

A1. Electrostatic and magnetostatic case

The present electromagnetic model assumes no temporal variation: images are static, stimuli are constant with respect to time. Hence the electrostatic differential Maxwell's equations [Ram84]:

$$\nabla\times \vec{E}= 0 \qquad \nabla\cdot \vec{D}= \rho \tag{12}$$

where $\vec{E}$ is the electrical field, $\vec{D}$ the electrical displacement and ρ the volumetric charge density. In a linear medium, $\vec{E}$ and $\vec{D}$ are linearly related by the medium permittivity ε:

$$\vec{D}= \varepsilon\, \vec{E} \tag{13}$$

Since the curl of a gradient is identically zero, Eq.12 implies that $\vec{E}$ is the gradient of some scalar field. This defines the electrical potential V:

$$\vec{E}= -\nabla V \tag{14}$$

A2. Electrostatic with charges: Poisson's equation

In the presence of space charges, Eq.12.b, Eq.13 and Eq.14 yield Poisson's equation:

$$\nabla^2 V = \Delta V = -\rho/\varepsilon \tag{15}$$

In the case of one point charge q (Fig.11), $\vec{E}$ has only a radial component E_r, where r is the distance to the charge:

$$E_r = q/(4\pi\varepsilon r^2) \tag{16}$$

The potential V is obtained by integration:

$$V = q/(4\pi\varepsilon r) \quad (+constant) \tag{17}$$

This result can be extended to a set of i point charges q_i, $i = 1..N$:

$$V(\vec{r}) = \frac{1}{4\pi\varepsilon}\sum_i \frac{q_i}{\left|\vec{r} - \vec{r_i}\right|} \tag{18}$$

where $\vec{r}$, $\vec{r_i}$ and $\left|\vec{r} - \vec{r_i}\right|$ respectively denote observer's position, charge i's position and the distance between observer and charge i, all in a common absolute coordinate system.

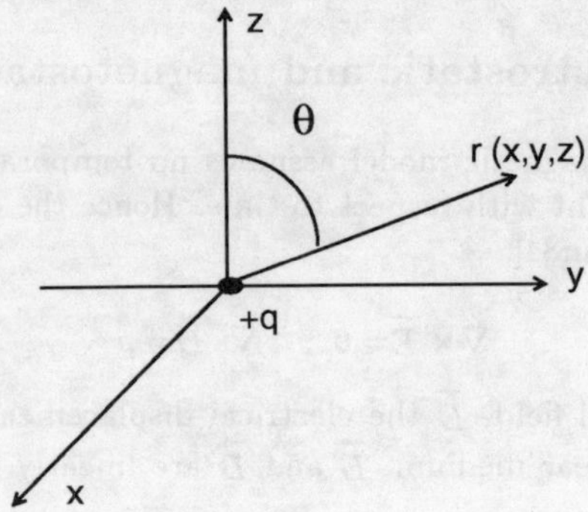

Figure 11: One electrical charge, with the 3D coordinate system centered on this charge.

$\vec{E}$ is obtained by expressing Eq.14 in cylindrical coordinates:

$$\vec{E} = -\nabla_{\vec{r}} V = \frac{1}{4\pi\varepsilon} \sum_i q_i \frac{(\vec{r} - \vec{r}_i)}{\left|\vec{r} - \vec{r}_i\right|^3} \tag{19}$$

V and $\vec{E}$ have singularities on the charges; this corresponds to the fact that $\vec{E}$ arrives radially on each on them. $V(x, y)$ and $\vec{E}\ (x, y)$ are directly computed by adding individual contributions (Eq.18, Eq.19). This approach is simpler than the one presented in [Pun89], where a relaxation mechanism is used to solve Laplace's equation $\Delta V = 0$ in order to obtain potential and field.

References

[Att54] F. Attneave, "Some informational aspects of visual perception", Psychological Review, 61, 1954, 183-193.

[Att82] F. Attneave, "Prägnanz and soap-bubble systems: a theoretical exploration", in: Organization And Representation In Perception, J. Beck, Ed., Lawrence Erlbaum Ass. Publ., 1982, 11-29.

[Bec82] J. Beck, Introduction of: Organization And Representation In Perception, J. Beck, Ed., Lawrence Erlbaum Ass. Publ., 1982, 1-10.

[Bec83] J. Beck, K. Prazdny and A. Rosenfeld, "A theory of textural segmentation", in: Human and Machine Vision, J. Beck, B. Hope and A. Rosenfeld, Eds., Academic Press, 1983, 1-38.

[Bru90] K. Brunnström, J.-O. Eklundh, T. Lindeberg, "On scale and resolution in the analysis of local image structure", in: Proc. First Eur. Conf. Comp. Vision, Antibes, France, 1990, 3-12.

[Bur91a]P.-Y. Burgi and T. Pun, "Temporal analysis of contrast and geometric selectivity in early human visual system", in: Channels In The Visual Nervous System: Neurophysiology, Psychophysics, Models, B. Blum, Ed., Freund, 1991.

[Bur91b] P.-Y. Burgi and T. Pun, "Figure-ground separation: evidences for asynchronous processing in visual perception?", Proc. Neural Networks for Vision and Image Processing (abstract), Boston, USA, May 10-12, 1991. In a different version: Proc. 13th Europ. Conf. on Visual Perception, Vilnius, Lithuania, August 25-29, 1991.

[Dod83] P.C. Dodwell, "The Lie transformation group model of visual perception", Perception and Psychophysics, 34, 1, 1983, 1-16.

[Ead88] P. Eades, "Symmetry finding algorithms", in: Computational Morphology, G.T. Toussaint, Ed., 1988, 41-51.

[Gla69] L. Glass, "Moiré effect from random dots", Nature, 223, Aug. 9, 1969, 578-580.

[Gro88] S. Grossberg, E. Mingolla, M.A. Cohen and coworkers, The Adaptive Brain II: Vision, Speech, Language, and Motor Control, S. Grossberg, Ed., North-Holland, 1988, Chaps. 1 to 4.

[Hof85] W.C. Hoffman and P.C. Dodwell," Geometric psychology generates the visual Gestalt", Canadian J. of Psychology, 39, 4, 1985, 491-528.

[Jul83] B. Julesz and J.R. Bergen, "Textons: the fundamental elements in preattentive vision and the perception of texture", Bell Systems Tech. Journal, 62, 6, Jul.-Aug., 1983, 1619-1644.

[Kas85] M. Kass and A. Witkin, "Analyzing oriented patterns", Proc. IJCAI, Los Angeles, 1985, 944-952.

[Kau84] L. Kaufman, Y.C. Okada, J. Tripp, H. Weinberg, "Evoked neuromagnetic fields", in: Brain and Information: Event-related Potentials, R. Karrer, J. Cohen, P. Tueting, Eds. Also: Annals of the New-York Academy of Sciences, 425, 1984, 722-742.

[Koe84] J.J. Koenderinck, "The structures of images", Biol. Cybernetics, 50, 363-370, 1984.

[Koh20] W. Köhler, "Die physichen Gestalten in Ruhe und im stationären Zustand", Eine naturphilosophische Untersuchung, Erlangen, 1920. Translated as: "Physical Gestalten", in: A Source Book in Gestalt Psychology, W.D. Ellis, Ed.,1938,17-54.

[Koh44] W. Kohler and H. Wallach, "Figural after-effects: an investigation of visual processes", Proc. Americ. Philos. Soc., 88, 1944, 269-357.

[Lee71] E. Leeuwenberg, "A perceptual coding language for visual and auditory patterns", Amer. J. of Psychology, 84, 1971, 307-349.

[Léo85] A. Léontiev, Théorie des Echanges de Chaleur et de Masse (Theory of Heat and Mass Exchanges), Editions Mir, Moscou, 1985.

[Low82] D.G. Lowe and T.O. Binford, "Segmentation and aggregation: an approach to figure-ground phenomena", Proc. DARPA Image Understanding Workshop, Palo-Alto, CA, Sept. 1982, 168-178.

[Mar82] D. Marr, Vision, Freeman and Co., New York, 1982.

[Mil90] R. Milanese, "Applying focus of attention to a computer vision system", AI and Vision Group, Computing Science Center, University of Geneva, Report 90.07, December 1990, 39p.

[Mil91] R. Milanese, J.-M. Bost and T. Pun, "Visual indexing with an attentive system", in: Trends in Artificial Intelligence, E. Ardizzone, S. Gaglio, F.Sorbello, Eds., Springer-Verlag, Lecture Notes in Artificial Intelligence, Vol. 549, 1991, 415–419.

[Pal83] S.E. Palmer, "The psychology of perceptual organization: a transformational approach", in: Human and Machine Vision, J. Beck, B. Hope and A. Rosenfeld, Eds., Academic Press, 1983, 269-339.

[Pom77] J.R. Pomerantz, L.C. Sager and R.J. Stoever, "Perception of wholes and their component parts: some configural superiority effects", Journal of Exp. Psychology: Human Perception and Performance, 3, 3, 1977, 422-435.

[Pun89] T. Pun and P.-Y. Burgi, "Perceptual grouping of image primitives using a potential and field approach", Proc. 6th Scand. Conf. Image Analysis, Oulu, Finland, 1989, 317-324.

[Pun90] T. Pun, "The Geneva Vision System: modules, integration and primal access", AI and Vision Group, Computing Science Center, University of Geneva, Report 90.06, December 12, 1990, 32p.

[Pun91] T. Pun, M. Regamey, P. Narbel and J. Mosig, "Perceptual grouping and directionality determination using an electromagnetic model", Proc. Vision Interface'91, Calgary, Canada, June 3-7, 1991, 81-88.

[Ram84] S. Ramo, J.R. Whinnery and T. van Duzer, Fields and Waves in Communication Electronics, 2nd. Ed., J. Wiley, 1984 (or any undergraduate textbook on electromagnetism).

[Res82] F. Restle, "Coding theory as an integration of Gestalt psychology and information processing theory", in: Organization And Representation In Perception, J. Beck, Ed., Lawrence Erlbaum Ass. Publ., 1982, 31-56.

[Sha90] A. Sha'ashua and S. Ullman, "Grouping contours by iterated pairing network", in: Advances in Neural Information Processing Systems 3 (NIPS 90), R.P. Lippman, J.E. Moody and Touretzky, Eds., 1991.

[Sch90] T.B. Schillen and P. König, "Coherency detection and response segregation by synchronizing and desynchronizing delay connections in a neuronal oscillator model", Proc. IJCNN, Int. Join Conf. on Neural Networks, San Diego, CA, June 17-21, 1990, II.387-II.395.

[Ste78] K.A. Stevens, "Computation of locally parallel structure", Biol. Cybernetics, 29, 1978, 19-28.

[Ull84] S. Ullman, "Visual routines", Cognition, 18, 1984, 97-159.

[Wer23] M. Wertheimer, "Laws of organization in perceptual forms", Psychologische Forschung, 1923, 4, 301-350. Translation in: A Source Book in Gestalt Psychology, W.D. Ellis, Ed., 1938, 71-88.

[Wit83] A.P. Witkin and J.M. Tenenbaum, "On the role of structure in vision", in: Human and Machine Vision, J. Beck, B. Hope and A. Rosenfeld, Eds., Academic Press, 1983, 269-339.

[Yue90] S.K. Yuen, "Shape from contours using symmetries", in: Proc. First Eur. Conf. Comp. Vision, Antibes, France, 1990, 437-453.

[Zuc83] S.W. Zucker, "Computational and psychophysical experiments in grouping: early orientation selection", in: Human and Machine Vision, J. Beck, B. Hope and A. Rosenfeld, Eds., Academic Press, 1983.

[Zuc85] S.W. Zucker, "Early orientation selection: tangent fields and the dimensionality of their support", Comp. Vision, Graphics and Image Processing, 32, 1, Oct. 1985, 74-103.

Using Complex Gabor Filters To Detect and Localize Edges and Bars

Zhengyan Wang and Michael Jenkin
Department of Computer Science
York University, North York
Ontario, Canada M3J 1P3

Abstract

Early vision process is the extraction of primitive measurements from the image, and this forms the first stage of computation in many vision systems. This paper presents a new approach to feature detection problem by transforming the problem to a complex space in which features may be more readily and robustly recognized. Aiming at providing a computational way of producing feature maps, we first define mathematically the feature of interest, then study phase, amplitude (local energy) and local frequency behavior of the feature to derive criteria for its detection.

This feature detection technique is applied to edges and bars, which are low level features used as primitives for complex visual tasks. We consider the use of phase, amplitude and local frequency information for edge and bar identification as an alternative to numerical differentiation, which is known to be unstable in the presence of noise. Algorithms for the detection and localization of one dimensional edges and bar-like targets of a particular width are developed and their application to real images indicate the promise of this approach.

1 Introduction

The gray level elements contain all the information about the image that is recorded by the camera, however, in order to obtain a symbolic description one has to extract structural information of the scene from the image first. Thus one of the early vision tasks is to obtain complete, meaningful, compact and robust measurements of structure in the scene which are well adapted to later stages, e.g., recognition.

Many vision systems utilize "edges" as this representation. In images, significant scene structures are often associated with sharp changes in image intensity (often called discontinuities) between neighboring pixels. These discontinuities are generally

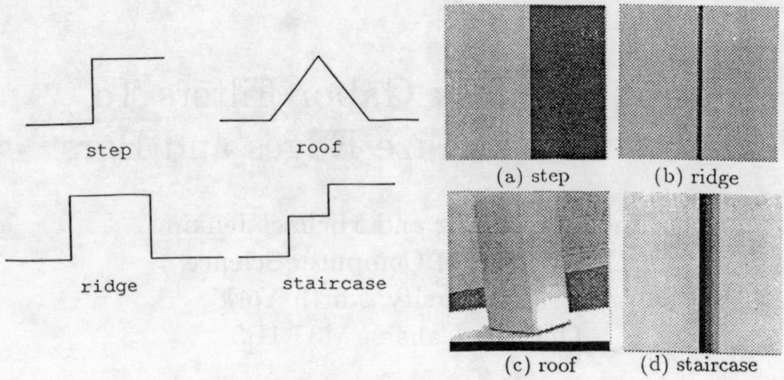

Figure 1: Edge Profiles and Corresponding Images

referred to as edges (more precisely, step edge, there are roof edges corresponding to discontinuities in the first derivative of the intensity). Edge detection is the process of localizing sharp changes of intensity in an image and linking them into line structures. This is essential in early vision because edges convey important structural information contained within the image. We can, for example, recognize a scene from its line drawing.

Consider the problem of identifying edges in the simpler case of a 1-D image. Even in the limited 1-D case, there are many possible types of edges. These edge types can be characterized by their local intensity profiles. Some examples of these are: (1) step, (2) roof, (3) ridge and (4) staircase which are shown in Figure 1 together with their corresponding images.

Although extensive results have been reported by many researchers on edge detection, except perhaps for the case of step edges, the results are still not satisfactory. One type of edge that has received less attention is the bar or ridge edge. The step edge may be a good primitive for some visual tasks, but the step edge is not the only available primitive. Other basic primitives are possible and may be better suited for particular applications. Bars, like edges, are low level features which can be used as primitives for complex visual tasks. Consider, for example, an application in which images consist of line drawings in which the width of the lines plays an important part in understanding the image. Such a situation occurs in maps or blueprints. In this type of application, a "bar" of a particular width might be a more suitable primitive representation of the image. In this paper, we will consider edge and bar-based representation as a special case of a more general symbolic phase-feature-based representation of an image. Within this representation, algorithms are developed for the detection and localization of edges and bars of a particular width. Other types of edges and other complex features could also be treated within a symbolic phase-feature-based representation but such treatments are beyond the scope of this paper.

For a detection mechanism to be successful it must possess certain properties. The mechanism must have good localization abilities (it should accurately position the feature), it should be stable and robust (the detection mechanism should be relatively insensitive to noise), and it should be reliable (the detection mechanism should label features so that they can be ranked in terms of goodness). Many previous algorithms (see [1] for a brief review) for edge detection are, in nature, differential processes. Differentiation processes amplify noise and are thus inherently unstable[2]. Following earlier phase based work[3, 4, 5, 6, 7], we consider the application of local phase measurements to the task of edge and bar detection. We formally define an edge and a bar and then introduce the notion of the use of local image phase as a localization measure, and we show how features identified from the local phase and instantaneous (local) frequency (defined as the derivative of phase) can be used to identify edges and bars in a scene. Algorithms to detect and localize one dimensional edges and bar-like targets of a particular width are developed using local phase information obtained from Gabor filtering.

2 Phase-Based Edge and Bar Detection, Localization

Recent advances in stereopsis[3, 5] and motion understanding[7] have approached the problem as one which utilizes the phase (or phase difference) of the underlying signal as the primitive upon which the more complex computation is to be based. Jepson and Fleet[8] have shown how local frequency information can be used to identify regions in which the local phase information can be trusted.

In this section we show how the problems of edge and bar detection can be reformulated as a phase measurement processes. In the 1-D case the process can be abstracted as follows:

i) define a mathematical model of the desired feature

ii) choose proper filters

iii) filter for the desired feature

iv) derive phase properties to identify the desired feature

v) phase may not uniquely characterize the desired feature, to detect the desired feature, other information may be needed, such as instantaneous frequency, amplitude or derivative of the amplitude.

Assume:

i) I is the space of local image,

ii) F is the model of the feature we are looking for and $F \subseteq I$,

we try to derive property P which is a subset of ordered $\mathcal{P}(F_\phi, F_{\phi'}, F_\rho, F_{\rho'})$ from the following transformations:

$$F \rightarrow F_{\phi}, \;\; F \rightarrow F_{\phi'}, \;\; F \rightarrow F_{\rho}, \;\; F \rightarrow F_{\rho'}$$

where $F_{\phi}, F_{\phi'}, F_{\rho}, F_{\rho'}$ represent the phase, local frequency, amplitude and derivative of the amplitude in terms of the feature respectively.

Although other filter choices are possible, we choose Gabor functions as our bandpass filters, as they achieve the theoretical lower bound of joint uncertainty in both space and spatial frequency.

2.1 Phase, Amplitude And Local Frequency From Gabor Kernels

Gabor functions are complex functions which take the form of : $e^{i\omega_0(x-x_0)}e^{-\frac{(x-x_0)^2}{2\sigma_0^2}}$. Take $x_0 = 0$ (without loss of generality) and consider the 1-D Gabor function defined to be:

$$Gab(x, \omega_0) = e^{i\omega_0 x} G(x, \sigma_0)$$

where $G(x, \sigma_0)$ is a normalized Gaussian with standard deviation σ_0. Gab is characterized by the following parameters:

- ω denotes the peak tuning spatial frequency which is a function of λ: $\omega \equiv 2\pi/\lambda$, where λ is wavelength

- σ represents the radius of support which we choose to be a function of λ: $\sigma = (2^{\beta}+1)/(2^{\beta}-1)\omega$, where β is the octave bandwidth and is usually set to one, so $\sigma = 3\lambda/2\pi$.

Gab(x) is a complex function. It can be written as the sum of real and imaginary parts: Gab(x)= $cos\omega_0 x\, G(x) + i \,\sin \omega_0 x\, G(x)$. The imaginary part is bandpass, but the real part has a dc component $\int_{-\infty}^{+\infty} cos\omega_0 x\, G(x)dx \neq 0$. In practice, we apply a slightly altered cosine-Gabor kernel in order to make it bandpass,

$$Gab_{cos}(x) = (cos\omega_0 x - \gamma)G(x)$$

where γ is a constant set to $e^{-\omega_0^2\sigma_0^2/2}$.

Filter a 1-D image with a Gabor kernel at scale λ. The response $R(x,\lambda)$ is: $R(x,\lambda) = R_{cos}(x,\lambda) + i\, R_{sin}(x,\lambda)$, where $R_{sin}(x,\lambda) = \sin(\omega x)G(x) * I(x)$ and $R_{cos}(x,\lambda) = (cos\omega x - \gamma)G(x) * I(x)$. This is a complex function and can be expressed in exponential form:

$$R(x,\lambda) = \rho(x,\lambda)e^{i\phi(x,\lambda)},$$

its amplitude and phase are given by:

$$\rho(x,\lambda) \equiv |R(x,\lambda)| = \sqrt{R_{sin}^2(x,\lambda) + R_{cos}^2(x,\lambda)} \tag{1}$$

$$\phi(x,\lambda) \equiv arg(R(x,\lambda)) = \tan^{-1}(\frac{R_{sin}(x,\lambda)}{R_{cos}(x,\lambda)}) \tag{2}$$

From equation 1, the amplitude derivative is given by:

$$\rho_x(x,\lambda) = \frac{R'_{sin}(x)R_{sin}(x,\lambda) + R'_{cos}(x,\lambda)R_{cos}(x,\lambda)}{\rho(x,\lambda)} \tag{3}$$

The local (instantaneous) frequency of R(x,λ) can be defined as the derivative of the phase: $\phi_x(x,\lambda)$[9], where

$$\phi_x(x,\lambda) = \frac{R'_{sin}(x,\lambda)R_{cos}(x,\lambda) - R_{sin}(x,\lambda)R'_{cos}(x,\lambda)}{R^2_{sin}(x,\lambda) + R^2_{cos}(x,\lambda)} \tag{4}$$

2.2 Finding Step Edges

In this section we introduce the model of an edge and show how the problem of step edge detection can be reduced to the problem of finding appropriate features in phase space.

2.2.1 Step Edge Model

Without loss of generality, we consider a step edge centered at the origin. An isolated step edge model is defined as follows:

$$I(x) = \begin{cases} A_1 & x < 0 \\ A_2 & x \geq 0 \end{cases} \tag{5}$$

where A_1 and A_2 are the gray-level of the two regions.

2.2.2 Phase and Related Properties of Step Edges

Consider the edge model defined in (5), and the convolution of the edge with the modified complex Gabor kernel. A number of useful properties can be shown based on the local phase and frequency of the result of the convolution.

Property I: At the center of a step edge, $R_{sin}(x)$ is nonzero.

$$R_{sin}(x) = \int_{-\infty}^{+\infty} \sin(\omega_0 z)\frac{e^{-\frac{z^2}{2\sigma_0^2}}}{\sqrt{2\pi}\sigma_0} I(x-z)dz$$

according to the definition of $I(x)$, we have:

$$R_{sin}(x) = A_2 \int_{-\infty}^{x} \sin(\omega_0 z)\frac{e^{-\frac{z^2}{2\sigma_0^2}}}{\sqrt{2\pi}\sigma_0} dz + A_1 \int_{x}^{+\infty} \sin(\omega_0 z)\frac{e^{-\frac{z^2}{2\sigma_0^2}}}{\sqrt{2\pi}\sigma_0} dz \tag{6}$$

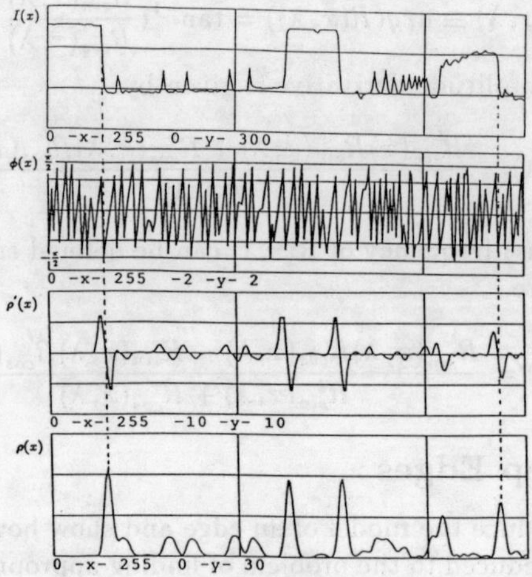

Figure 2: A scan line of an image, its phase from Gabor filtering, derivative of amplitude and amplitude of Gabor response

since $\sin x$ is odd, we have

$$R_{sin}(0) = (A_2 - A_1)\int_0^{+\infty} \sin(\omega_0 z)\frac{e^{-\frac{z^2}{2\sigma_0^2}}}{\sqrt{2\pi}\sigma_0}dz$$

$A_1 \neq A_2$ and the integral $\int_0^{+\infty} \sin(\omega_0 x)e^{-\frac{x^2}{2\sigma_0^2}}dx$ is a non-zero constant, therefore $R_{sin}(0) \neq 0$.

Property II: At the center of an edge, $R_{cos}(x)$ is zero.

$$\begin{aligned} R_{cos}(x) &= \int_{-\infty}^{+\infty} [\cos(\omega_0 z) - \gamma]\frac{e^{-\frac{z^2}{2\sigma_0^2}}}{\sqrt{2\pi}\sigma_0} I(x-z)dz \quad (where\ \ \gamma = e^{-\frac{\omega_0^2\sigma_0^2}{2}}) \\ &= A_2\int_{-\infty}^{x} [\cos(\omega_0 z) - \gamma]\frac{e^{-\frac{z^2}{2\sigma_0^2}}}{\sqrt{2\pi}\sigma_0}dz + A_1\int_{x}^{+\infty} [\cos(\omega_0 z) - \gamma]\frac{e^{-\frac{z^2}{2\sigma_0^2}}}{\sqrt{2\pi}\sigma_0}dz \end{aligned}$$

$\cos x$ is even, thus

$$R_{cos}(0) = (A_2 + A_1)\int_0^{+\infty} [\cos(\omega_0 z) - \gamma]\frac{e^{-\frac{z^2}{2\sigma_0^2}}}{\sqrt{2\pi}\sigma_0}dz$$

We know that $\frac{1}{\sqrt{2\pi}\sigma}\int_{-\infty}^{+\infty}\cos(\omega x)\,e^{-\frac{x^2}{2\sigma^2}}dx = e^{-\omega^2\sigma^2/2}$ *and* $\frac{1}{\sqrt{2\pi}\sigma}\int_{-\infty}^{+\infty}e^{-\frac{x^2}{2\sigma^2}}dx = 1.$

Hence $\frac{1}{\sqrt{2\pi}\sigma_0}\int_0^{+\infty}[\cos(\omega_0 z)-\gamma]e^{-\frac{z^2}{2\sigma_0^2}}dz = \frac{1}{2}\{\frac{1}{\sqrt{2\pi}\sigma_0}\int_{-\infty}^{+\infty}[\cos(\omega_0 z)-\gamma]e^{-\frac{z^2}{2\sigma_0^2}}dz\} = 0,$ $R_{cos}(0)=0$

Property III: At the center of an edge, $|\phi(x)|$ is $\frac{\pi}{2}$.

This is a straightforward result of the above two properties.

In practice, the constraint is relaxed to phase near $\pm\pi/2$. The type of an edge can be further classified by the sign of the phase which indicates a positive- or negative-going edge.

Property IV: At the center of an edge, $R'_{sin}(x)$ is zero.

Differentiating (6) results in

$$R'_{sin}(x) = (A_2 - A_1)\sin(\omega_0 x)\frac{e^{-\frac{x^2}{2\sigma_0^2}}}{\sqrt{2\pi}\sigma_0}$$

Clearly $R'_{sin}(0) = 0$.

Property V: At the center of an edge, $\rho_x(x)$ is zero.

Recall equation 3:

$$\rho_x(x) = \frac{R'_{sin}(x)R_{sin}(x) + R'_{cos}(x)R_{cos}(x)}{\rho(x)}$$

substituting value for $R_{cos}(0)$, $R'_{sin}(0)$ and from the above properties $\rho(0) \neq 0$ gives $\rho_x(0) = 0$, i.e., its amplitude is either a local maximum or a local minimum.

Figure 2 shows the intensity profile of a scan line of an image containing step transactions, its phase from Gabor filtering, the amplitude of Gabor response and the derivative of the amplitude.

2.2.3 Edge Detection And Localization

Based on the above properties, we can identify step edge candidates by searching for a phase response near (ideal step edges rarely exist in real images) $\pm\pi/2$. The distance between the recovered phase and $\pi/2$ can be measured as $||\phi(x)| - \pi/2|$ and is bounded by ϵ. ϵ is used to control the sharpness of edges detected. Candidate edges are then checked to see if they violate constraints on amplitude ($\rho(x)$) or on the derivative of the amplitude ($\rho_x(x)$). A threshold is set to make sure the strength of the amplitude of the response is high enough in the presence of the edge. Many spurious edges will be eliminated as they violate these amplitude constraints. Edges are only labelled for those locations at which all three constraints are satisfied.

2.3 Finding Bar-Like Targets Of Particular Width

A ridge edge or bar is a strip lying on a background with a relatively uniform grey level. Although the line and bar detectors reviewed in [1] can be applied to reveal the presence of bars, it is not really a suitable application of these techniques; we are interested in finding bars of particular width rather than any bar-like targets. Edge techniques are not suitable for finding bars of particular width. If the bar width is relatively wide with respect to the size of operator, its two edge pairs can be located separately, the width may be approximately measured as the distance between the edge pairs and the center of the bar can be taken as the midpoint of the edge pairs (although the process of matching the edges is non-trivial). However, operators usually have a large region of support because small operators can be sensitive to noise. Two step edges may show up together in the step operator's "window" which will result in the dislocalization or merging of the edge pairs. It is more proper to develop a mechanism that is sensitive to the width of bar-like targets. In this section, we define a bar model and show that this problem can also be reduced to phase measurement.

2.3.1 Bar Model

Assuming a bar with uniform intensity on a uniform gray-level background, we define an isolated bar model centered at x=0 as:

$$I(x) = \begin{cases} A_2, & |x| \leq \lambda/2 \\ A_1, & |x| > \lambda/2, \quad A_1 \neq A_2 \end{cases} \tag{7}$$

where A_1 and A_2 are the gray-level of background and bar respectively, and λ is the width of the bar.

2.3.2 Phase and Related Properties of Bar

Consider the bar model defined in (7), and the convolution of the bar with the modified complex Gabor kernel with $\sigma = 3\lambda/2\pi$ and $\lambda = \lambda_0$ the bar width. A number of useful properties can be derived based on the local phase and frequency of the result of the convolution.

Property I: At the center of a bar of width λ, $R_{sin}(x)$ is zero.

$$R_{sin}(x) = \int_{-\infty}^{+\infty} \sin(\omega_0 z) \frac{e^{-\frac{z^2}{2\sigma_0^2}}}{\sqrt{2\pi}\sigma_0} I(x-z)dz$$

since $I(x-z)$ is A_2 for z between $x - \lambda/2$ and $x + \lambda/2$ and A_1 elsewhere, we have:

$$R_{sin}(x) \;=\; A_2 \int_{x-\lambda/2}^{x+\lambda/2} \sin(\omega_0 z) \frac{e^{-\frac{z^2}{2\sigma_0^2}}}{\sqrt{2\pi}\sigma_0} dz + A_1 \int_{-\infty}^{+\infty} \sin(\omega_0 z) \frac{e^{-\frac{z^2}{2\sigma_0^2}}}{\sqrt{2\pi}\sigma_0} dz$$

$$-A_1 \int_{x-\lambda/2}^{x+\lambda/2} \sin(\omega_0 z) \frac{e^{-\frac{z^2}{2\sigma_0^2}}}{\sqrt{2\pi}\sigma_0} dz$$

$$= (A_2 - A_1) \int_{x-\lambda/2}^{x+\lambda/2} \sin(\omega_0 z) \frac{e^{-\frac{z^2}{2\sigma_0^2}}}{\sqrt{2\pi}\sigma_0} dz + A_1 \int_{-\infty}^{+\infty} \sin(\omega_0 z) \frac{e^{-\frac{z^2}{2\sigma_0^2}}}{\sqrt{2\pi}\sigma_0} dz$$

$\int_{-\infty}^{+\infty} \sin(\omega x) e^{-\frac{x^2}{2\sigma_0^2}} dx = 0$, therefore

$$R_{sin}(x) = (A_2 - A_1) \int_{x-\lambda/2}^{x+\lambda/2} \sin(\omega_0 z) \frac{e^{-\frac{z^2}{2\sigma_0^2}}}{\sqrt{2\pi}\sigma_0} dz \tag{8}$$

Since $\sin(\omega_0 z)\, e^{-\frac{z^2}{2\sigma_0^2}}$ is odd, $R_{sin}(0) = 0$.

Property II: At the center of a bar of width $\lambda = \lambda_0$, $R_{cos}(x)$ is nonzero.

$$R_{cos}(x) = \int_{-\infty}^{+\infty} [\cos(\omega_0 z) - \gamma] \frac{e^{-\frac{z^2}{2\sigma_0^2}}}{\sqrt{2\pi}\sigma_0} I(x-z) dz \quad (where\ \ \gamma = e^{-\frac{\omega_0^2 \sigma_0^2}{2}})$$

$$= (A_2 - A_1) \int_{x-\lambda_0/2}^{x+\lambda_0/2} [\cos(\omega_0 z) - \gamma] \frac{e^{-\frac{z^2}{2\sigma_0^2}}}{\sqrt{2\pi}\sigma_0} dz + A_1 \int_{-\infty}^{+\infty} [\cos(\omega_0 z) - \gamma] \frac{e^{-\frac{z^2}{2\sigma_0^2}}}{\sqrt{2\pi}\sigma_0} dz$$

We know that $\frac{1}{\sqrt{2\pi}\sigma_0} \int_{-\infty}^{+\infty} [\cos(\omega_0 z) - \gamma] e^{-\frac{z^2}{2\sigma_0^2}} dz = 0$, thus

$$R_{cos}(x) = (A_2 - A_1) \int_{x-\lambda_0/2}^{x+\lambda_0/2} [\cos(\omega_0 z) - \gamma] \frac{e^{-\frac{z^2}{2\sigma_0^2}}}{\sqrt{2\pi}\sigma_0} dz \tag{9}$$

recall that $\omega_0 = 2\pi/\lambda_0$, $\sigma_0 = 3\lambda_0/2\pi$, and that $(\cos x - \gamma) e^{-\frac{x^2}{2\sigma_0^2}}$ is even, with a change of variable, we get:

$$R_{cos}(0) = \frac{2(A_2 - A_1)}{\sqrt{2\pi}\sigma_0} \int_0^{\lambda_0/2} (\cos \frac{2\pi}{\lambda_0} x - \gamma) e^{-\frac{(2x\pi)^2}{18\lambda_0^2}} dx$$

The integral is a non-zero constant, as $A_1 \neq A_2$, so $R_{cos}(0) \neq 0$.

Property III: At the center of a bar of width $\lambda = \lambda_0$, $\rho(x)$ is nonzero.
This is a direct consequence of the above two properties.

Property IV: At the center of a bar of width $\lambda = \lambda_0$, $\phi(x)$ is zero.
$\phi(x)$ is defined in (2) so $\phi(0) = \tan^{-1}(R_{sin}(0)/R_{cos}(0)) = \tan^{-1}(0/K)$, where $K \neq 0$, thus $\phi(0) = 0$.

Property V: At the center of a bar of width $\lambda = \lambda_0$, $R'_{sin}(x)$ is zero.

Differentiating (8) results in

$$R'_{sin}(x) = \frac{A_2 - A_1}{\sqrt{2\pi}\sigma_0}\{\sin(\omega_0(x+\lambda_0/2))e^{-\frac{(x+\lambda_0/2)^2}{2\sigma_0^2}} - \sin(\omega_0(x-\lambda_0/2))e^{-\frac{(x-\lambda_0/2)^2}{2\sigma_0^2}}\}$$

Clearly $R'_{sin}(0) = 0$.

Property VI: At the center of a bar of width λ, $R'_{cos}(x)$ is zero.

Differentiating (9) results in

$$R'_{cos}(x) = \frac{A_2 - A_1}{\sqrt{2\pi}\sigma_0}\{[\cos(\omega_0(x+\lambda/2)) - \gamma]e^{-\frac{(x+\lambda/2)^2}{2\sigma_0^2}} - [\cos(\omega_0(x-\lambda/2)) - \gamma]e^{-\frac{(x-\lambda/2)^2}{2\sigma_0^2}}\}$$

$R'_{cos}(x)$ is odd, so $R'_{cos}(0) = 0$.

Property VII: At the center of a bar of width $\lambda = \lambda_0$, $\phi_x(x)$ is zero.

From (4) and using the above results of $R_{sin}(0), R_{cos}(0), R'_{sin}(0)$ and $R'_{cos}(0)$, then clearly $\phi_x(0) = 0$.

Property VIII: At the center of a bar of width $\lambda = \lambda_0$, $\rho_x(x)$ is zero.

In equation 3:

$$\rho_x(x) = \frac{R'_{sin}(x)R_{sin}(x) + R'_{cos}(x)R_{cos}(x)}{\rho(x)}$$

substituting value for $R_{sin}(0), R_{cos}(0), R'_{sin}(0)$ and $R'_{cos}(0)$ results in $\rho_x(0) = 0$.

Observation from experiments shows that at the center of a bar, the amplitude may achieve a local maxima for some widths and a local minima for other widths, in either event, it must be an extreme point.

2.3.3 Bar Detection

In[8], phase and the behavior of its singularity neighborhoods were considered in a Gabor scale-space expansion of a 1-D signal. Let (x_0, λ_0) denote the location of a singularity. The neighborhoods above (below) singular points (above:$\lambda > \lambda_0$, below:$\lambda < \lambda_0$) are characterized by local frequencies that are significantly below (above) the corresponding peak tuning frequency. There exist retrograde regions along which the local frequency is zero, i.e. $\phi_x(x, \lambda) = 0$. In[8], neighbors above or below the singular points (which are characterized by local frequencies significantly below or above the peak tuning frequency) were detected by limiting the distance between the local frequency of response and the peak tuning frequency. Neighborhoods spatially adjacent to singular points were detected by applying constraints on local

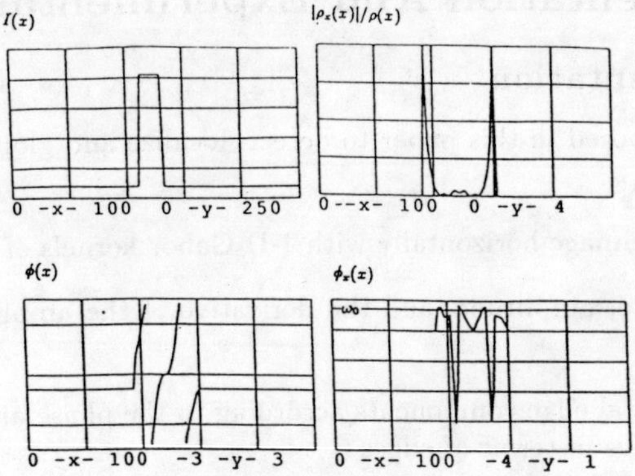

Figure 3: A 1-D bar, relative amplitude derivative, phase and local frequency

amplitude and its variation. The local phase structure for an ideal bar is as shown in Figure 3. The local frequency $\phi_x(x)$ is roughly ω_0 with wrap around effects as well as retrograde regions (places where $\phi_x(x) = 0$). For the ideal situation ($\lambda = \lambda_0$) a retrograde region corresponds to the exact center of the bar. For more practical situations in which $\lambda \approx \lambda_0$, then the retrograde region may vanish. Nevertheless, the center of the bar is characterized by $\phi(x) = 0$.

The constraints for the candidates being the intended bar targets can be expressed as:

$$|\phi_x(x) - \omega_0| < \tau_1, \quad |\rho_x(x)|/\rho(x) < \tau_2$$

where τ_1 and τ_2 are constants related to λ_0. The first constraint forces the instantaneous frequency of the output of the Gabor filter to agree with the pass frequency of the Gabor. The second constraint requires that (a) there is sufficient energy in the pass region, and (b) the energy is relatively constant over a local region spatially. We can identify potential candidate bars as zero-crossings in $\phi(x)$. If at a position $\phi(x) = 0$, we find that the amplitude constraint is violated, then this candidate bar is ignored. If on the other hand the local frequency constraint is violated then a check is made to see if the retrograde region lies within a region of consistent frequency ($\omega \approx \omega_0$). This corresponds to a bar of exactly the correct width as shown in Figure 3.

Note that this approach generates a necessary (but not sufficient) technique for identifying bars in an image. The measurements can be rated by how well the sinusoidal constraints are satisfied by the data, i.e. $e = \alpha|\phi_x(x) - \omega_0| + (1-\alpha)|\rho_x(x)|/\rho(x)$, where α is a parameter weighting the two constraints.

3 Implementation And Experimental Results

3.1 Implementation

The algorithm proposed in this paper to detect, localize and globalize edges consists of the following steps:

- Convolve the image horizontally with 1-D Gabor kernels of different λ
- Measure phase, amplitude, and the derivative of the amplitude and the local frequency
- Label horizontal edge components according to the phase and related measurement properties in terms of edges
- Produce analogously a vertical edge component map
- Combine the horizontal and vertical edge map to obtain an edge map with edge location and local edge orientation
- Apply Hough transform on the edge map and utilize orientation information to aggregate edges approximately lying on straight lines.

Similarly, the algorithm to detect, localize and globalize bars of certain width is as follows:

- Convolve the image horizontally with 1-D Gabor kernels (λ is set to the intended width λ_0)
- Measure phase, amplitude, and the derivative of the amplitude and the local frequency
- Register the presence of a bar with horizontal width near λ_0 according to the phase and related measurement properties in terms of a bar of certain width
- Produce analogously a vertical bar map
- Combine the horizontal and vertical bar map — there is a valid bar center if it is registered in horizontal or vertical bar map.
- Apply the Hough transform on the bar map to globalize discrete bars as connected bars.

The filter we use is sampled from the Gabor function with a limited support size. As $Gab_{sin}(x)$ and $Gab_{cos}(x)$ become negligible with $x > 3\sigma$, the insignificant parts can be truncated. In finding edges, we use varied λ (hence peak tuning frequency) to yield varied resolution. As required by the sampling theorem, the sampling interval should be less than $\lambda/2$. As the regular grid is of interval 1, λ should be greater than 2 to avoid aliasing. For simplicity, the octave bandwidth β is set to 1 in the proof given in section 2. In order to reduce the bandwidth and distortion of local phase, we take β to be 0.8 in the implementation.

3.2 Experiments

The algorithm as described in this paper is a one dimensional process. By applying the algorithm in horizontal and vertical directions separately, a feature map is obtained by combining these results. In the case of edges, directional information is available from the strength of horizontal and vertical components of an edge. Figure 4 (a)-(f) shows input images and the edge maps produced by our phase-based edge finding algorithm. It can be seen that most of the intensity discontinuities are labelled but there exist aliasing phenomenon in the edge map in Figure 4 (d) (left of the bear) which contains multiple edges one next to the other. This is due to our isolated edge model and the constraint of Nyquist frequency. To improve the output, multi-edge models have to be defined and the behaviors of interactions between adjacent edges have to be studied.

3.3 Comparison With Other Approaches

Gradient-based techniques or zero-crossing approaches are almost standard, thus comparisons are usually made against this type of detector. Since zero-crossings of $\nabla^2 G$ is a kind of standard gauge among edge detectors, we apply it to the test image and compare it against our result. The Canny edge operator[10] is commonly adopted in some sense as the best operator, so test images are also processed using the Canny operator. Figure 5 shows input images and the result of these different edge finding schemes. Our result is obviously better than the $\nabla^2 G$ while it is hard to tell between Canny's and ours. A noteworthy feature of $\nabla^2 G$ is that it generates closed contours, while edge maps produced by other techniques do not possess such a property. Instead the output is a collection of "line" segments. Figure 6 shows input images which contain a number of bar like features and also shows the outputs of combining the result of the algorithm applied in both the horizontal and vertical directions. Most of the bar like structures in the image have been identified and note that certain structures, such as crossings of horizontal and vertical bars, have not been selected, however, they can be filled in by extending the bars' ends. In addition, bars of widths relatively far from the tuning width have not been reported. As the target feature is bars of a particular width, edges should not be detected and Figure 6 (b) shows that only bars of valid widths are present in the bar map. Besides, we have to mention a limitation of our scheme that it cannot be applied to detect bars of less than two pixels width, in these cases, frequencies are equal to or above the Nyquist limit and aliasing will be present in the resultant image (Figure 6 (b)).

3.4 Stability in the Presence of Noise

To illustrate the robustness of the phase-based technique, a real image was corrupted with additive white Gaussian noise. The resultant edge maps produced by phase measurement, $\nabla^2 G$ and the Canny operator are shown in Figure 7. There are several tunable parameters in Canny's algorithm and they are non-trivial to tune to get a clean result while our algorithm is simple and effective. Compared with the edge maps produced by the differential technique, we noticed that our results are not significantly affected by the noise in the image. In fact, less false edges are shown in our edge map and these can be lessened by linking (a line segment must be of certain length). However, differential techniques are by their very nature sensitive to noise and a post-process is necessary to eliminate spurious measures (usually by threshold tuning). The stability of the phase-based approach is due to the fact that the detection of the feature, say, a step edge, is based on the model's properties manifested in the phase space.

4 Discussion

Phase based techniques play an important role in measuring stereo disparity[3, 4, 5, 6], and image velocity[7]. Noting that phase contains rich information, we applied a phase technique to the problem of detecting and localizing image features, in particular, edges and bar-like targets of particular width. We should point out that although Gabor kernels are employed in our scheme other kernels are possible. In particular, complex wavelets with the simultaneous concentration in space and frequency domain close to the lower bound and possessing other desirable properties are good candidates. Through the analysis of phase and related information in terms of intended features, it is observed that edges and bars can be characterized by phase, instantaneous frequency, amplitude and derivative of amplitude from Gabor filtering. The relationship between candidates of the two features in phase space is orthogonal: one is characterized by a phase of $\pm\pi/2$, the other by a phase of 0. Other information, such as amplitude, is needed to further discriminate the validity of the feature. Since phase determines the location of a feature and amplitude signals the presence of a feature, we first use phase information to localize the feature and utilize amplitude and other information to reveal the desired features and discard undesired ones.

The problem of edge or bar detection can be viewed in an abstract way, i.e., the problem can be taken as reducing the feature finding in image space to another space in which the representation of corresponding features may be built more readily. Such is the approach of differential technique and phase-based techniques. Both techniques serve as transforms from one space to another. The detection scheme we developed is to reduce the feature finding in image space to feature finding in a complex space. Comparing phase-based technique with other approaches, such as zero-crossing or Canny's detection technique, it turns out that our approach is more robust, because more information (characterizations) are utilized in terms of information theory. However of them has its advantages and pitfalls. Phase-based feature detection is robust and reliable in the sense that features are first detected using phase information which is relatively insensitive to noise and further judgments involving digital differentiation are measured in feature candidates regions in which data are much less noisy. Experimental results show that the phase detecting scheme is less sensitive to noise than the differential scheme (e.g., $\nabla^2 G$). Criteria for finding features are derived from the phase behavior in terms of the features. Features are first localized by satisfying the phase criterion. Further measurements like amplitude and local frequency are concentrated in regions with valid phase measurements, and are applied to rule out undesired regions. The detection and localization of bar-like targets using phase information is independent of the intensity of the feature being looked for and is independent of the intensity difference between the target and its surroundings. As with other feature detection approaches which define a model and derive operators to optimize a set of criteria, the detection mechanism provides a necessary, but not sufficient process for identifying bars. Thus noise may also be present in the edge map while non-bar structures may appear in the bar map. For example, in the test image, blobs and text are identified as bars. Some subsequent process must be used to remove this extra non-bar structure. Also we have a problem which confronted by every approach, i.e., the scale problem, in order to get a satisfiable result, a proper scale has to be chosen.

The spirit of the technique proposed in this paper can be applied to detect other well-defined features. First, choose a complex filter, its real and imaginary parts form a pair of quadrature band-pass filters. Then, a complex response is obtained by band-pass filtering the original signal. The phase and amplitude of the complex response can be computed as:

$$\phi(x) = \tan^{-1}\left(\frac{f_I * I}{f_R * I}\right)$$

$$\rho(x) = \sqrt{(f_I * I)^2 + (f_R * I)^2}$$

where f_R and f_I denote the real and imaginary part of the complex filter respectively. The ordered property set can be extended according to the application, say $P(F_\phi, F_{\phi'}, F_{\phi''}, F_\rho, F_{\rho'}, F_{\rho''}, ...)$, and properties should be derived at positions which characterize the feature.

In this paper we investigate feature detection problem in complex space for 1-D case, however, images are two-dimensional and so are most features. Thus, it is more appropriate to develop 2-D operators. In principle 1-D algorithms can be applied horizontally and vertically to reveal the presence of 2-D features. On the other hand, many directional operators have been developed and their results are more robust than collections of 1-D operators. Utilizing orientation tuned filters to provide direction selectivity, 2-D extensions of our schemes could be devised.

References

[1] Z. Wang, "Using phase information to detect and localize edges and bars", Masters Thesis, Department of Computer Science, York University, 1991

[2] V. Torre and T. Poggio, "On edge detection," *IEEE Trans. on PAMI*, 8(2), pp. 147–163, 1986.

[3] M. R. M. Jenkin, A. D. Jepson, and J. K. Tsotsos, "Techniques of Disparity Measurement," *Computer Vision, Graphics and Image Processing*, 53(1), pp. 14–30, 1991.

[4] T. D. Sanger, "Stereo disparity computation using Gabor filters", *Biol. Cybern.*, vol. 59, pp. 405–418, 1988.

[5] A. Jepson and M. R. M. Jenkin, "The Fast Computation of Disparity from Phase Differences," *Proc. CVPR 89*, pp. 398–403 San Diego, California, 1989.

[6] K. Langley, M.H.E. Larcombe, T.J. Atherton, and R.G. Wilson, "Vertical and horizontal disparities from phase", *Proc. 1st ECCV*, pp. 315–325, Antibes, France, 1990.

[7] D. J. Fleet and A. D. Jepson, "Computation of normal velocity from local phase information," Research in Biological and Computational Vision, RBCV-TR-89-27, University of Toronto, 1989.

[8] A.D. Jepson and D.J. Fleet , "Scale-space singularities," *Proc. 1st ECCV* , pp. 50–55, Antibes, France, 1990.

[9] G.B. Whitham, *Linear and Nonlinear Waves*, John Wiley and Sons, New York, 1974.

[10] J.F. Canny , "A Computational Approach To Edge Detection" In *Readings in Computer Vision*, pp. 591–604, M.A. Fischler and O. Firschein, eds., Morgan Kaufmann Publishers Inc., New York, 1986.

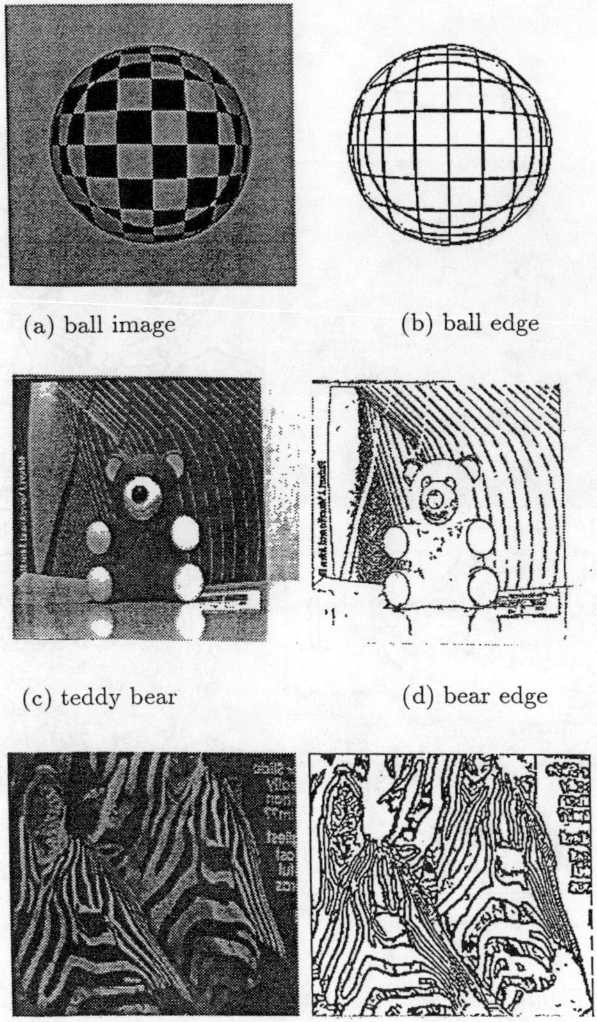

(a) ball image (b) ball edge

(c) teddy bear (d) bear edge

(e) zebra image (f) zebra edge

Figure 4: 2-D scenes and their edge maps produced by phase measurements

(a) original image
(b) phase edge map
(c) $\nabla^2 G$ edge map
(d) Canny edge map

Figure 5: Edge maps: phase-based finding, $\nabla^2 G$, Canny

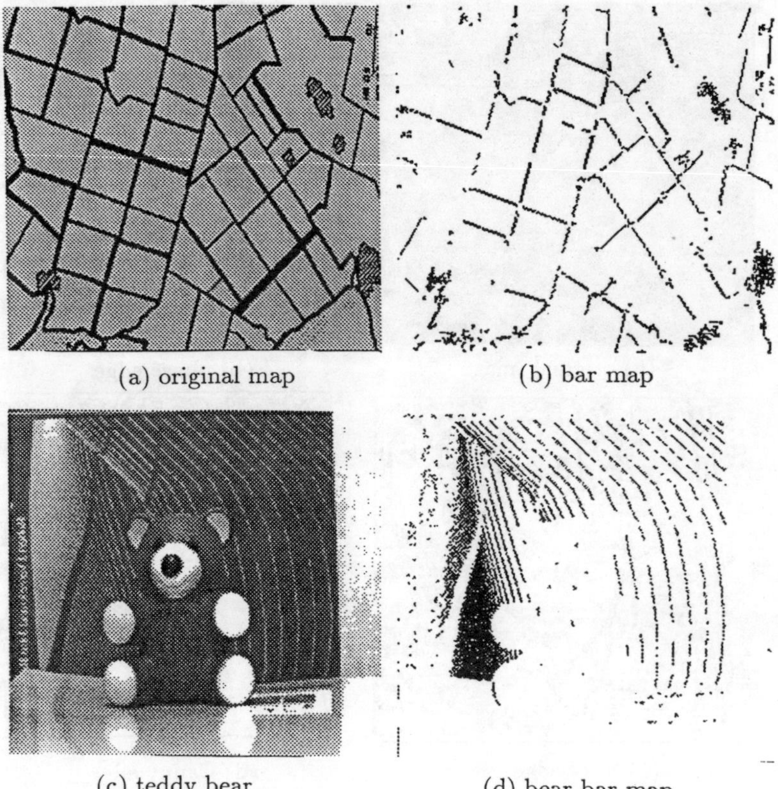

(a) original map (b) bar map

(c) teddy bear (d) bear bar map

Figure 6: 2-D scenes and their bar maps generated by our algorithm

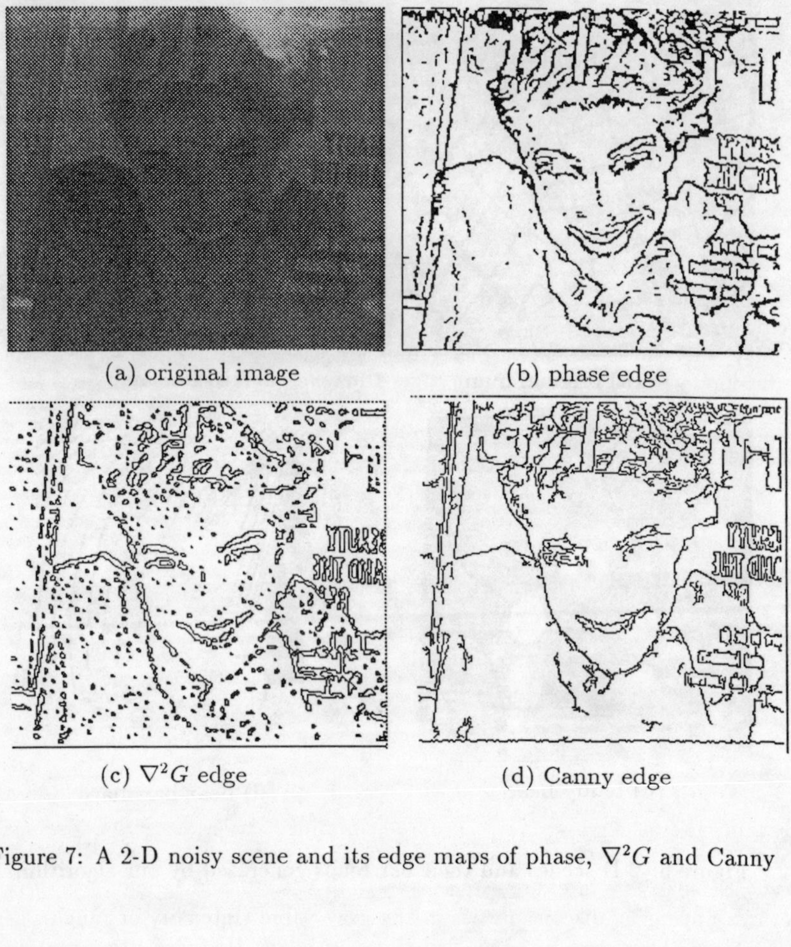

(a) original image (b) phase edge

(c) $\nabla^2 G$ edge (d) Canny edge

Figure 7: A 2-D noisy scene and its edge maps of phase, $\nabla^2 G$ and Canny

Roof Edge Detection: A Morphological Skeleton Approach*

Jin-Chang Cheng† and Hon-Son Don
Department of Electrical Engineering
State University of New York
Stony Brook, NY 11794

Abstract

This paper presents a new method for roof edge detection. In practice, digital images are noise contaminated. A simple method of roof edge detection is to smooth the noisy image, take its derivative, and then use step edge detection techniques to detect roof edges. However, reducing noise by smoothing will also blur the edges since edges correspond to the high frequencies as noises do. Our method is to classify roof peaks from noise peaks by morphological testing. Roof edges on digital images are found by looking for the leaf nodes of the skeletal tree. A leaf node represents a local maximum (or peak). The size of the subtree containing the leaf node is tested to distinguish the noise from the roof edge. The skeletal trees of gray scale images can be easily found by gray scale morphological skeleton operations.

1 Introduction

The segmentation problem is defined as partitioning the image into segments. It is an intrinsic part of an image understanding system. Boundaries are one of the basic sorts of segments to extract. If the boundary of an object can be extracted successfully then the scene analysis can be simplified and recognition becomes highly feasible. The boundary is composed of edge segments and thus edge detection plays an important role in low-level image processing. In images, any type of significant scene structures appear as discontinuities in the gray value (intensity or range), generally called edges. There are at least two types of edges that are important in image analysis: jump (or step) edges and roof edges. In Figure 1, we observe 2-D instances of jump edges and roof edges. For intensity images, a jump edge is defined as the

*This work was supported by the U.S. National Science Foundation under Grant IRI-8710856 and U.S. Army Research Office under Contract DAAL 0338K0033.

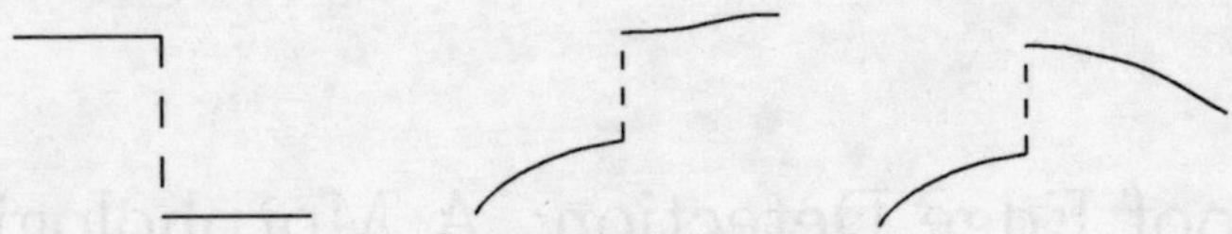

(a)

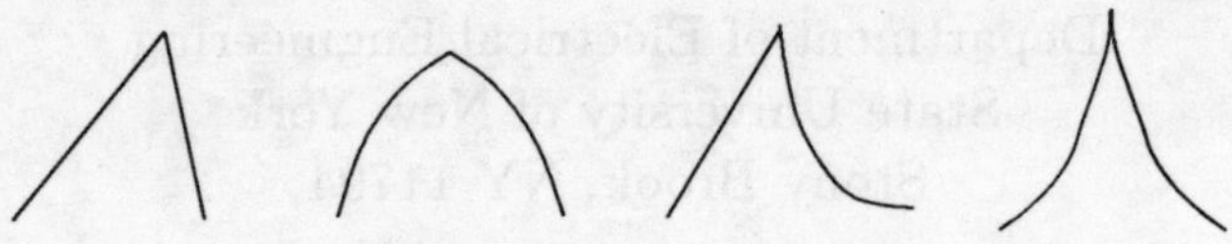

(b)

Figure 1: (a) Jump edges, (b) roof edges.

boundary between two regions whose brightness values are significantly different, and usually corresponding to occluding boundaries of objects in a scene. Jump edges in range images are formed where depth values are discontinuous. Such edges occur when an object occludes another object or when part of an object occludes itself. Here, roof edges include crease edges and ridge edges. A crease edge occurs on a crease line which is the intersection of two planes. A crease line is a straight line consisting of a series of crease points. As we walk along the crease line the points to the right and left of us are lower than the ones we are on. A crease edge occurs where there is a local maximum in one direction. A ridge edge is identical to a crease edge except it is the intersection of two non-planar surfaces. The ridge line may be flat, sloped upward, sloped downward, curving upward, or curving downward.

Most work on edge detection has been focused on jump edge detection. Roof edge detection has received relatively less attention. One possible reason may be due to the fact that identification of roof edges is a particularly more difficult problem than that of jump edges. Our main interest in this paper is with the detection of roof edges using morphological operations.

Theoretically, any jump edge detectors can be used to detect roof edges by taking the smoothed derivative (or gradient) of the image. However, a problem with the gradient vector estimation is the need to define a neighborhood. There will be distortion of the "true" gradients in the vicinity of roof edges, so that gradients will not be wholly discontinuous in traveling over an edge but will gradually change. Also smoothing may destroy edge details. Consider the Laplacian-Gaussian jump edge detector which takes the derivative twice. Then the gradient method needs to take the derivative three times in total. For each derivative operation, the operating window will be enlarged and thus the locality of the roof edge detector will be inaccurate.

Moreover, taking the derivative too many times may enhance the noise, resulting in many spurious edges.

Most approaches involve applying a local operator to the neighborhood of every pixel to determine which pixels have a "roof" quality [1-4]. These approaches are usually very sensitive to noise. Inokuchi, et al [2] proposed a pixel classification method, called a ring operator, to classify pixels into jump edge, roof edge and plane in range images. It transforms a set of pixels on a ring placed on an observed point into the frequency domain. A pixel is classified as a roof edge pixel if the amplitude of the second component is sufficiently "large" (using a preset threshold t_2) and the first and the third components are fairly small (using two more preset thresholds t_1 and t_3). Haralick [1] proposed fitting the image with a two-variable cubic polynomial. Ridges are marked at pixels by looking for zero crossings of the first directional derivative taken in a direction which extremizes the second directional derivative.

Mathematical morphology is an approach to image processing based on set theoretic concepts of shape. Serra [5] defined a ridge detector by upper thinnings. The upper thinnings method is complicated and does not have an explicit representation. Basically, it is derived from the Hit-Miss transformation. The Hit-Miss transformation required two disjoint structuring elements. Serra provided a guide to the selection of structuring elements; however, hexagonal grids are employed, which makes the presentation difficult to interpret. Recently, Noble [6] proposed a morphological ridge detector by the set difference of the closing and opening of the image surface by a spherical structuring element. Noble's detector is very simple but very sensitive to noise.

In this paper, we propose a roof edge detector by using mathematical morphology. Roof edges on digital images are found by looking for the leaf nodes of the skeletal tree. A leaf node represents a local maximum (or peak) which may be caused by noise, a roof edge or a top flat plane. A top flat is a simple, horizontal surface which is a local maximum. The size of the subtree containing the leaf node is measured to distinguish the noise from the roof edge or top flat. The skeletal tree can be easily found by gray scale morphological skeleton operations [7]. Results are shown for two simulated simple edges and some range images. Our proposed method is less noise sensitive and the true edge locations are more preserved. The computation speed is also much faster than the Laplacian-Gaussian step-edge detector, ring operator and Haralick's ridge detector.

2 Review of Simple Morphological Edge Detectors

In this section, we review some simple morphological edge detectors. Roof edges, jump edges and sharp corners are similar entities as far as a morphological operator is concerned, for in all cases the operator responds to the sudden discontinuities of

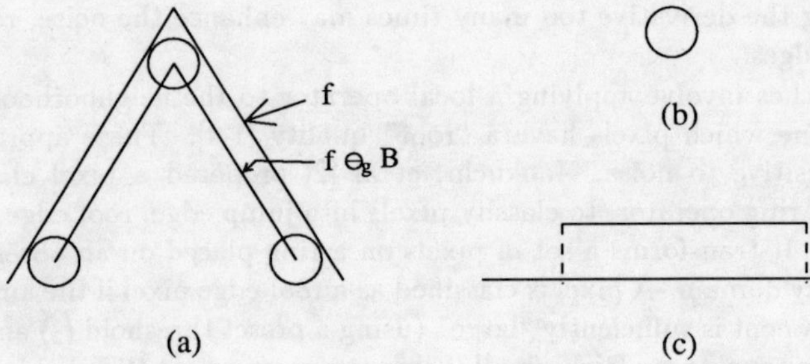

Figure 2: Application of the erosion residue operator on (a) the ideal roof edge by using (b) spherical structuring element, (c) result of the operation.

gray scale values in the image. Several simple morphological corner and step edge detectors are proposed in [8,9], which include:

1. Erosion residue operator ($f - f \ominus_g B$);
2. Dilation residue operator ($f \oplus_g B - f$);
3. Combinations (maximum, minimum, or summation) of (1) and (2); and
4. Blur and minimum operator ($\min\{f_1 - f_1 \ominus_g B, f_1 \oplus_g B - f_1\}$), where $\oplus_g$ and $\ominus_g$ represent gray scale morphological dilation and erosion, respectively, and f_1 is the result of the input image f with a blurring operation.

The results of the above operators are the images of edge strength. An appropriate threshold value is then selected to threshold the edge strength image into a binary edge image. Although these detectors are simple, (1)–(3) are shown to be noise sensitive and/or position biased [9]. Despite their noise sensitivity, there are some other reasons that (1)–(4) are not suitable for the roof edge detection. For example, given an ideal roof edge shown in Figure 2a and a spherical structuring element in Figure 2b. The application of the erosion residue operator on the ideal roof edge results in the edge strength shown in Figure 2c. Obviously, Figure 2c shows that this operator fails to detect the ideal roof edge.

Recently, Nobel [6] proposed a morphological roof and corner detector by the set difference of the gray scale morphological closing and opening of the image:

$$(f \bullet_g B - f \circ_g B)$$

The opening operation can be visualized as rolling the structuring element under the top surface and only keeping points where the structuring element touches the surface. The dual operation of closing rolls the structuring element over the complement of the umbra of the surface. To test its performance, we apply Nobel's roof edge detector to a computer-generated roof edge profile, shown in Figure 3a, which consists of two

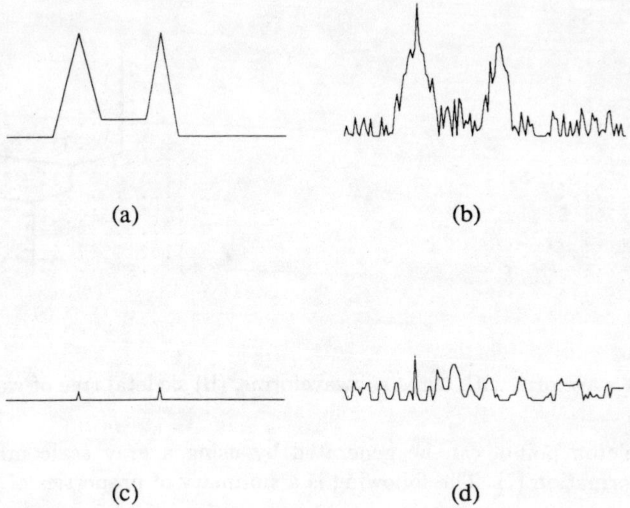

Figure 3: An example of Nobel's roof edge detector: (a) an ideal roof edge profile, (b) noise contaminant of (a), (c) and (d) show respective results of Nobel's operator on (a) and (b).

ideal roof edges of different "sizes". We see that this detector has better response on sharper corners. Figure 3b is obtained by adding zero-mean Gaussian noise with standard deviation of 15 to the ideal roof edges shown in Figure 3a. The result of Nobel's operator applied on Figure 3b is shown in Figure 3d. As we can see from Figure 3d, this detector is also sensitive to noise.

3 A New Morphological Ridge Detector

As discussed in the previous section, the simple morphological detectors based on erosion residue, dilation residue or difference of closing and opening are either sensitive to noise or have poor localization. In this section an algorithm is proposed for ridge edge detection, which is less noise sensitive. The algorithm is based on the skeletal trees of the gray scale image which is provided by the gray scale morphological skeleton transformation.

3.1 Skeletal Tree

The skeletal trees of a waveform are composed of the skeleton points of the waveform. A node on the skeletal tree is a skeleton point associated with its radius as the

Figure 4: (a) A waveform with three subwaveforms, (b) skeletal tree of waveform (a).

attribute. Skeleton points can be generated by using a gray scale morphological skeleton transformation [7]. The following is a summary of properties of the skeletal tree:

1. Each peak on the waveform will correspond to a leaf node which is the smallest radius in the subwaveform. A subwaveform is defined as the subset of the waveform containing a peak and the portion between its two neighboring valleys. A subwaveform is labeled with its corresponding peak. Figure 4a shows three subwaveforms S_1, S_2 and S_3 .

2. Every node has a single parent. Most nodes have only one son, but some nodes may have more than one sons.

3. Parent nodes always have a radius greater than or equal to its son nodes. Numbers shown in Figure 4b indicate the corresponding radius at each node. Node Q is, of radius 9, a parent node of nodes P_1 and P_2 which are of radius 4 and 2, respectively.

4. The nearest common predecessor of two nodes corresponds to a valley on the waveform. Figure 4b shows node Q, which has two son nodes P_1 and P_2, corresponding to a valley between subwaveform S_1 and subwaveform S_2.

5. The original waveform can be easily reconstructed from its skeletal tree by using inverse gray scale morphological skeleton transformation [7].

3.2 The Construction of the Skeletal Tree

The Construction of skeletal tree is briefly described in this section. We apply the one-dimensional gray scale skeleton transformation to the waveform with a fixed unit sized structuring element. The domain of the structuring element B of radius 1 is defined by the set

$$D_B = \{-1, 0, 1\}$$

and $B(t)$, $t \in D_B$, is zero. The gray scale skeleton transformation [7] needs two kinds of morphological operations: erosion and opening. Morphological gray scale opening is performed by gray scale erosion followed by gray scale dilation. The erosion of a gray scale waveform $f(t)$ by the structuring element B of radius 1 can be carried out by:

$$f \ominus_g B(t) = \min_{i \in D_B} \{f(t+i) - B(i)\},$$

which, in the case of a zero height structuring element, becomes:

$$f \ominus_g B(t) = \min_{i \in D_B} \{f(t+i)\} = \min\{f(t-1), f(t), f(t+1)\}$$

So, the erosion operation can be reduced to finding the minimum of three numbers. The dilation of a gray scale waveform $f(t)$ by the structuring element B of radius 1 can be carried out by:

$$f \oplus_g B(t) = \max_{i \in D_B} \{f(t-i) + B(i)\},$$

which, in the case of a zero height structuring element, becomes:

$$f \oplus_g B(t) = \max_{i \in D_B} \{f(t-i)\} = \max\{f(t-1), f(t), f(t+1)\}$$

The dilation operation can be achieved by finding the maximum of three numbers.

We find $f \ominus_g B$ first, and save it as $g(t)$ for later use, then the set S_0 of all skeleton points of zero radius are:

$$g = f \ominus_g B$$

$$S_0 = U[f] - U[g \bigcirc_g B]$$

Where $U[f]$ is the umbra of f. Copy $g(t)$ to $f(t)$ and repeat the above steps for the set S_1 of all skeleton points of radius 1 are:

$$f = g$$

$$g = f \ominus_g B$$

$$S_1 = U[f] - U[g \bigcirc_g B]$$

Recursively apply the above steps. All sets $S_2, S_3, \ldots, S_{maxr}$ of all skeleton points of radii from 2 to maxr can be found.

Note that each step generates all skeleton points of the same radius. Two skeleton points of the same radius may be close to each other. That means the skeleton may be, at most, of width two. This is due to the shape of the structuring element we choose. Also, skeleton points of the same radius within a subwaveform must have the same t value (or t and $t+1$ in the case of width two) and therefore, their plots form one (or two, at most) vertical line segment. In the real implementation, those skeleton points with same t value can be saved as only one tree node, in the case of width 1. In the case of width two, we combine these two close skeleton points into one with radius one half more. That is, let $P_1 = (r, t_1)$ and $P_2 = (r, t_1 + 1)$ be two close skeleton points of radius r at $t = t_1$ and $t_1 + 1$, respectively, then we combine them into one point $P = (r + 0.5, t_1 + 0.5)$ with radius $r + 0.5$ at $t = t_1 + 0.5$.

After all skeleton sets S_i are found, the construction of the skeletal tree is simple. The parent-son relationship is built as follows. Remember that the nodes of the skeletal tree are the skeleton points of different sizes. Let $P = (r, t_1) \in S_r$ and $Q = (r+1, t_2) \in S_{r+1}$ be any two skeleton points of radius r and $r+1$, respectively. (Note that, in the case of skeleton points of radius $r + 0.5$ exist, we consider skeleton points of radii r and $r + 0.5$ first, then those of $r + 0.5$ and r.) Then node P is a son node of Q, if and only if:

$$|t_2 - t_1| \leq r + 1$$

Similarly, every point in S_0 is naturally a leaf node and all nodes which have no sons are leaf nodes. Leaf nodes are particularly interesting since they are candidates of ridge points. Ridge point detection is our major concern in this paper.

3.3 Roof Edge Detection

Every leaf node in the skeletal tree corresponds to a peak (local maximum). Roof edges and top flats are local maxima; so they are peaks in the waveform. They can be located from the leaf nodes of the skeletal tree. The difference between them is the radius associated with the leaf node. Roof edges associate with a small radius and top flat with a large radius. The size depends on the resolution of the picture. However, not every leaf node corresponds to a roof edge or a top flat, since noise may generate peaks too. The noise subwaveforms must be small compared with the roof or top flat subwaveforms. To distinguish between small subwaveforms and large subwaveforms, we set a threshold value T. The decision then becomes very simple since most nodes have only one son. Those nodes, which have more than one sons, must be corresponding to more than two subwaveforms (or two peaks). If one son node P_1 is of radius less than T and the other son node P_2 is of radius greater than or equal to T, we claim that node P_1 corresponds to a noise peak. Therefore the whole subtree rooted by node P_1 must be removed. On the other hand, if both node P_1 and node P_2 are of radius less than T, then both nodes may correspond to two noise peaks or one node corresponds to a roof peak and the other node a noise peak but close to the roof peak. In such case, we check the difference between the radius of node P_1 (or P_2) and the radius of their parent node Q. If this difference is large (more than 3 in our implementation), then both node P_1 and node P_2 must be noises. Otherwise, one of them is a roof peak and the other is a noise peak which is close to the roof. In such case, a heuristic decision is made by removing the one with smaller

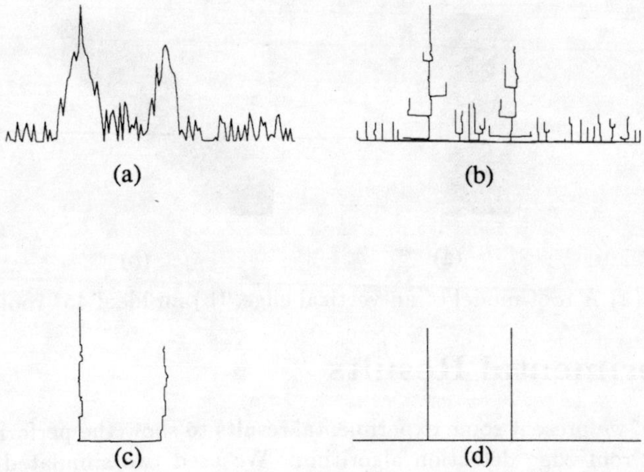

Figure 5: (a) Same as Figure 3b, (b) skeletal tree of (a), (c) result after removing noise peaks, (d) two roof edge locations are found.

gray scale value. Figure 5 demonstrates steps of this roof edge detector. Figure 5b shows the skeletal tree of Figure 5a, which is the same as Figure 3b. Figure 5c shows the result after the small branches, which correspond to noise peaks, are trimmed. From Figure 5c, we see that there are only two leaf nodes left, which correspond to the original roof edge pixels and the result is shown in Figure 5d.

3.4 Computational Complexity

Since the threshold T determines the acceptance of a subwaveform as a roof subwaveform or the rejection as a noise subwaveform. Any subwaveform has size greater than or equal to T will not be a noise subwaveform. Therefore, the skeleton set $\{S_i\}, 0 \leq i \leq maxr$, only needs to be calculated up to $maxr = T$. In all the experiments, which will be described in next section, $T = 10$ is large enough. The time complexity analysis is similar to that described in [7], since the computation time in this roof edge detection algorithm is basically dominated by the skeletonization process. As analyzed in Section 3.3.1, the skeletonization process of raster scan fashion (1-D) takes $(T^2 + 5T + 2)$ parallel min/max operations on a parallel machine, or, about $(T^2 + 5T + 2) \cdot N^2$ operations on a conventional sequential machine for image of size $N \times N$ pixels.

(a) (b)

Figure 6: (a) A roof model of an vertical edge, (b) an ideal 45° roof model.

4 Experimental Results

In this section, we present some experimental results to show the performance of our morphological roof edge detection algorithm. We used two simulated roof images and four range images for the experiments. For the purpose of comparison, the images in all experiments were processed by applying 1) our algorithm, 2) Differential-Laplacian-of-Gaussian (DLOG) edge detector, (3) cubic facet ridge detector (Haralick [1]), and (4) ring operator (Inokuchi, et al [2]). The Differential-Laplacian-of-Gaussian (DLOG) edge detector is implemented by taking the first order difference along its gradient direction and then apply the Laplacian-of-Gaussian (LOG) edge detector. The ring operator can classify pixels into jump edge, roof edge and plane in range images. We use it for the detection of only roof edges. Before running the ring operator, three thresholds must be set. We have run the ring operator for several different sets of thresholds and only the best results will be shown.

The first simulated image is an ideal vertical roof edge, shown in Figure 6(a), which has slope +10 (in pixel units) in one side and −10 in the other side. The size of the image is 128 × 128 pixels. The second image, shown in Figure 6(b), simulates an ideal 45° roof edge which is obtained by 45° rotation of Figure 6(a). If we take the differential operation on the first image, the result will have an ideal major jump edge with edge contrast 20 plus two minor jump edges with edge contrast 10 on both sides of the major jump edge, which correspond to two concave ridges. The signal to noise ratio (S/N) for the jump edge is defined as edge contrast divided by noise standard deviation. For these two simulated images, we add to them the various Gaussian noise with zero mean and standard deviation 2.5, 5.0, 10.0, 15.0 and 20.0. Thus the S/N of the major jump edges are 8.0, 4.0, 2.0, 1.33 and 1.0, respectively.

To evaluate the performance of different roof edge operators, we use:

1. the conditional probability of the label edge [1] given the true edge $P_0 = Prob(L|E)$, where L is the set of the label edges and E is the set of the true edges, [2]

[1] Label edge is referred to as the pixel acclaimed to be edge pixel by the edge detector.

[2] P_0 is approximated by $n(E \cap L)/n(E)$, where $n(S)$ is the cardinal number of the set S.

2. the conditional probability of the true edge given the label edge $P_1 = Prob(E|L)$, and [3]

3. mean distance (Δ) of label edge pixels to the center of the true edge.

The measurements of P_0 and P_1 indicate how good the detection is and the measurement of Δ indicates how good the localization is. The higher the P_0 and P_1, the better the performance. Since the higher the P_0, the lower the probability of falsely marking nonedge pixels; and the higher the P_1, the lower the probability of failing to mark real edge pixels. The measurement of Δ should be small for good localization. The adjustable parameters for each edge operator are chosen to equalize these two conditional probabilities P_0 and P_1.

The DLOG detector responds to convex roof edges as well as concave roof edges. Therefore, when calculating the probabilities P_0 and P_1, the set E of true edges for this detector will include both convex roof edge pixels and concave roof edge pixels. Our morphological roof edge detector responds to convex roof edges only. The set E only includes convex roof edge pixels in the evaluation of our algorithm. To detect the concave roof edges, our algorithm can be easily modified based on the dual properties of morphological operations (erosion duals to dilation and opening duals to closing).

Figure 7 shows a sequence of edge images which are the results of applying our algorithm to the two simulated images with zero mean Gaussian white noise of standard deviations 0.0, 2.5, 5.0, 10.0, 15.0 and 20.0. Figure 8, 9, and 10 show the corresponding parts of DLOG edge detector, cubic facet ridge detector, and ring operator, respectively. Those edge detectors are implemented in the C language on VAX-11/780. The computation time (user time plus system time) of our algorithm for each image is about 18 seconds. That of DLOG algorithm is about 1030 seconds, cubic facet algorithm 263 seconds, and ring operator 42 seconds. Figure 11 and Figure 12 show the performance plots of those detectors. To suppress image border effects, the performance probabilities P_0 and P_1 are measured in a window of size 100×100. The locality (Δ) are measured in the neighborhood of the ideal roof edge with distance half way to the two minor edges. The results show that our algorithm have much better performance than the other edge detectors. The DLOG operator produces thick edge lines and lots of false edge pixels. Cubic facet ridge detector and ring operator have good results when the noise is small.

Finally, we illustrate by using examples of applying our algorithm to four range images. The images are shown in Figure 13 (a), Figure 14 (a), Figure 15 (a), and Figure 16 (a), respectively. We apply our algorithm, the cubic facet ridge detector and the ring operator to these images. The resulting edge images are shown in Figure 13 (b)-(d), Figure 14 (b)-(d), Figure 15 (b)-(d) and Figure 16 (b)-(d), respectively. A visual evaluation leaves the impression that our algorithm has the best performance.

[3]Similarly, P_1 is approximated by $n(E \cap L)/n(L)$.

5 Conclusions

We have proposed a roof edge detection algorithm based on mathematical morphology. Roof edges in digital images are detected by looking for the leaf nodes of the skeletal tree. A leaf node represents a local maximum (peak) in the image. The size of subtree containing the leaf node is tested to distinguish the noise from the roof edge. The skeletal trees of gray scale images can be obtained by applying the gray scale morphological skeleton operations. Experimental results show that this algorithm is faster and has better performance than many other methods.

References

[1] R. M. Haralick, "Ridges and valleys on digital images", *Computer Vision, Graphics, and Image Processing*, vol.2, no.1, pp.28-38, 1983.

[2] S. Inokuchi, T. Nita, F. Matsuda, and Y. Sakurai, "A three-dimensional edge-region operator for range pictures", *Proc. 6th ICPR*, pp.918-920, 1982.

[3] A. Mitiche and J. K. Aggarwal, "Detection of edges using range information", *IEEE Trans. Pattern Anal. Machine Intell.*, vol. PAMI-5, no.2, pp.174-178, 1983.

[4] J. Ponce and M. Brady, "Toward a surface primal sketch", *Proc. 2nd IEEE Int. Conf. Robotics*, pp.420-425, 1985.

[5] J. Serra, *"Image analysis and mathematical morphology"*, Academic Press, New York, 1982.

[6] J. Alison Noble, "Morphological feature detection", *Proc. 2nd IEEE Int. Conf. Computer Vision*, Florida, pp.112-116, 1988.

[7] J. C. Cheng and H. S. Don, "Segmentation of Bilevel Images: A Morphological Approach", *Pattern Recognition: Algorithms, Architectures and Applications.* ed. R. Plamondon and H. D. Cheng, World Scientific Publishing, Co., New Jersey, 1991.

[8] V. Goetcherian, "From binary to grey tone image processing using fuzzy logic concepts", *Pattern Recognition*, vol.12, no.1, pp.7-15, 1979.

[9] J. S. J. Lee, R. M. Haralick and L. G. Shapiro, "Morphologic edge detection", *IEEE J. Robotics and Automation*, vol. RA-3, no.2, April 1987.

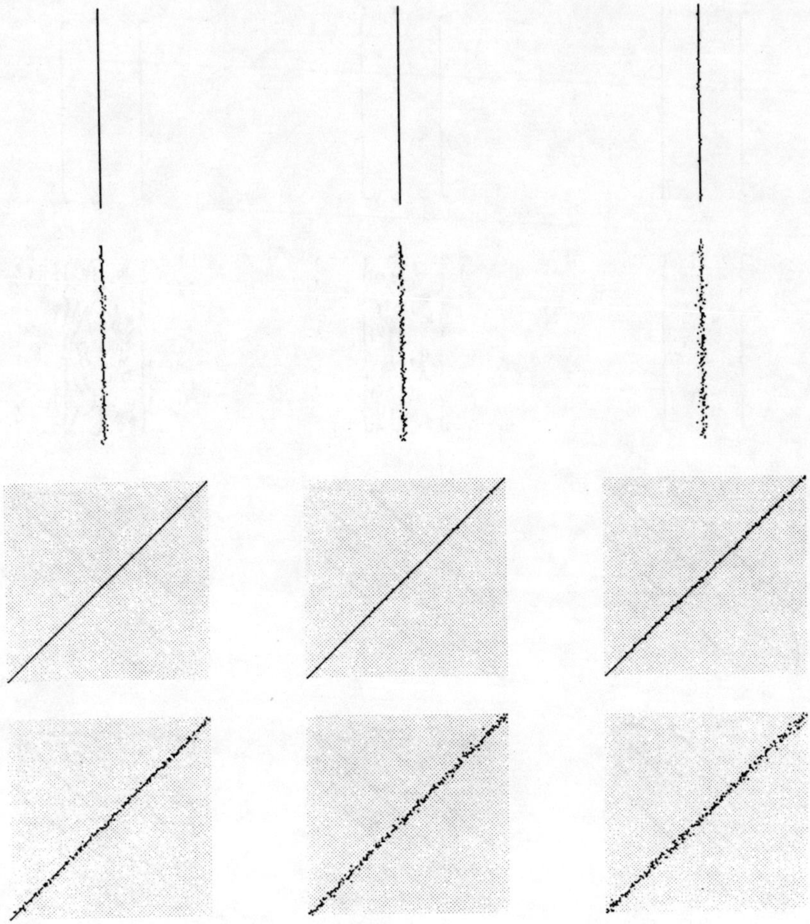

Figure 7: Results of our roof edge detection algorithm applied on images shown in Figure 6 (a) and (b) with zero mean Gaussian white noise of standard deviations 0.0, 2.5, 5.0, 10.0, 15.0 and 20.0, respectively (cpu time $\approx$ 18 sec).

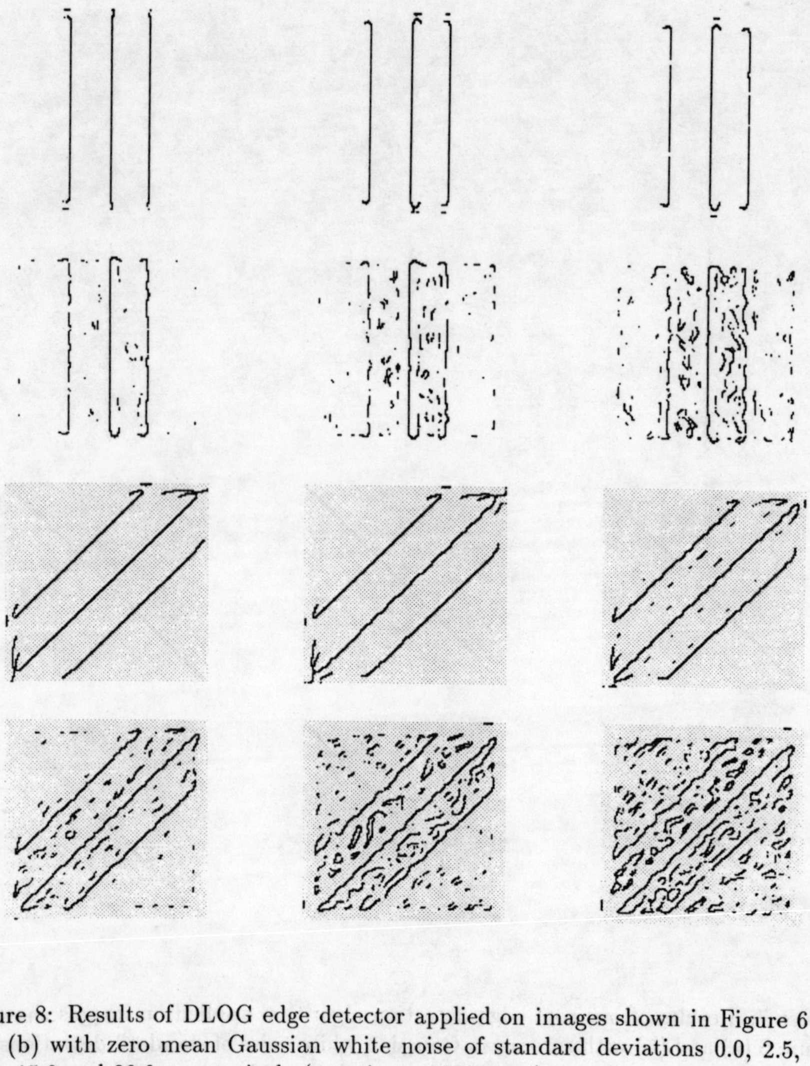

Figure 8: Results of DLOG edge detector applied on images shown in Figure 6 (a) and (b) with zero mean Gaussian white noise of standard deviations 0.0, 2.5, 5.0, 10.0, 15.0 and 20.0, respectively (cpu time $\approx$ 1030 sec.)

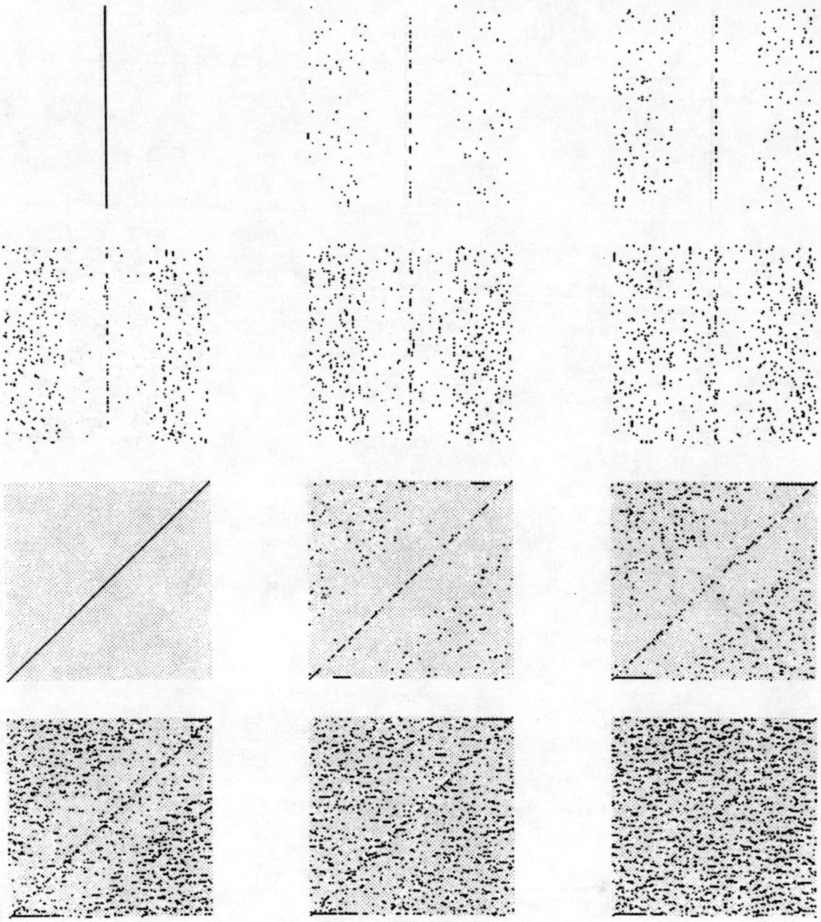

Figure 9: Results of cubic facet ridge detector (Haralick [1]) applied on images shown in Figure 6 (a) and (b) with zero mean Gaussian white noise of standard deviations 0.0, 2.5, 5.0, 10.0, 15.0 and 20.0, respectively (cpu time ≈ 263 sec.)

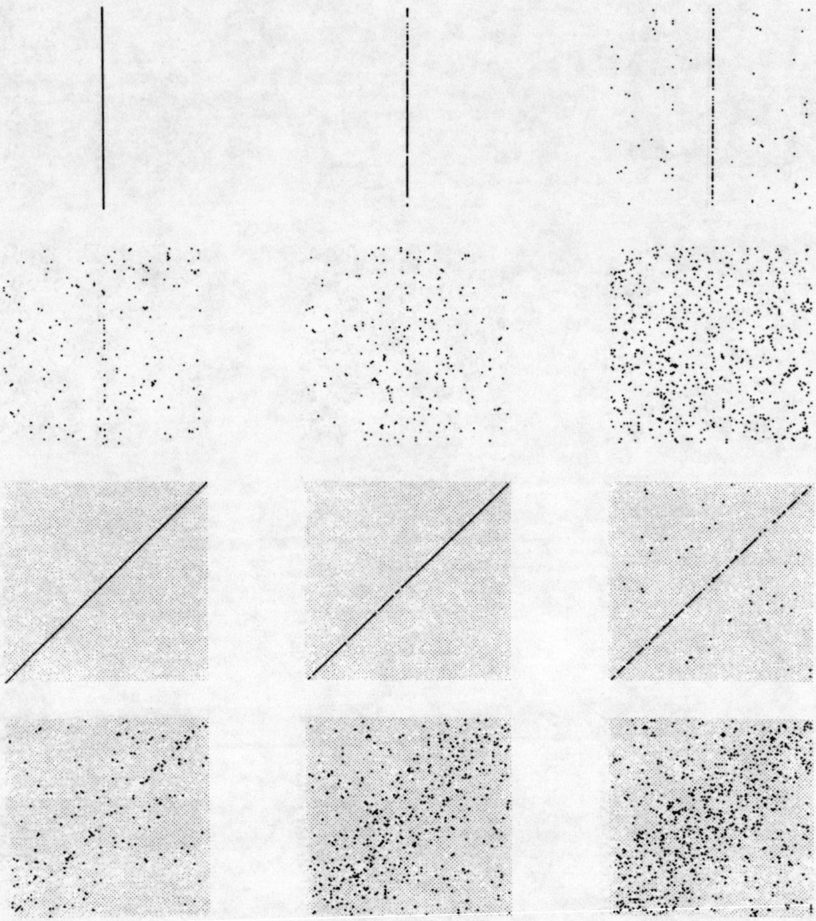

Figure 10: Results of ring operator (Inokuchi, et al [2]) applied on images shown in Figure 6 (a) and (b) with zero mean Gaussian white noise of standard deviations 0.0, 2.5, 5.0, 10.0, 15.0 and 20.0, respectively (cpu time $\approx$ 42 sec.)

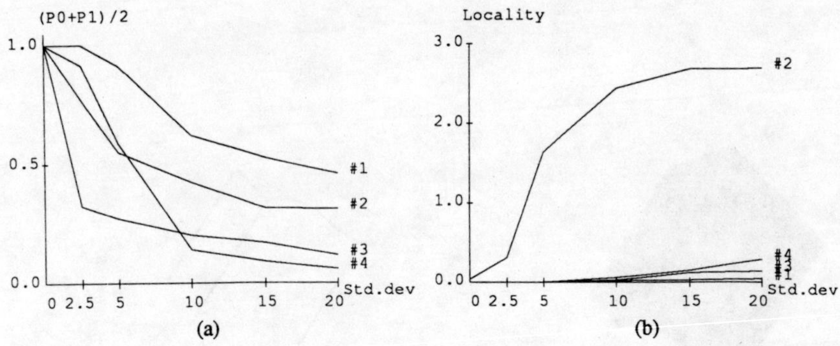

Figure 11: (a) Performance probabilities, and (b) locality measure (Δ) of roof edge detectors on noisy vertical roof edge images. Curve 1: our morphological roof edge detection algorithm, Curve 2: DLOG detector, Curve 3: cubic facet ridge detector, Curve 4: ring operator.

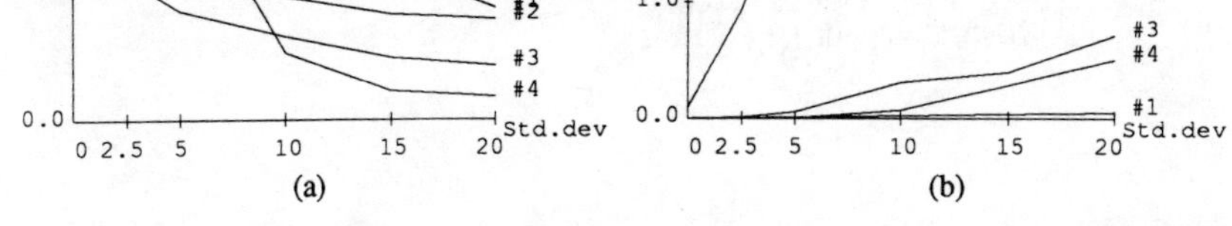

Figure 12: (a) Performance probabilities, and (b) locality measure (Δ) of roof edge detectors on noisy diagonal 45° roof edge images. Curve 1: our morphological roof edge detection algorithm, Curve 2: DLOG detector, Curve 3: cubic facet ridge detector, Curve 4: ring operator.

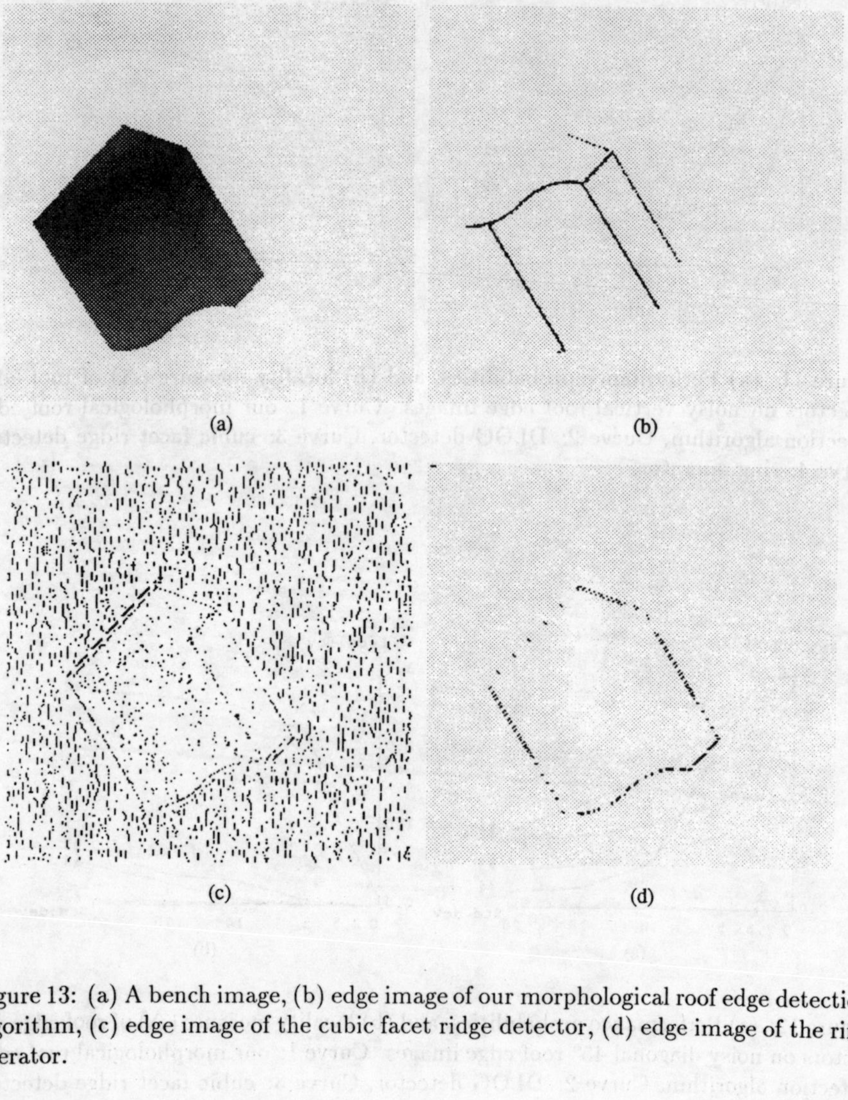

Figure 13: (a) A bench image, (b) edge image of our morphological roof edge detection algorithm, (c) edge image of the cubic facet ridge detector, (d) edge image of the ring operator.

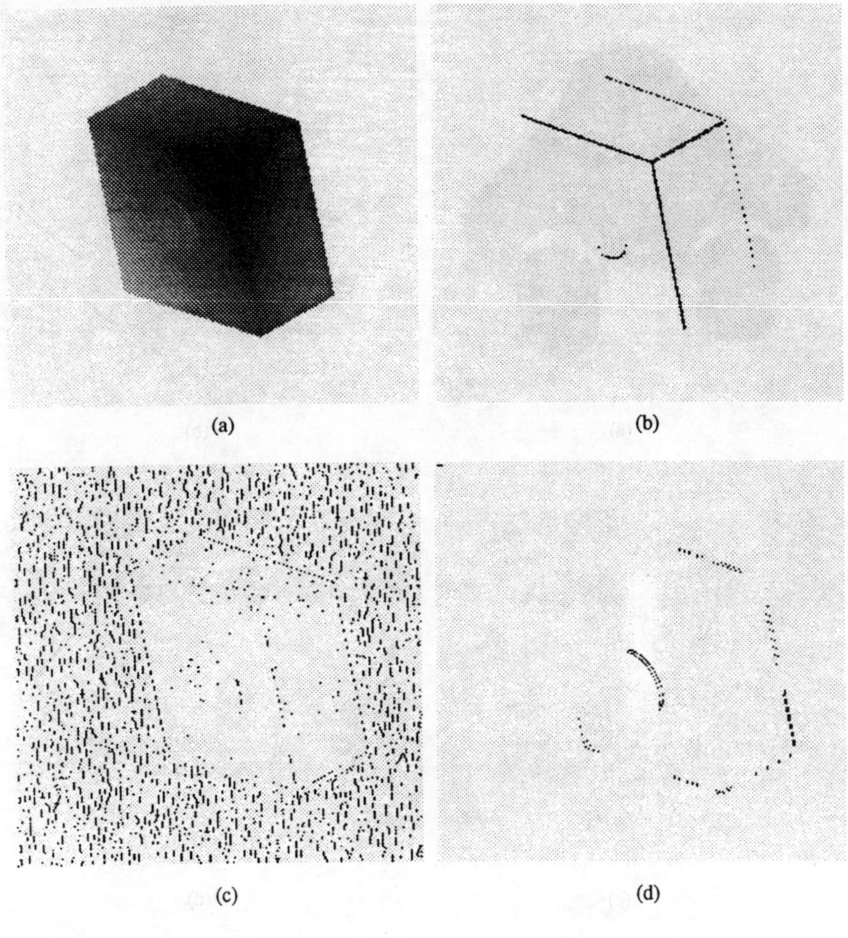

Figure 14: (a) A dice image, (b) edge image of our morphological roof edge detection algorithm, (c) edge image of the cubic facet ridge detector, (d) edge image of the ring operator.

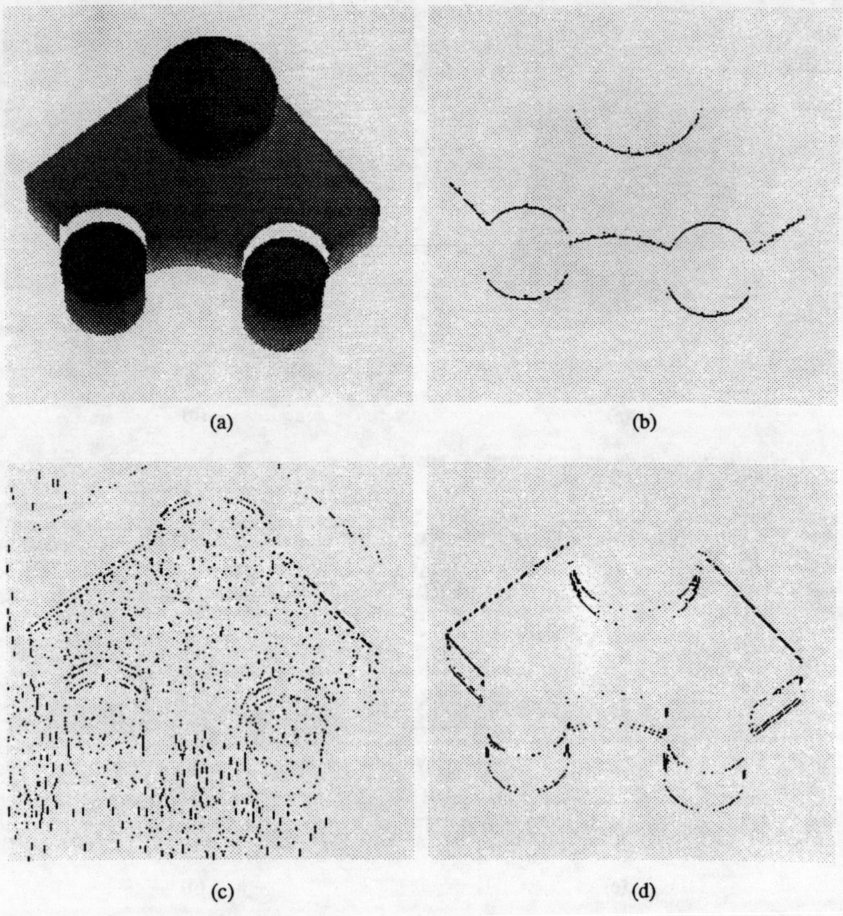

(a) (b) (c) (d)

Figure 15: (a) Third range image, (b) edge image of our morphological roof edge detection algorithm, (c) edge image of the cubic facet ridge detector, (d) edge image of the ring operator.

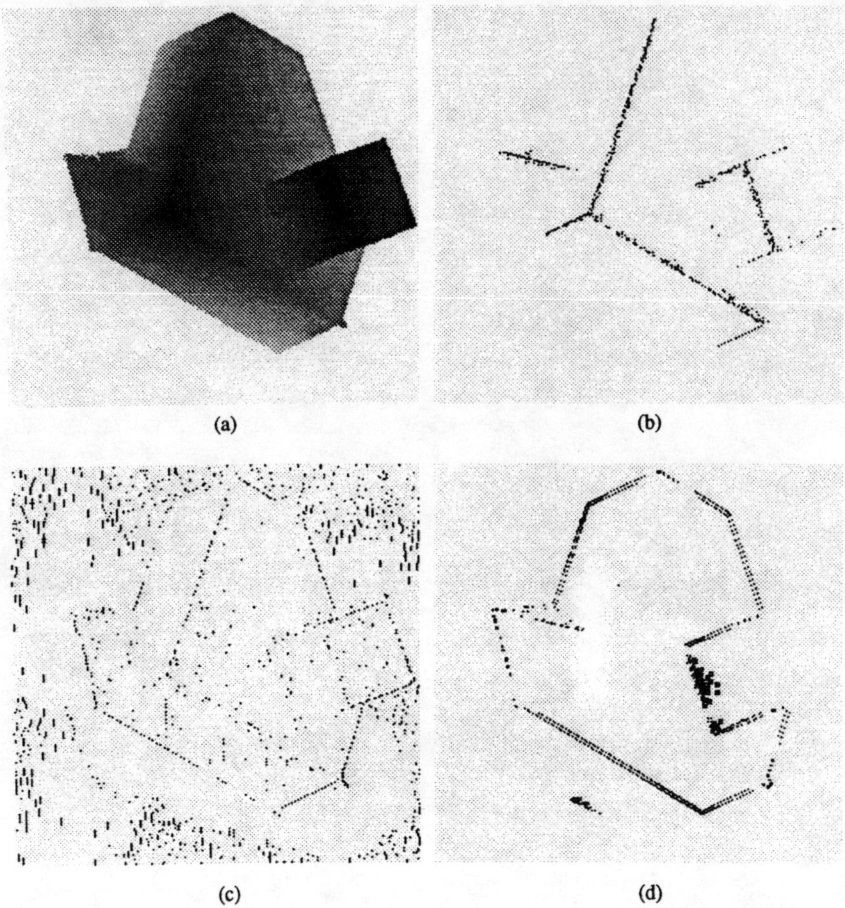

Figure 16: (a) Fourth range image, (b) edge image of our morphological roof edge detection algorithm, (c) edge image of the cubic facet ridge detector, (d) edge image of the ring operator.

Representation of Visual Textural Properties

H.C. Shen
C.Y.C. Bie
Department of Systems Design Engineering
University of Waterloo
Waterloo, Ontario
Canada N2L 3G1

Abstract

This paper presents a set of feature frequency matrices(FFM) to reflect four visual properties of a texture image. The visual properties considered are **coarseness, contrast, directionality** and **line-likeness**. We shall demonstrate the efficacy of the proposed approach by simulated as well as natural textures.

1 Introduction

Texture is a very important property of an image. it is closely related to aeronautic remote sense image, biomedical image processing, objects recognition, industrial quality inspection and many others. Texture analysis is a very important research area. Theoretically, texture can be measured on many aspects. Practically, it is mainly measured based on its statistical and structural aspects. Texture is usually described by its visual aspects such as coarseness, contrast, directionality, line-likeness, regularity, and others. Over the years, researchers have attempted to quantify texture using various features [1, 6, 8, 10, 15, 11, 14, 16]. The 'success' of a set of features is usually measured by two criteria, namely, how good is the set of features for texture classification and segmentation and how closely does the set of features resemble human visual perception. In [12, 13], we have proposed a new approach to represent textures. A set of feature frequency matrices (FFM) of different dimensions at various resolution levels is defined. Its efficacy is demonstrated by the high rate of correct classification when compared with some existing models.

In this paper, based on the feature frequency matrices, we shall define a meaningful feature set to reflect four most common visual textural properties, namely, **coarseness, contrast, directionality** and **line-likeness**. The efficacy of this proposed approach will be demonstrated by simulated as well as natural texture images.

2 The FFM Representation

In [12, 13], we have proposed a scheme of representation of textures, i.e. a set of feature frequency matrices (FFM) of different dimensions at various resolution levels. In this section, we shall briefly give the definitions of feature frequency matrices. It is believed that images at different resolution levels include different information for texture differentiation [11]. Given an image at certain resolution level, a set of operators can be used to extract the corresponding features. By applying one operator to the image, a feature image can be obtained. From the feature images, feature frequency matrices in different dimensions can be extracted and they are defined as follows:

Definition 1: Feature Image, (F)

Given an $(N1 \times N2)$ image S at a certain resolution level and an $n \times n$ operator OP, a feature image F is obtained by convolving OP over the image S, $F = \{0 \leq f(i,j) \leq L-1 \mid \frac{n-1}{2} \leq i \leq N1 - \frac{n-1}{2}; \frac{n-1}{2} \leq j \leq N2 - \frac{n-1}{2}\}$, where L is the number of feature values in the feature image.

Definition 2: 1-D Feature Frequency Matrix, (FFM_1)

Given a feature image F, the 1-D feature frequency matrix is defined by a vector FFM_1, where the element $ffm(i)$ is the number of pixels with feature value i in F.

Definition 3: 2-D Feature Frequency Matrix, (FFM_2)

Given two feature images, $F1$ and $F2$, the 2-D feature frequency matrix for $F1$ and $F2$ is defined as a matrix FFM_2 where the element $ffm(i,j)$ is the number of pixels with feature value i in $F1$ **and** feature value j in $F2$.

Definition 4: Generalized n-D Feature Frequency Matrix, (FFM_n)

Given n feature images, $F1$, $F2$, ..., Fn, the generalized n-D feature frequency matrix for these feature images is defined as an n-D array (matrix), FFM_n, where the element $ffm(k_1, k_2, \ldots, k_n)$ is the number of pixels with feature value k_j in the feature image Fj, $j = 1, \ldots, n$.

The representation scheme is shown in Figure 1. From the definitions, it can be seen that the low dimension feature frequency matrix can be considered as the 'marginal frequency matrix' of higher level frequency matrix. The higher dimension feature frequency matrix include more joint structural information. However, within the context of this paper, only 2-D feature frequency matrices are explored to measure visual textural properties.

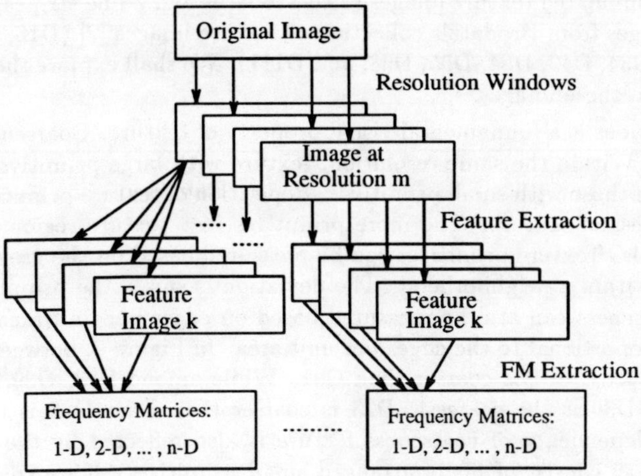

Figure 1: Representation Scheme

3 Texture Features Corresponding to Visual Properties

Researchers have carried out investigations on how human beings describe and differentiate textured images for a long time. Julesz [9] found that human beings are very sensitive to the second order statistics. From this view, many texture image feature representations were developed. The second order co-occurrence matrix [7] is one representative of them. Campbell [3] proposed a multi-channel filtering theory for processing visual information in the early stages of human visual system. According to that theory, the human visual system decomposes the retinal image into a number of filtered images which contains intensity variations over a narrow range of frequency and orientation. Based on this theory, Faugeras [5] and Coggins [4] have developed various techniques to extract texture features. Visual textural properties are reflected on the distributions of grey levels and the changes of grey levels. When human being looks at a textured image, the most important features captured are the texture pattern (locally) and how the patterns are presented in the image. The pattern and its presentation can be described by some visual concepts like coarseness, contrast, directionality, line-likeness, regularity, complexity, etc. which are relatively easy to measure by statistical methods. From these properties, many researchers developed different methods to measure them [1, 14]. These visual properties are called statistical visual properties. For a homogeneous texture, there are many visual textural properties. Four of them are considered most important for visual perception [1, 14]. They are: **Coarseness, Contrast, Directionality, and Line-likeness.** Figure 2 shows four simulated texture images (2 chessboards and 2 line stripes) and 12 natural texture images from Brodatz's collection of texture images [2] (D16, D20, D21, D23, D29, D30, D34, D49, D54, D55, D68, and D111). We shall explore the visual textural properties in these images.

Coarseness is a fundamental visual property of texture. Coarseness is resolution dependent. Within the same resolution, texture with large primitives or particles is coarser than those with small primitives. Repetition of texture primitives also reflects the coarseness of texture. The more primitives in a texture region, the less coarse the texture is. Texture primitives can be measured based on the grey level deviation of texture within a neighborhood. The deviation is small, the primitive is relatively large. Coarseness can also be measured based on edges per unit area. Coarseness is inversely proportional to the edges per unit area. In Figure 2, between the simulated two chessboards, C01 is coarser than C02. While among D23, D29, D30, and D54, we can see D30 is the coarsest; D23 is coarser than D54; D29 is the finest. The resolution dependence of coarseness feature is also reflected by the fact that large primitive may consist of finely arranged small primitives. If we consider the small primitives, originally coarse texture may become fine. An edge threshold may be set to suppress the edges from the small primitives.

Contrast measures the degree of differences and spatial frequency changes of grey levels in a texture region. High contrast texture has large grey level differences and high spatial frequency changes in a region and low contrast texture has relatively small differences and low frequency changes. Tamura [14] assumes four factors for contrast: dynamic range of grey-levels, polarization of the distribution of black and white on the grey-level histogram or ratio of black and white areas, sharpness of edges, and period of repeating patterns. The fourth factor reflects the connection between coarseness and contrast. For example, when we see two chessboard patterns which only differ in scale, the one with large scale looks both coarse and high contrast. In [14], Tamura assumed the smaller primitives has a high contrast. It is not consistent with the human visual perception. Contrast can be considered in two aspects: locally and globally. Locally high contrast texture image may be low contrast globally. Using the two texture image with stripes in Figure 2 as an example, although both images consist of same grey levels. S02 with large stripes has a globally (also visually) higher contrast than S01 with smaller stripes while locally the latter has a higher contrast. We will consider grey level distribution and local dynamic range of grey levels as the measurement of texture contrast. The local dynamic range will measure the local contrast and the grey level distribution will reflect the global contrast. Usually with the same grey level distribution, the high local contrast will have a low global contrast. The relationship between the global and local contrast justifies the resolution dependence of contrast. In Figure 2, between C01 and C02, C01 has a higher contrast. Also D20 has a higher contrast than D16.

Directionality measures the directional property of grey level changes. Two aspects are important with the directionality of texture. One is the direction of

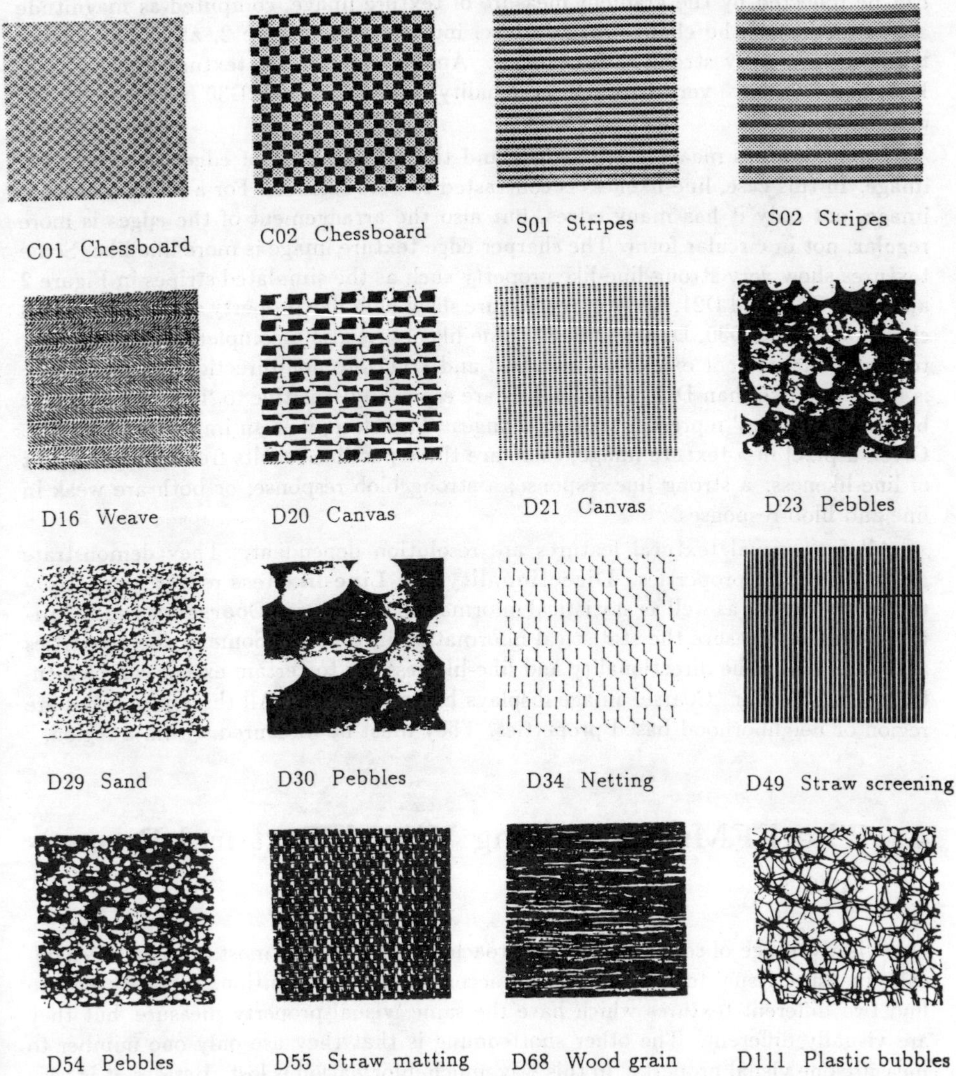

Figure 2: Simulated and Natural Textures

grey level changes and the other is the extent of the changes. Thus, directionality can be reflected by the gradient measure of texture image, computed as magnitude and direction of the change in grey level intensity. In Figure 2, all the simulated textures have very strong directionality. Among the natural texture images, D49, D55, and D68 have very strong directionality while D29, D23, D30 have shown non-directionality.

Line-likeness measures the edges and the arrangement of edges in a textured image. In this case, line-likeness is contrasted to blob-likeness. For a line-like texture image, not only it has many edges, but also the arrangement of the edges is more regular, not in circular form. The sharper edge texture image is more line-like. Some textures show very strong line-like property such as the simulated stripes in Figure 2 and D49, D68, and D21, while other texture show blob-like property such as simulated chessboards and D30, D23, and D29 . Line-likeness is also a complementary property to directionality. For example, both D23 and D111 are non-directionality, but D111 is more line-like than D23. Human eyes are especially sensitive to 'lines' in an image because the 'lines' represent sudden changes. Therefore, it is an important measure. Given a pixel in a texture image, there are three possible results from the viewpoint of line-likeness: a strong line response; a strong blob response; or both are weak in line and blob responses.

All four visual textural features are resolution dependent. They demonstrate global and local properties. **Directionality** and **Line-likeness** reflect some structural information as well as statistical information in texture. **Coarseness** and **Contrast** mainly measure the statistical information in texture. Some of the properties are correlated. The directionality and line-likeness are to certain extent complementary to each other. Coarser image displays higher contrast. All these properties are region or neighborhood based properties. They must be measured within a region.

4 The FFMs Measuring Visual Textural Properties

One disadvantage of some previous approaches [1, 14] is that most of them mechanically measure visual textural properties according to the definition. It is very easy to find two different textures which have the same 'visual property measure' but they are visually different. The other shortcoming is that they use only one number to measure one visual property. In this way, much information is lost. Besides, it is very difficult to use only one number to measure both global and local properties. We propose to represent visual properties by 2-D feature frequency matrices (FFM) [12, 13]. Not only the visual properties are properly represented, some structural information related to these properties are also measured. Therefore, much information can be retained by the FFM representation.

Based upon the concept of 2-D FFM, we proceed to define the features and the corresponding frequency matrices for each of the visual properties.

4.1 Coarseness

We define the deviation over a local window and the edges per unit area to measure the coarseness. One deviation operator and four edge operators are used to extract coarseness information.

Deviation operator calculates the grey value deviation over an $m \times m$ window on the original image. By applying the operator to the original image, deviation feature image D_σ is obtained. The deviation at pixel (i, j) is computed by:

$$\sigma_{i,j} = \frac{1}{m^2} \sum_{k=i-(m-1)/2}^{i+(m-1)/2} \sum_{l=j-(m-1)/2}^{j+(m-1)/2} |g(k,l) - \mu_{i,j}| \quad (1)$$

where

$$\mu_{i,j} = \frac{1}{m^2} \sum_{k=i-(m-1)/2}^{i+(m-1)/2} \sum_{l=j-(m-1)/2}^{j+(m-1)/2} g(k,l) \quad (2)$$

Four edge detectors for 0, 45, 90, 135 degree directions shown in Figure 3 are used to detect edge response in every pixel of the original image. The maximum response of the four edge detectors in a pixel is the edge response in the pixel. One threshold is set for edge response. If the edge response is larger than the threshold, the output of edge detector is '1'; otherwise the output is '0'. Therefore, by applying the four edge detectors to the original image, we obtain an edge feature image with pixel value either '1' for large edge response or '0' for small edge response. Then we apply an edge-per-unit-area operator on the edge feature image. The operator simply counts the '1's in an $m \times m$ region. Thus, the output of the edge-per-unit-area operator $epa_{i,j}$ is between 0 and m^2. From the output of the edge-per-unit-area operator, we obtain another feature image of edge-per-unit-area D_{epa}. From D_{epa} and D_σ, the 2-D FFM F_{crs} can be obtained.

Independently, $(\sigma_{i,j})$ and $(epa_{i,j})$ measure the coarseness information. Jointly, i.e. 2-D FFM, they provide the statistical-structural information about coarseness $(epa_{i,j}, \sigma_{i,j})$. $(\sigma_{i,j})$ to some extent depends on $(epa_{i,j})$. That two pixels have the same *epa* does not mean they have the same σ, and vice versa. If texture is coarse, $\sigma_{i,j}$ should be relatively small indicating large primitives. Edge-per-unit-area value should be small too. Therefore, the peaks of the FFM will occur closer to the origin of the axes.

$$\begin{matrix} 1 & 1 & 1 \\ 0 & 0 & 0 \\ -1 & -1 & -1 \end{matrix} \qquad \begin{matrix} 1 & 0 & -1 \\ 1 & 0 & -1 \\ 1 & 0 & -1 \end{matrix}$$

(a) (b)

$$\begin{matrix} 0 & 1 & 1 \\ -1 & 0 & 1 \\ -1 & -1 & 0 \end{matrix} \qquad \begin{matrix} 1 & 1 & 0 \\ 1 & 0 & -1 \\ 0 & -1 & -1 \end{matrix}$$

(c) (d)

Figure 3: Four edge detectors

4.2 Contrast

Two operators are used to extract contrast information. One operator extracts information about the local dynamic range of grey levels D_{drg}, the other extracts the grey level information at each pixel D_{gl}. From the 2 feature images, a 2-D FFM for contrast measure F_{con} can be obtained.

Let $s_1, s_2, ..., s_{m^2}$ represent the $m \times m$ grey level values in observation neighborhood of m-window at (i, j). The grey value is between 0 and $L-1$. The output of dynamic range operator is defined as:

$$drg_{i,j} = \frac{1}{m^4} \sum_{k=1}^{m^2} \sum_{p=1}^{m^2} |s_k - s_p| \tag{3}$$

where $0 \leq drg_{i,j} \leq L/2 - 1$. The 2-D FFM for contrast is a $L \times L/2$ matrix. While the dynamic range operator measures the local contrast, the grey level distribution reflects the global contrast. If the texture has a high local contrast, the dynamic grey level range have high values. If the texture has a high global contrast, the dynamic range of grey values will be relatively low and the peaks of grey level distribution will be far apart.

4.3 Directionality

Directionality can be measured based on the gradient operators, e.g. the Sobel operators as shown in Figure 4. In general, two $(m \times m)$ gradient operators can be defined to extract gradient information (magnitude and directionality) for a m-window. Then the magnitude and directionality are normalized. Therefore, from the magnitude feature image and directionality feature image, 2-D directionality FFM F_{dir} can be formed.

If texture image has a high directionality, F_{dir} should have some "banded" areas

-1	0	1		1	2	1
-2	0	2		0	0	0
-1	0	1		-1	-2	-1
	(b) Sx				(a) Sy	

Figure 4: Sobel Operators

corresponding to the normalized directionality values. The extent of grey level changes in certain direction is reflected by the normalized magnitude.

4.4 Line-likeness

Line detectors and line-likeness detector can be used to detect line-likeness. Line detectors will find how sharp is the edges for the texture and line-likeness detector will find how edges are distributed in the texture. Four line detectors are used to detect lines: 0° , 90°, 45° and 135°. For (3×3) window, the four operators are defined in Figure 3. The outputs of four line detectors at (i, j) are recorded as $E_0(i, j)$, $E_{90}(i, j)$, $E_{45}(i, j)$, $E_{135}(i, j)$. Traditional methods use one edge detector to obtain one edge image, we combine the four edge feature images as one edge feature image, $E(i, j)$, and together with a line-like feature image, $B(i, j)$, to form a 2-D FFM. The edge feature image, $E(i, j)$, is defined as the maximum of $\{E_0(i, j),\ E_{90}(i, j),\ E_{45}(i, j),\ E_{135}(i, j)\}$ and is normalized to between 0 and $(L - 1)$. From the edge feature image, we obtain an edge register image, $R(i, j)$, which is defined as :

$$R(i,j) = \begin{cases} 0 & E(i,j) < T \\ 1 & E(i,j) = E_0(i,j) \\ 2 & E(i,j) = E_{45}(i,j) \\ 3 & E(i,j) = E_{90}(i,j) \\ 4 & E(i,j) = E_{135}(i,j) \end{cases}$$

where T is a threshold for edge response. Line-like operator will reveal how the edges are distributed in a $W \times W$ local observation window over $R(i, j)$. If pixel (i, j) is very line-like, the directionality of edges in the window will be relatively consistent. If the edges in the observation window are in different directions, then it is not very line-like at pixel (i, j) since the edges may 'cancel' each other and tend to form blobs.

To account for the cancelling effect within a window, we define the line-likeness detector, $B(i,j)$, at pixel (i,j) in the following way:
Let K be the maximum value in the edge register image;
r_k be the number of pixels in the $W \times W$ window with edge register value k; and a function $\beta(k)$ be defined as

$$\beta(k) = \begin{cases} 0 & k = 0; \\ |\beta(k-1) - r_k| & k > 0 \end{cases}$$

Then,

$$B(i,j) = \beta(K).$$

$B(i,j)$ is between 0 and W^2. If the texture within the window centred at (i,j) is more line-like, the directionality will be relatively consistent and $B(i,j)$ will be relatively large. If the texture is more blob-like, the directionality will be relatively evenly distributed and $B(i,j)$ will be relatively small. Thus, the 2-D FFM formed based on $E(i,j)$ and $B(i,j)$ is a $L \times (W^2+1)$ matrix.

If an image is not line-like, line-like response will be relatively low, F_{lin} will spread out along the low value line-likeness frequency axis. If an image has strong edge response, F_{lin} will spread out over high edge value areas. Usually, line-likeness response will be more spread out if there are more edges or if the edge responses are high.

5 Experiments

Four simulated textures (chessboards and stripes) and 12 natural textures from Brodatz's texture album [2] as shown in Figure 2 are used to demonstrate the effectiveness of the proposed approach.

The grey level of the original texture image ranges from 0 to 255. For the purpose of faster computation, we compress the range to between 0 and 63, i.e. $L = 64$. Window size of 5 by 5 is used to extract the local grey level deviation, edge per unit area and line-likeness features. The Sobel gradient operators (Figure 4) are used to extract the magnitude and direction of the gradient measure. The magnitude of the gradient is normalized to values between 0 and 63 while the direction of the gradient is normalized to values between 0 and 32 with 32 indicating non-directional. Thus, the sizes of the feature frequency matrices for coarseness, contrast, directionality and line-likeness are (26×32), (64×32), (64×33) and (64×26) respectively. The FFM of the 16 texture images shown in Figure 2 are extracted.

In the following subsections, we will discuss how the 2D feature frequency matrices proposed reflect the visual texture properties. In general, the location and number of peaks, the spread of the distribution along each axes in the FFM provide information to describe the visual properties of the corresponding texture image.

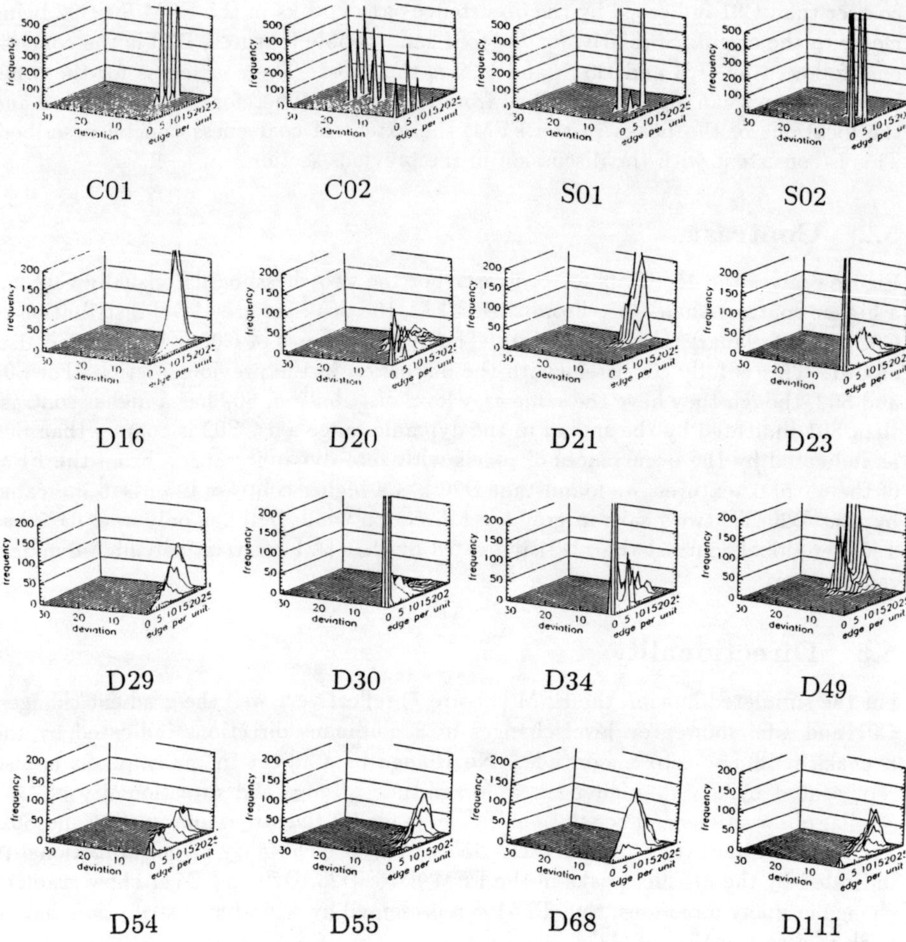

Figure 5: Feature Frequency Matrices (FFM) for Coarseness

5.1 Coarseness

Figure 5 gives the 16 FFM for coarseness. For the two simulated textures, C02 is coarser than C01 indicated by the occurrence of the peaks in the FFM for C02 being closer to the origin. Similarly, for the four sand pebble textures, D30 is the coarsest one, followed by D23 and D54 , and D29 is the finest. The coarseness for the other texture images can also be identified from their FFM. Therefore, by the spread and the locations of the peaks in the FFM, the extent of coarseness can be described. This is consistent with the discussion in the previous section.

5.2 Contrast

Figure 6 gives the 16 FFMs for contrast. For the two chessboards, visually C02 has a higher contrast than C01. From their FFM, the peaks of grey level distribution of C02 is farther apart than that of C01. The dynamic range of C02 is higher than that of C01. This is fully consistent with the discussion in the previous section. For S01 and S02, though they have the same grey level distribution, S02 has a higher contrast than S01 indicated by the spread in the dynamic range axis. S02 is coarser than S01 as indicated by the occurrences of pixels with zero dynamic range. From the FFM of the natural textures, we found that D20 has a higher contrast than D16 indicated by that D20 has two peaks in grey level direction while D16 has only one. D20 has a higher global contrast than D21 indicated by that D21 has relatively high dynamic range values.

5.3 Directionality

For the simulated images, the FFM (Figure 7) reflect very well the gradient changes. C01 and C02 show grey level changes in 8 dominant directions indicated by the 8 peaks with non-zero magnitude. No change in gradient in many pixels is also represented by a very high peak at magnitude of zero and directionality of '32'. Similar observations and conclusions can be made for the stripe images, S01 and S02. For the natural textures, D49, D21, D55 and D68 have fairly strong directionality indicated by the distinct peaks in the FFM. D16, D23, D29 and D111 show gradient change in many directions, thus FFM is represented by a 'flatter' distribution and no peak at directionality of '32'.

5.4 Line-likeness

Figure 8 shows the 16 FFM for line-likeness. Comparing the chessboard and stripe images, obviously, the stripe images are more line-like than the chessboard images. We can infer this characteristic from the FFM. The locations of the peaks of the S01 FFM is away from the origin and occur at the high values of the line-likeness axis. For the natural textures, D49 demonstrates the most line-like indicated by the high

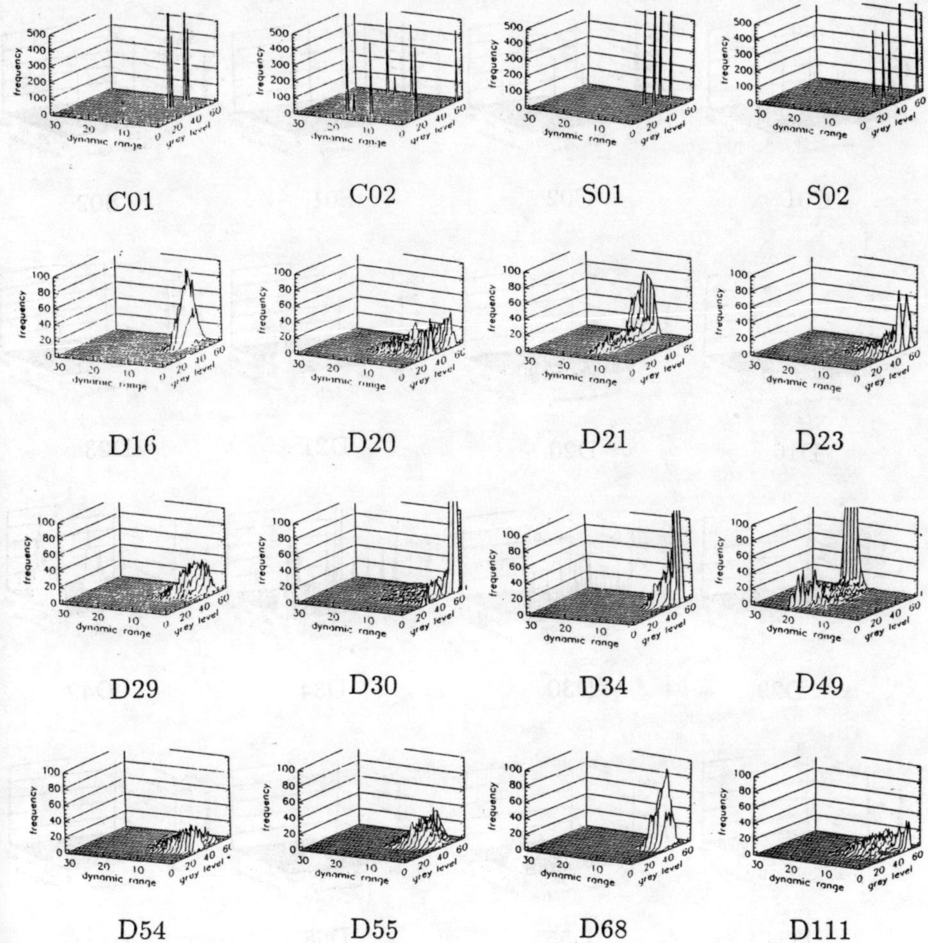

Figure 6: Feature Frequency Matrices (FFM) for Contrast

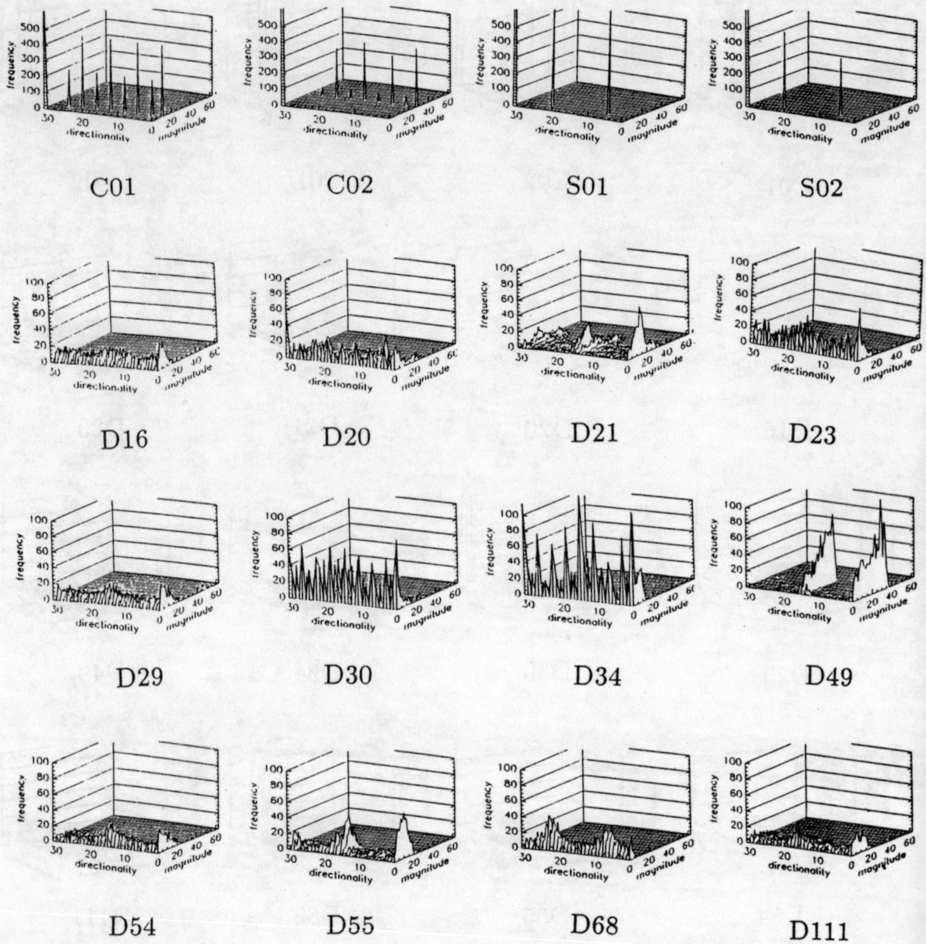

Figure 7: Feature Frequency Matrices (FFM) for Directionality

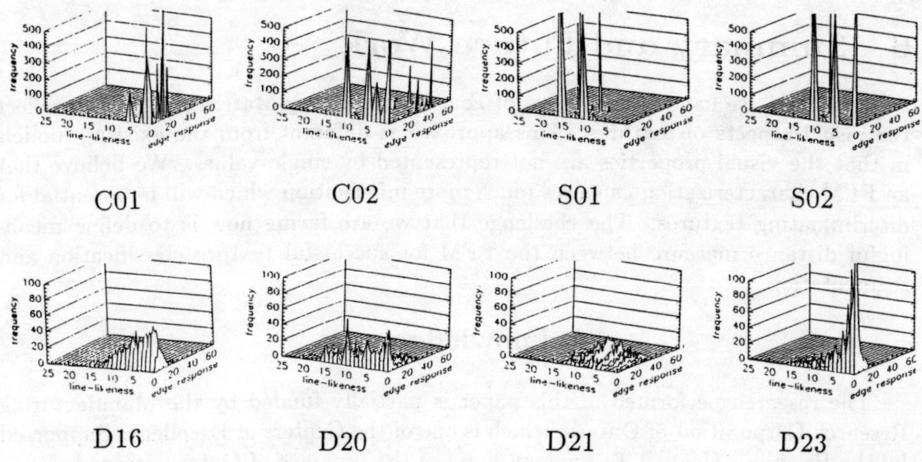

Figure 8: Feature Frequency Matrices (FFM) for Line-likeness

response along both axes. D23, D30, D34 are not very line-like reflected by the peaks occurred close to the origin. Other textures vary in between.

6 Summary and Future Work

In this paper, we have defined a set of Feature Frequency Matrices (FFM) to reflect the visual aspects of textures. This approach is different from the existing models in that the visual properties are not represented by single values. We believe that an FFM characterization captures much more information which will be essential for discriminating textures. The challenge that we are facing now is to define meaningful distance measure between the FFM for successful texture classification and segmentation.

Acknowledgement

The research performed in this paper is partially funded by the Manufacturing Research Corporation of Ontario which is one of the Centers of Excellence Supported by the Premier's Council Technology Fund of the province of Ontario, Canada.

References

[1] Amadasun, M. and R. King (1989). Textural Features Corresponding to Textural Properties. *IEEE Trans. on System, Man, and Cybernectics* Vol. 19, No. 5.

[2] Brodatz, P. (1968). *Textures*, Reinhold, New York.

[3] Campbell, F.W. and J.G. Robson (1968). Application of Fourier Analysis to the Visibility of Gratings, *J. Physiology* 197, 551-556.

[4] Coggins, J. M. and A. K. Jain (1985). A Spatial Filtering Approach to Texture Analysis. *Pattern Recognition Letters 3(3)*, 195-203.

[5] Faugeras, O.D. (1978) Texture Analysis and Classification Using a Human Visual Model. *Proc. Int. Conf. on Pattern Recognition*, Tokyo, Japan, 549-552.

[6] Gool, L.V., P. Dewaele, and A. Oosterlinck (1985). Survey, Texture Analysis Anno 1983. *Computer Vision, Graphics and Image Processing* 29, 336-357.

[7] Haralick, R.M., K. Shanmugam, and I. Dinstein (1973). Textural Features for Image Classification, *IEEE Trans. Syst., Man, Cybern.*, Vol. SMC-3, 610-621.

[8] Haralick, R.M. (1979). Statistical and Structural Approaches to Texture. *Proc. IEEE* 67, 786-804.

[9] Julesz, B. (1965). Texture and Visual Perception, *Scientific American*, Vol. 212, 38-54.

[10] Rao, A.R. (Ed.) (1990). *A Taxonomy for Texture Description and Identification.* Springer-Verlag.

[11] Shen, H.C. and Wong, A.K.C. (1983). Generalized Texture Representation and Metric. *Computer Vision, Graphics and Image Processing* 23, 187-206.

[12] Shen, H.C. and C.Y.C. Bie (1991). Feature Frequency Matrices as Texture Image Representation. To appear *Pattern Recognition Letters.*

[13] Shen, H.C. and C.Y.C. Bie (1991). Visual Properties of Textures Based on Feature Frequency Matrices. *Proc. of Vision Interface '91*, Calgary, Alberta, Canada, 170-175.

[14] Tamura, H., S. Mori, and T. Yamawaki (1978) Textural Features Corresponding to Visual Perception. *IEEE Trans. on System, Man and Cybernetics*, Vol. SMC-8, No. 6.

[15] Tomita, F. and S. Tsuji (1990). *Computer Analysis of Visual Textures.* Kluwer Academic Publilshers.

[16] Wechsler, H. (1980). Texture Analysis - A Survey. *Signal Processing* 2, 271-282.

Multi-font Alpha-numeric Recognition using Multi-layer Neural Networks with a Rejection Function

Keiji YAMADA
C&C Information Technology Research Laboratories, NEC Corporation
1-1, Miyazaki 4chome, Miyamae-ku, Kawasaki, 216, Japan
yamada@pat.cl.nec.co.jp

Abstract

A new method for pattern recognition using multi-layer neural networks with a rejection function is described. The system has two components: (1) STELA (STandstill Evading Learning Algorithm), a new learning algorithm that extends back propagation by adding a technique to avoid learning standstill states, and (2) the derivation of a rejection function which estimates the uncertainty of the classification, and rejects the pattern as unclassifiable if the network response is too uncertain. The ability of the method to recognize deformed and noisy patterns is demonstrated in the domain of multi-font alphanumeric character recognition. STELA achieves perfect recognition for the training data. Without using rejection function, the back propagation makes 12 times as many errors on a recognition task using 20 different fonts as STELA does. Furthermore, a conventional rejection method is compared to the one derived here, and found to require twice as many rejections to achieve the same error rate.

1 Introduction

There has been considerable recent interest in the use of Multi-layer Neural Networks (MLNN) and back propagation learning for pattern recognition[1-8], in particular for character recognition. In some cases, the image data is presented without preprocessing. In others, features are extracted from raw image data and presented to the network. In both cases, the MLNN works as a template matching classifier that has good pattern recognition accuracy when pattern variation is smaller within classes than between classes.

Pattern recognition performance should be discussed from two view points, tolerance for unlearned geometrical deformation and tolerance for meaningless noise superimposed on patterns. Handwritten digit recognition is often discussed as an example where new deformations should be correctly classified. Usually, many kinds of deformed digits are learned and then a digit pattern with a shape which has never been learned must be recognized. As discussed in [8], a simple MLNN often does not perform correctly on novel shapes, even when all of the training patterns can be recognized. For this task, an MLNN should have a special structure so that spatial fluctuation of local features might be absorbed.

Multi-font character recognition can be viewed as an example of meaningless noise superimposed upon normal patterns. Here, character patterns of all the fonts are learned previously and noisy input patterns should be recognized accurately. An MLNN is expected to have good performance for such a task.

For most practical applications, however, it is important for the classifier to signal when it cannot classify an input with certainty. The goal of this paper is to show how a network can achieve perfect performance on a training set and then achieve a 0% error rate (misclassification rate) on novel patterns by rejecting some as unclassifiable. Back propagation learning cannot be used to achieve perfect performance on the training set, since it often reaches what may be termed a "learning standstill state" (this is not equivalent to a local minima - see below). We have proposed an improved learning algorithm called STELA (STandstill Evading Learning Algorithm) which avoids this state. Second, when an input pattern is too noisy to be recognized exactly, the recognition process should not be forced to output an answer. The answer is uncertain. In the case where the answer is too uncertain, the recognition process should output a rejection signal rather than a category name. This rejection function increases the reliability of the pattern recognition system.

This paper describes multi-font alpha-numeric recognition performance of an MLNN with consideration of the two kinds of functions mentioned above. At first, the STELA method is briefly described. Next an uncertainty expression is defined and criteria for consistency of the behavior of this expression are proposed. Finally, experimental results of using the learning algorithm and the rejection function for pattern recognition of multi-font alpha-numerics are presented which support the utility of these mechanisms.

2 Improved Back Propagation Algorithm

2.1 Multi-layer neural network

The typical neural network used for pattern recognition is a three-layer network as shown in Figure 1. The pattern to be classified is presented as input, activation propagates forward through the network, activating the output units. There is an output unit for each category. The output unit with the highest activation is taken

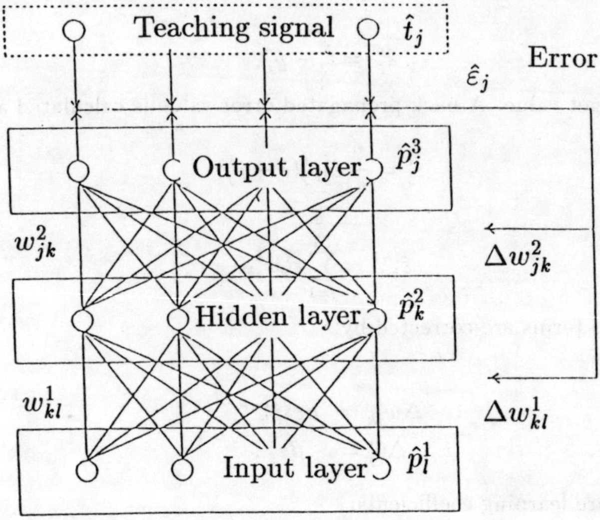

Figure 1: Multi-layer neural network conceptual diagram

as the category to which the pattern belongs.

The feedforward process and the learning process are described below. A unit's activation value is the result of passing the weighted sum of its inputs through a nonlinear sigmoid function. Thus the output value $\hat{p}_j^i$ on the j-th unit on the i-th layer is given as

$$\hat{p}_j^i = f(\sum_{k=1}^{N_{i-1}} w_{jk}^i \hat{p}_k^{i-1} + \theta_j^i) , \tag{1}$$

where $\hat{p}_k^{i-1}$ is an output value on the k-th unit on the $(i-1)$-th layer and w_{jk}^i is a weighting value between $\hat{p}_j^i$ and $\hat{p}_k^{i-1}$. θ_j^i is a bias term for the j-th unit on the i-th layer. $f(x)$ is a sigmoid function, denoted as

$$f(x) = \frac{1}{2}(1 + \tanh \frac{x}{u_0}) , \tag{2}$$

where u_0 is a parameter for controlling the sigmoid function slant. An output value from a sigmoid function is between one and zero.

During learning, weights are changed according to the back propagation learning algorithm [10], which minimizes mean-squared error. Output unit values are calculated according to the feedforward process. The error at the output is computed as a difference of presented and expected signal for each unit on the output layer (the m-th layer):

$$\hat{\varepsilon}_j^m = \hat{t}_j - \hat{p}_j^m \,, \tag{3}$$

where $\hat{t}_j$ is a target value. A back propagated error value is calculated as

$$\varepsilon_j^i = \frac{2}{u_0}\hat{\varepsilon}_j^i \hat{p}_j^i (1 - \hat{p}_j^i) \tag{4}$$

and

$$\hat{\varepsilon}_k^{i-1} = \sum_{j=1}^{N_i} \varepsilon_j^i w_{jk}^i \,. \tag{5}$$

Weights and bias terms are corrected by:

$$\begin{aligned} \Delta w_{jk}^i &= \alpha \varepsilon_j^i \hat{p}_k^{i-1} \\ \Delta \theta_j^i &= \beta \varepsilon_j^i \,, \end{aligned} \tag{6}$$

where α and β are learning coefficients.

Using equations (4),(5),(6) repeatedly, weights at lower layers are updated.

2.2 The STELA Algorithm

In preliminary experiments using back propagation learning, the network often reached a learning standstill state [9]. In this state, weights do not change, as if the network were in a local minimum, even when the $|\hat{\varepsilon}_j^m|$'s are large. This state, however, is not a local minimum. The standstill state is due to the slope term in Eq.(4). When a unit is near its maximum or minimum value, ($\hat{p}_j^i$ is near one or zero), the slope is near zero, so the error value ε_j^i in Eq.(4) is near zero and the weight change term Δw_{jk}^i is near zero. Weights are not corrected in this state. This also occurs if all $\hat{p}_k^{i-1}$'s are near zero. We term such a state a learning standstill.

The learning standstill state is very serious, especially in digital neuro-computers, owing to underflow. In order to evade the learning standstill state, the slant parameter is controlled. That is, u_0 in Eq.(2) is enlarged so that output values are far from one and zero, as shown Figure 2. This is done during the presentation of a single training pattern.

The algorithm is described below(see also [8]).

(1) Detecting a learning standstill state

A1. A "serious" error is judged using the condition

$$|\hat{\varepsilon}_j^m| > \delta \tag{7}$$

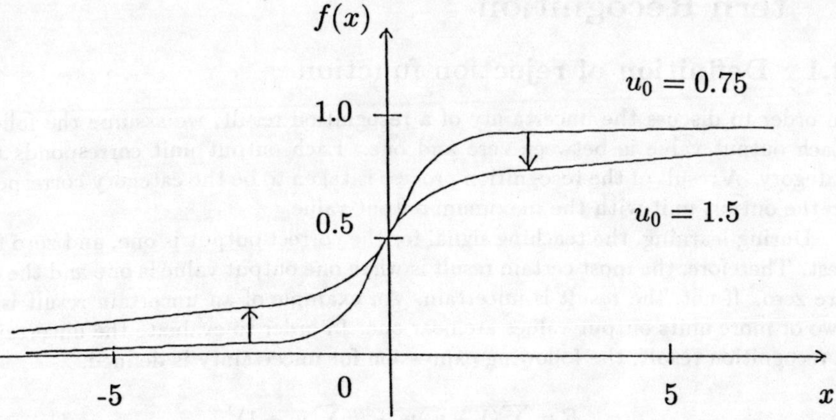

Figure 2: Slant control using sigmoid slant parameter, u_0, in $f(x) = \frac{1}{2}(1 + \tanh(\frac{x}{u_0}))$ (In order to evade the learning standstill state, the slant parameter, u_0, is temporarily enlarged.)

A2. A network weight change is too small if:

$$\sum_{j,k} |\Delta w^i_{jk}| < \eta \tag{8}$$

In order to reduce the amount of computation, we actually use the following condition instead of A2:

$$\sum_{j,k} |\hat{p}^{i-1}_k \hat{p}^i_j (1 - \hat{p}^i_j)| < \xi \tag{9}$$

This clearly captures the zero-slope and the zero-output conditions characteristic of the learning standstill state.

(2) Evading a learning standstill state

A3. If the conditions in Eqs.(7) and (9) are satisfied, u_0 is enlarged and all output values are calculated again. Otherwise, go to A5.

A4. The judgements are again carried out from A1.

A5. Weighting values are corrected according to Eqs.(4),(5), and (6). After that, the initial value for the slant is set again to u_0.

3 Rejection Function for Neural Network Pattern Recognition

3.1 Definition of rejection function

In order to discuss the uncertainty of a recognition result, we assume the following: Each output value is between zero and one. Each output unit corresponds to one category. A result of the recognition process is taken to be the category corresponding to the output unit with the maximum output value.

During learning, the teaching signal for the correct output is one, and zero for the rest. Therefore, the most certain result is when one output value is one and the others are zero. If not, the result is uncertain. An example of an uncertain result is when two or more units output values are near one. In order to evaluate the uncertainty of a recognition result, the following expression for uncertainty is defined.

$$R = \sum_i (1 - p_i)p_i + \beta(\sum_i p_i - 1)^2 \ , \tag{10}$$

where p_i is the value of the i-th output unit.

This function is non-negative when the p_i's are between 0 and 1, and when β is non-negative. The left-hand term is minimized when all units are near values of 1 or 0. The right hand term is minimized when all of the output units' activations sum to 1.

In the next section, Constraints are derived on approximate threshold values.

3.2 Consistency of the uncertainty

The stability of the uncertainty degree defined above is discussed, appropriate parameter values of β in Eq.(10) are derived, and a rejection threshold is estimated.

We would like the parameter β to be such that the following consistency criteria are obeyed.

Criterion 1

Uncertainty should decrease as the maximum-valued output, p_M, increases:

$$R(p_1, \cdots, p_M, \cdots, p_N) > R(p_1, \cdots, p_M + \Delta p, \cdots, p_N)$$

Criterion 2

When p_j is not the maximum value, uncertainty should decrease as p_j decreases:

$$R(p_1, \cdots, p_j, \cdots, p_N) > R(p_1, \cdots, p_j - \Delta p, \cdots, p_N)$$

where Δp is a small positive value.

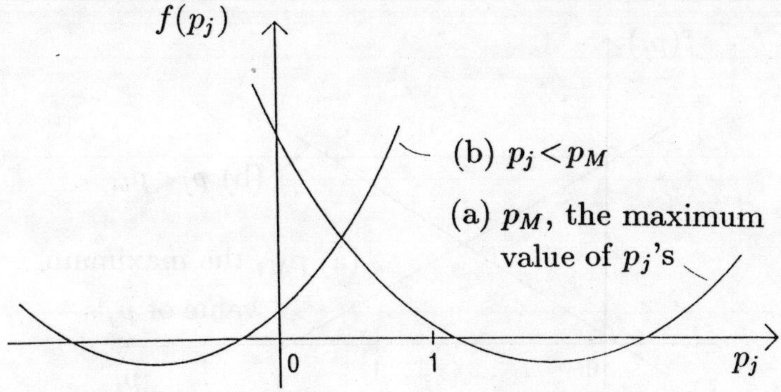

Figure 3: Uncertainty degree $f(p_j)$ in the case that $\beta > 1$ ($f(p_M)$ should be monotonically decreasing in [0, 1]. $f(p_j)(p_j < p_M)$ should be monotonically increasing in [0,1].)

The consistency of the uncertainty defined in Eq.(10) is assured if it satisfies these criteria.

In order to consider the behavior of the uncertainty degree R concerning an output value, p_i, Eq.(10) is arranged as follows:

$$R = f(p_j) = (\beta - 1)p_j^2 + (1 + 2\beta D)p_j + C + \beta D^2, \tag{11}$$

where $C = \sum_{i \neq j}(1 - p_i)p_i$ and $D = \sum_{i \neq j} p_i - 1$.

This equation will be discussed in the following three cases.

i) $\beta > 1$

In this case, $f(p_j)$ is a concave upward quadratic function of p_j. If $j = M$, that is, p_j is the maximum value, $f(p_M)$ should decrease as p_M increases. Therefore, the x-coordinate of the minimum of the quadratic function must be larger than one, as shown in Figure 3(a). By taking the partial derivative of f with respect to p_j, and imposing the condition that it be negative in the [0,1] interval, we can derive the following condition:

$$\sum_{i \neq M} p_i < \frac{1}{2\beta}\,. \tag{12}$$

If $j \neq M$, the following condition is derived from Criterion 2:

$$\sum_{i \neq j} p_i > 1 - \frac{1}{2\beta}\,. \tag{13}$$

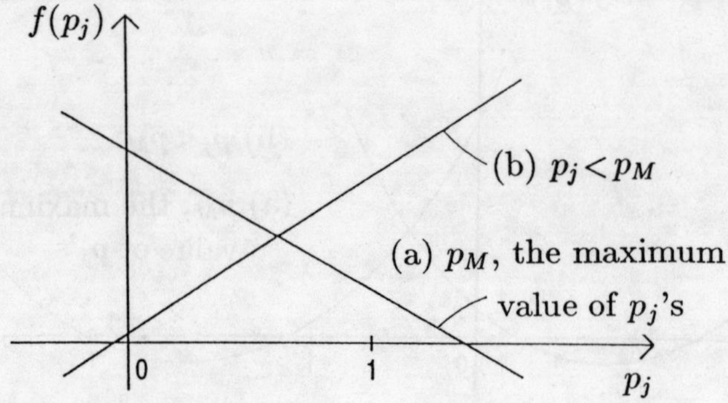

Figure 4: Uncertainty degree $f(p_j)$ in the case that $\beta = 1$ ($f(p_M)$ should be monotonically decreasing in [0,1]. $f(p_j)(p_j < p_M)$ should be monotonically increasing in [0,1].)

ii) $\beta = 1$

In this case, f is linear, (see Figure 4). If $j = M$, the following condition is derived as above:

$$\sum_{j \neq M} p_i < \frac{1}{2}. \tag{14}$$

And the condition for $p_j(\neq p_M)$ is :

$$\sum_{i \neq j} p_i > \frac{1}{2} \quad (j \neq M). \tag{15}$$

iii) $0 < \beta < 1$

The preceding analysis is difficult to carry over to this case. $f(p_j)$ is shown in Figure 5. The coordinate q of the peak along the p_j axis is :

$$q(\beta, j) = \frac{1 + 2\beta(\sum_{i \neq j} p_i - 1)}{2(1-\beta)}. \tag{16}$$

Satisfaction of the criteria depends on whether the p_j value is larger than $q(\beta, j)$. When $q(\beta, j)$ is regarded as a function of β, rough shapes of $q(\beta, j)$ are shown in Figure 6. Criteria 1 and 2 yield the conditions: $p_M > q(\beta, M)$ and $p_j < q(\beta, j)$ $(j \neq M)$, respectively. In Figure 6, p_M should be above the $q(\beta, M)$ curve and p_j should be below the $q(\beta, j)$ curve.

Let us discuss p_M with respect to β and $\sum_{i \neq M} p_i$(see Figure 6).

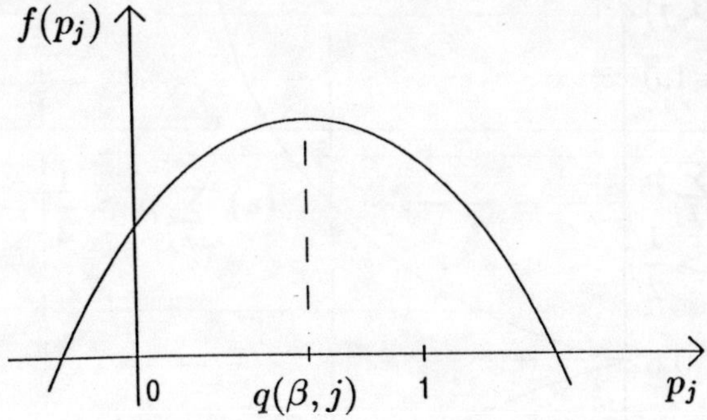

Figure 5: Uncertainty degree $f(p_j)$ in the case that $0 < \beta < 1$
(p_M should be larger than $q(\beta, M)$, so that $f(p_M)$ should be a decreasing function. p_j ($p_j < p_M$) should be less than $q(\beta, j)$, so that $f(p_j)$ should be a increasing function.)

If β is near zero, p_M must be larger than 0.5. If β is near one, the condition is $\sum_{i \neq M} p_i < \frac{1}{2}$. This includes the case that $\beta \geq 1$. If β is about 0.5, p_M must satisfy the following condition:

$$p_M > \sum_{i \neq M} p_i \ . \tag{17}$$

On the other hand, p_j ($j \neq M$) must satisfy $p_j < q(\beta, j)$. If β is near zero, p_j must be smaller than 0.5 independently of the value of $\sum_{i \neq j} p_i$. If β is about one, the condition for satisfying Criterion 2 is that $\sum_{i \neq j} p_i > \frac{1}{2}$. If β is about 0.5, the condition is that:

$$p_j < \sum_{i \neq j} p_i \qquad (j \neq M) \ . \tag{18}$$

This is always satisfied because $p_M \in \{p_i | i \neq j (\neq M)\}$ and $p_j < p_M \leq \sum_{i \neq j} p_i$.

Based on the above discussion, the case that β is about 0.5 is better than the cases that β is near zero or larger than about one, because the consistency criteria are satisfied under less strict conditions than in the other cases. In the following sections, β is set to be 0.5.

The boundary condition in the case that $\beta = 0.5$ is given from Eq.(17) as:

$$p_M = \sum_{j \neq M} p_j \ . \tag{19}$$

The substitution of Eq.(19) into Eq.(10) yields

$$BC(R) = \frac{1}{2} + {p_M}^2 - \sum_{j \neq M} {p_j}^2 \ . \tag{20}$$

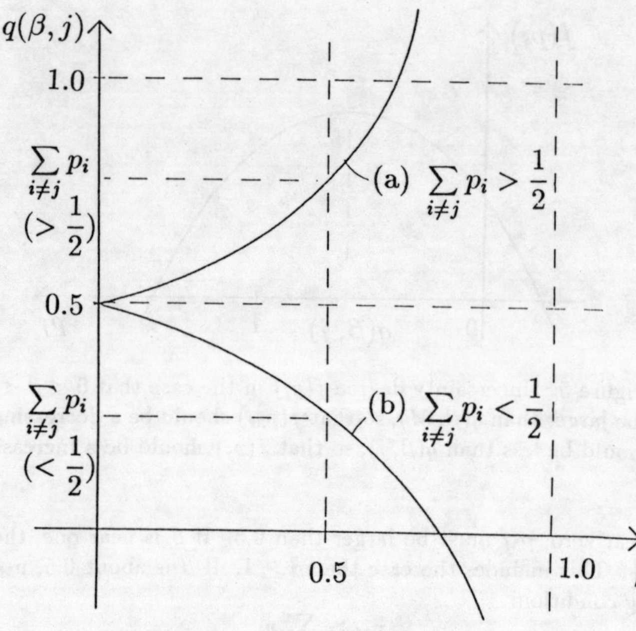

Figure 6: Boundary condition $q(\beta, j)$ by β and $\sum_{i \neq j} p_i$
(p_M should be larger than $q(\beta, M)$. p_j should be less than $q(\beta, j)$.)

The sum of the second term and the third term gives a small positive value. Any value of R below 0.5 is thus guaranteed to be consistent. Therefore, R is constrained to be smaller than 0.5 and the rejection threshold of R is set to be smaller than 0.5 .

4 Experiments of Multi-font Alpha-numeric Recognition

4.1 Alpha-numeric characters and neural network model for experiments

In order to test our approach on a realistic data set, we obtained twenty kinds of alpha-numeric fonts, as shown in Figure 7. The number of categories is 74. The same category name is given to pairs that are identical in some fonts, such as O(oh) and 0(zero), l(el) and 1(one), l(el) and I(Capital letter without serif), and '(apostrophe) and ,(comma). Fonts 1-12 are used for twelve font recognition experiments. Fonts 1-16 are used for the sixteen font recognition, and Fonts 1-20 for twenty font recognition.

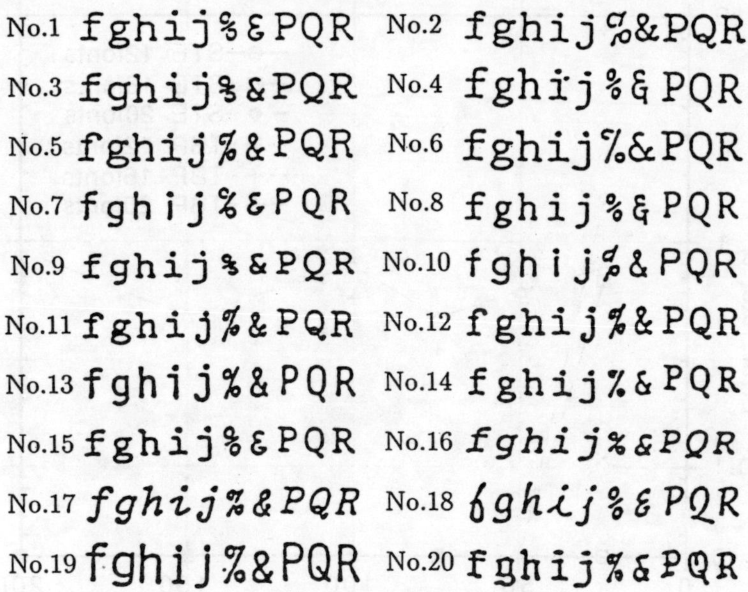

Figure 7: Original samples of 20 fonts
(No.1 ~ No.12 are used for 12 font recognition)
(No.1 ~ No.16 are used for 16 font recognition)
(No.1 ~ No.20 are used for 20 font recognition)

Ten characters of each category and each font are typed and scanned with a 240dpi density scanner in order to obtain realistically noisy binary patterns. Half of them are used for training and the other half are used for recognition tests. The scanned binary character image is transformed into a 9 by 11 pixel gray level image using a Gaussian filter. In these experiments, we did not use size normalization.

We use a fully connected feedforward network with one hidden layer. The number of input units is 99 and the number of output units is 74. Locally connected models, with hidden units connected to a local area of the input layer [9, 1], have been found to be more effective in handwritten numeral recognition than fully connected networks [8]. Preliminary experiments on our multi-font character data set showed similar performance for both kinds of networks.

In this paper, we describe experimental results of a fully connected network. The number of hidden units is 25, 50, 75, 100, 150, and 200. TBP (Typical Back Propagation) and STELA (STandstill Evading Learning Algorithm) are used for training.

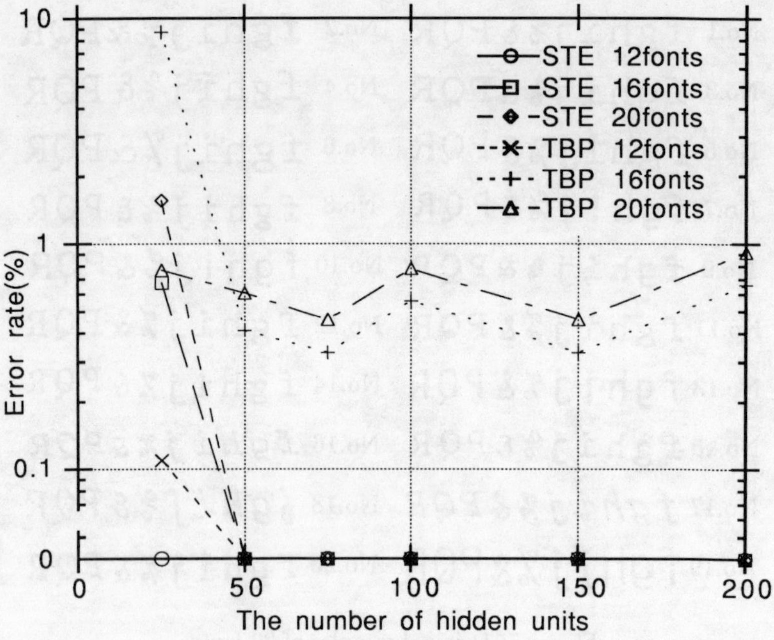

Figure 8: Error rates for training data by the number of hidden units and the number of fonts when using TBP and STE(STELA) without rejection
(In 20fonts recognition, TBP:0.5% error rate. STELA:no error.)

4.2 Recognition experiments without rejection

The recognition rates for the training data and the test data are investigated in each case of twelve fonts, sixteen fonts, and twenty fonts.

The results are shown in Figure 8 and Figure 9. When twenty-five hidden units are used, neither TBP or STELA can recognize all of the training data. In this case, the recognition rate by STELA is worse than that by TBP. This is because STELA will not let a neural network state stay in a local minimum and will continue to work at finding a solution where all the training data can be recognized.

In the case of twelve font recognition trained by TBP, all of the training data are exactly recognized. However, for the larger number of fonts, TBP does not achieve 100 % accuracy on the training set. STELA achieves 100% accuracy for all 20 fonts when the network has at least 50 hidden units.

In the case of TBP, error rates for test data of the 20 font sets are about 0.6%. In

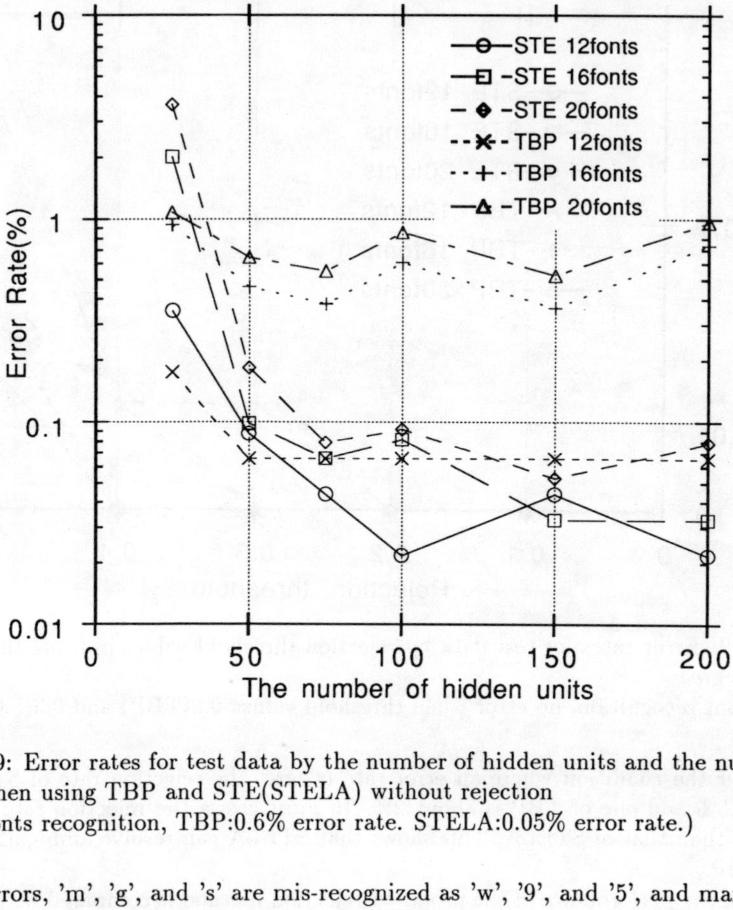

Figure 9: Error rates for test data by the number of hidden units and the number of fonts when using TBP and STE(STELA) without rejection
(In 20fonts recognition, TBP:0.6% error rate. STELA:0.05% error rate.)

these errors, 'm', 'g', and 's' are mis-recognized as 'w', '9', and '5', and many other characters are not recognized. In the case of STELA, the error rates are about 0.05%. The character 'i' is recognized as '1', because the dot in the 'i' is connected to the line due to noise.

4.3 Recognition experiments with rejection

The rejection rates are examined using Eq.(10) with $\beta = 0.5$. Rejection threshold values between 0.1 and 0.5 are used. The error rates by the threshold values and the rejection rates by the threshold values are shown in Figure 10 and Figure 11.

In sixteen and twenty font recognition, STELA and TBP achieve no-error recognition when threshold values are 0.3 and 0.2, respectively.

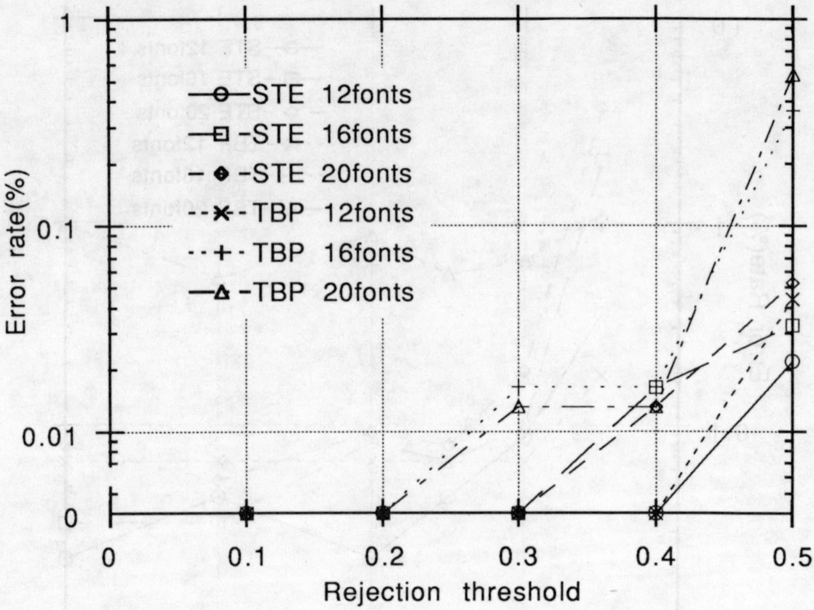

Figure 10: Error rates for test data by rejection threshold values in using the uncertainty degree
(In 20 font recognition, no error when threshold values 0.2(TBP) and 0.3(STELA).)

Under the condition where an error rate is zero, the rejection rate of STELA is about 0.5% and one of TBP is about 2%. In most cases, the rejection rate of TBP is larger than that of STELA. This shows that STELA can resolve ambiguity better than TBP.

Next, the performance of the proposed rejection method is compared with that of a conventional method. As a conventional rejection method, the difference between the maximum value and the next highest value is used. If the difference is smaller than a threshold value, the input pattern is rejected.

Figure 12 shows error rates and rejection rates for each font set using STELA. Under the zero error rate condition, the rejection rates for sixteen fonts and twenty fonts are about 1% , twice as large as those by the proposed rejection method.

The conventional method uses only the primary candidate and the secondary candidate and ignores the others. However, the third and fourth candidates must influence the confidence of a recognition result. For example, assume that the maximum value is 0.8 and the second value is 0.5, the confidence in the case where the third value is 0.2 should be better than in the case where it is 0.5. By using all of the

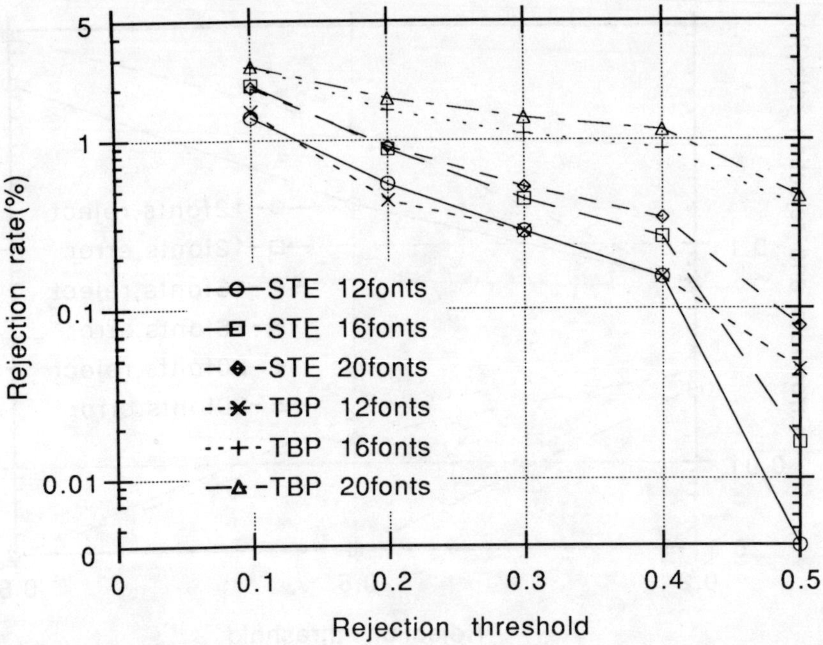

Figure 11: Rejection rates for test data by rejection threshold values when using uncertainty degree
(Under the no error condition, TBP:2% rejection (threshold value: 0.2) and STELA: 0.5% rejection (threshold value: 0.3).)

output values, our method significantly reduces the rejection rate.

5 Concluding Remarks

This paper has described a new method for pattern recognition using multi-layer neural networks with a rejection function. We demonstrated the superiority of STELA over TBP (Typical Back Propagation). We also proposed an expression for the uncertainty of recognition results, and a rejection method using the degree of uncertainty.

The rejection method was compared with a conventional rejection method, and shown to be superior.

We tested the method on the multi-font character recognition task and found that

1) STELA can learn 100% of the training set when TBP fails.
2) STELA achieves better generalization than TBP. Hence it has not over learned

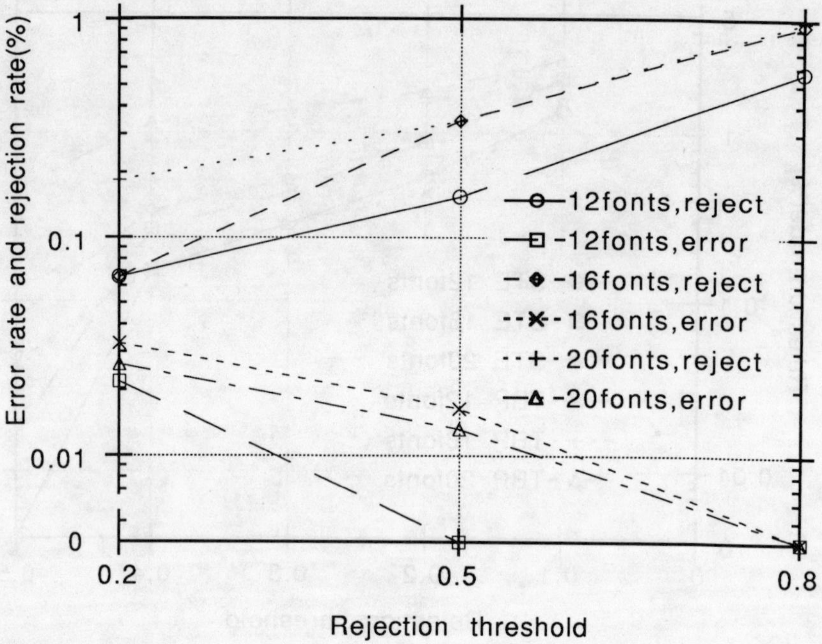

Figure 12: Error rates and rejection rates by threshold values when using training by STELA and a conventional rejection method
(Under the no error condition, the rejection rate: about 1%.
The rejection rate by the proposed method: about 0.5% (ref. Fig.11).)

the training set.
3) The rejection rate required to achieve 0 errors for STELA is one fourth as large as that for TBP.
4) The rejection rate required to achieve 0 errors is half as big as that for a conventional rejection method.

In future work, we plan to apply this technique to handwritten digit recognition, where error rates need to be as near 0 as possible.

Acknowledgements

The author is indebted to their colleagues in Pattern Recognition Research Laboratory at NEC for their helpful discussion. He also thanks Mr. Tsutomu Temma and Mr. Jun Tsukumo of the same laboratory for their continuous encouragement. Finally, he thanks Dr. Garrison W. Cottrell of Univ. of California, San Diego for improvements to this paper.

References

[1] Y.LeCun et al.: Backpropagation Applied to Handwritten Zip Code Recognition, Neural Computation Vol.1, pp.541-551 (1989)

[2] T.F.Pawlicki et al.: Neural Network Models and their Application to Handwritten Digit Recognition, Proc. IEEE Second Int. Conf. on Neural Networks, San Diego, p.II-63 (1988)

[3] D.J.Burr: Experiments with a Connectionist Text Reader, Proc. IEEE First Int. Conf. on Neural Networks, San Diego, p.IV-717 (1987)

[4] E.Gullichsen et al: Pattern classification by Neural Network: An Experimental System for Icon Recognition, Proc. IEEE First Int. Conf. on Neural Networks, San Diego, p.IV-725 (1987)

[5] T.Kohonen: The Neural Phonetic Typewriter, IEEE Computer, vol.21, No.3, pp.11-22 (1988)

[6] S.E.Troxel et al.: The Use of Neural Networks in PSRI Target Recognition, Proc. IEEE Second Int. Conf. on Neural Networks, San Diego, p.I-593 (1988)

[7] A.Rajavelu, et al.: A neural Network Approach to Character Recognition, Neural Networks, Vol.2, pp.387-393 (1989)

[8] K.Yamada, et al.: Handwritten Numeral Recognition by Multi-layered Neural Network with Improved learning Algorithm, Proc. IJCNN 89, pp.II259-II266 (1989)

[9] K.Yamada: Improved Learning Algorithm for Multi-layer Neural Networks and Handwritten Numeral Recognition, NEC R&D, No.98, pp.81-88 (1990)

[10] D.E.Rumelhart et al.: Parallel Distributed Processing, Vol.1, p.322, The MIT Press (1986)

ARisTotle: A Neural Net Architecture for the Auto-classification, Grouping and Recognition of Partially Distorted Binary Images

D. Boulanger and D. Poussart
Laboratoire de Vision et Systèmes Numériques
Département de génie électrique
Université Laval
Québec, Canada

Abstract

In this paper a classification network is presented for the recognition and hierarchical grouping of binary patterns based on an ART 1 neural classifier. The network is first presented with a series of primitive patterns which are memorized. After each presentation of a complex pattern the network automatically selects the classes of the previously learned primitives that match the parts of the image shown. For each primitive selected, the network spatially isolates its counterpart in the image and computes the Hamming distance between the two parts. After recognizing all the parts of the complex image, a grouping of these constituent classes is made to a new class for the complex image. This hierarchical representation is subsequently used for the recognition of partially distorted complex patterns.

The network was trained for recognizing handwritten numerals. After the presentation of a few primitives and only one example of the numbers from 0 to 9, the network was able to identify many distorted and noisy patterns of numbers.

1 Introduction

Aristotle thought that things in the universe could be classified in five groups. These groups, structured in an ascending hierarchical order, extended from minerals to pure thought. Man, a class by itself, was the highest form of being on this planet and was placed just below pure thought. A class was thought superior to another if it possessed the characteristics of the lower classes and some of its own [1].

It was evident to Aristotle that man's knowledge was also driven by this hierarchical structuralism, as in nature. Man's knowledge is the result of his experiences acquired through his senses.

This philosophy of man's knowledge structure still has an influence on today's comprehension of the complex structure of our visual system. In spite of the complex world surrounding us, our visual system, from the retina to the cortex, is able to capture this world with a certain ease.

Primary information is first extracted from a complex scene, and is then classified at different levels of representation helping us later, to recognize things of the same nature [14].

When analysing a complex scene, our visual system seems to search the most dominant features. These features, like edges, color, texture and others, excite in parallel the different cognitive levels and induce images or models of things that are compared with the inputs. This hierarchical search is generally non-oriented, meaning that the initiative of this search originated from external stimulation and not from internal thinking like the search for a particular object in a complex scene.

In our approach we have tried to keep these principles in perspective. The parallel algorithm described in this paper is able to make a hierarchical classification of binary patterns. When a complex pattern is presented to the net, a fast and direct search of its constituent parts is made from the previously learned ones. The classes for these recognized parts are then grouped together with the class for the complex image, resulting in an ascendingly complex representation. These complex groupings, now part of the knowledge base, are then used subsequently for the identification of still more complex patterns.

This system, mainly based on an ART 1 neural classifier, is capable of identifying partially distorted patterns by means of this hierarchical representation. In the first part of this paper the ART 1 system is briefly described with an emphasis on the emergent properties of this neural architecture. The proposed system, named ARisTotle, is then described and compared with an ART network. Finally the properties of ARisTotle are demonstrated by applying it to the task of handwriting recognition.

2 ART 1 system

Carpenter and Grossberg [3,8,9] described a number of classification models in their adaptive resonance theory. The ART 1 system [2] implements a massively parallel algorithm for the non-supervised classification of binary patterns. The network is made of two layers of units, or neurons. Units of the first layer called F1, serve as input as well as comparison units. These input units are tied to the classification units by a series of weighted links which adapt slowly in order to memorize the input patterns. The classification units, call F2, select the unit receiving the maximum excitation from F1 layer ("winner-take-all"). The network also has a reset unit for the validation of the selected class and a gain control unit for switching the F1 units from the input mode to the comparison mode.

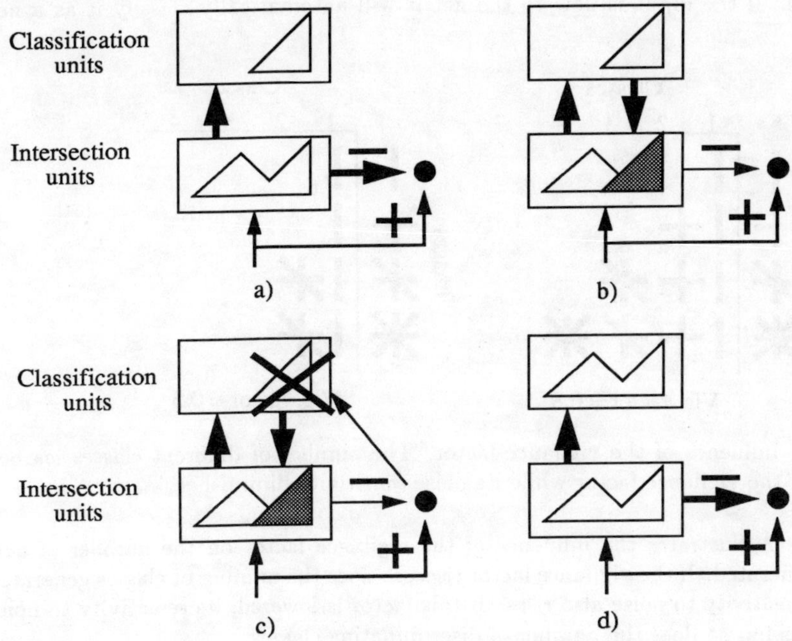

Figure 1: Search process in an ART 1 system: a) The network first selects a class through the bottom-up links b) An image of the selected class is then fed back to the input layer. c) If the relative difference between the input pattern and the class is lower than a certain threshold, the selected unit is reset. d) The process continues until the selected class matches the input.

The search process, illustrated by figure 1, is as follows: The network begins by comparing the inputs in parallel with all the memorized models using the bottom-up interconnections between the two layers. This generates a series of excitations to the classification units. The unit receiving the highest excitation is then enhanced and selected. This selected unit then generates an image of the memorized model through the top-down links to the first layer. The intersection of the generated model and the input is then computed at the first layer. If the relative difference between the intersection and the input is lower than a certain factor called the vigilance factor, the selected classification unit is deactivated temporarily by the reset unit. In other words: if the Euclidian distance between the presented image and the selected model is too large, forget this selected class and try another one.

After the reset, the process is re-activated and another classification unit is selected and validated. This continues until a match is found which gives a relative difference higher than the vigilance factor. If there is no reset, then the net is said to be in resonance. If the input is new to the net it will automatically classify it as a new unit.

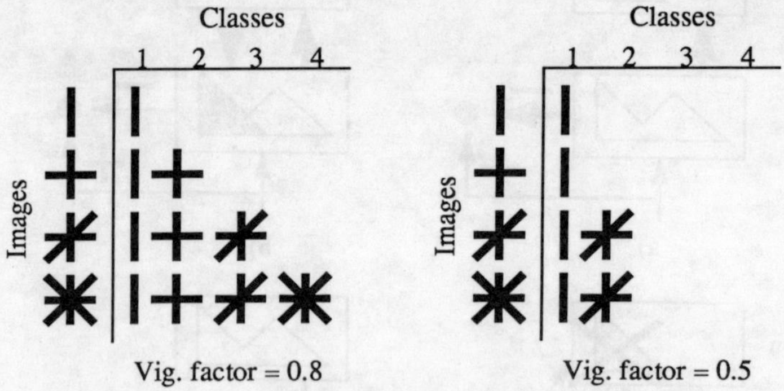

Figure 2: Influence of the vigilance factor: The number of different classes learned rises with the vigilance factor while its noise sensitivity diminishes.

Figure 2 illustrates the influence of the vigilance factor on the number of new classes generated. If the vigilance factor rises, so does the number of classes generated but its sensitivity to noise also rises. If this factor is lowered, its sensitivity to noise diminishes but so does the number of discriminating classes.

Every time the net is in resonance, the bottom-up and the top-down links between the two layers are gradually modified to adapt to the variations between the input and the model.

The values of the weights for these links are chosen in such a way that inputs of the same nature will have a direct access to this class without searching through all the others.

3 ARisTotle: A hierarchical binary classifier

One of the problems with an ART classifier is that it cannot identify the constituent parts of a complex image. This global approach does not provide a good symbolic representation needed for recognition. The proposed system has the same properties as the ART system, such as direct access, but it can also cope with the parts of a complex image by using a spatial delimitation network. Every object in a complex image generates, as in an ART system, a series of excitations to the classification units which are selected one after another to be compared with the whole image or parts of it. With the spatial delimitation network, the selected models can be classified as being a part of the presented image or as the whole.

After every part is recognized by the net, the whole image is classified with a different unit. A grouping of this last class with all the sub-classes found previously, is then made. This symbolic representation of complex images is used later on for the recognition of partially distorted images of the same nature.

3.1 General description of ARisTotle network

The net has six main sections. A spatial delimitation network which is able to focus on parts of the image. A classification network which memorizes the patterns. Another section is used to temporarily memorize and make hierarchical groupings of the validated classes. Finally, units are used to validate the selected classes while others make an asynchronous update of the possible net.

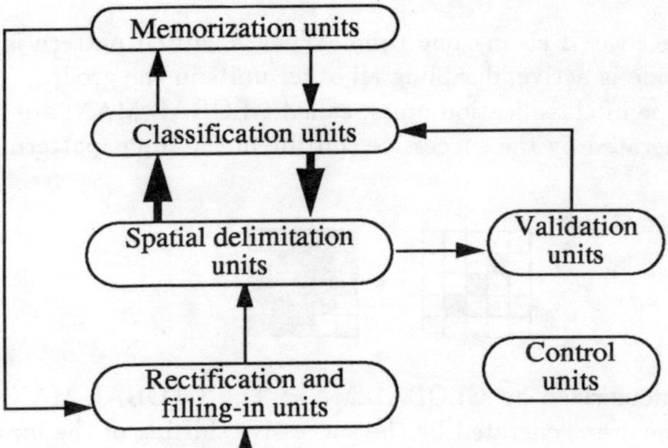

Figure 3: The ARisTotle network: The network consists of six sections: the control units, the validation units, the spatial delimitation units, the classification units, the memorizing units and the rectification and filling-in units.

3.1.1 Class selection and image focussing

ARisTotle has two types of classification units. First, the units called LOCAL_MAX are used to classify the images at nine shifted positions around the input image.

This type of representation will enable the recognition of distorted patterns. Every unit in each group of nine are linked to one another in such a way that only one unit

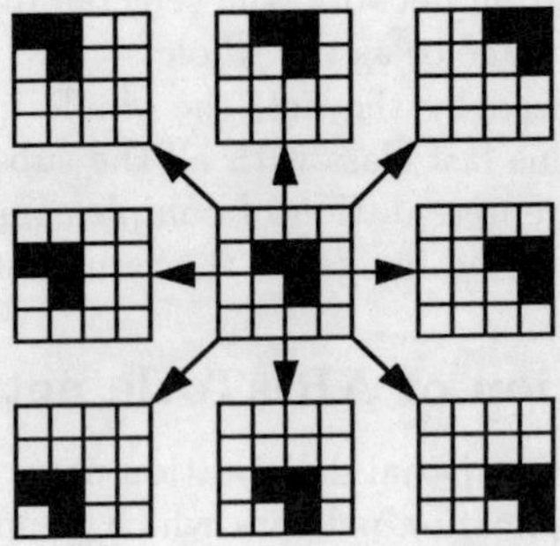

Figure 4: Nine positions memorized by LOCAL_MAX: Each new pattern is memorized at LOCAL_MAX by shifting by one bit every bit in the input pattern in each direction.

of each group is activated at any one time. When a shifted pattern is presented to these units, only one is active, disabling all other units in the group.

The second type of classification units, called GLOBAL_MAX, are used to memorize the area generated by the successive shifting of the input pattern.

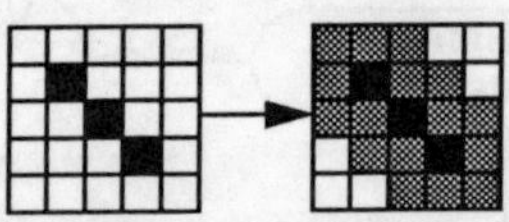

Figure 5: Area memorized by GLOBAL_MAX: The GLOBAL_MAX classification units memorize the area generated by the successive shifting of the input.

Each unit represents one pattern. Units are linked together in such a way as to limit activation to a single cell. The selected class will be the one corresponding most closely to the input, like in the ART system. Even though the input pattern is distorted, the corresponding class is still activated because it falls in the memorized area.

The network sequence of execution is as follows. After the presentation and learning of the primitives and prototypes of complex images, the input units process a complex image made of those previously learned patterns. This image generates a series of excitations through the links of the long term memory to the classification units LOCAL_MAX and GLOBAL_MAX. The unit in GLOBAL_MAX receiving the highest excitation is activated while all others are disabled. One unit in each group of nine from LOCAL_MAX is also activated. Those activated units correspond to the possible sub-patterns of the input image.

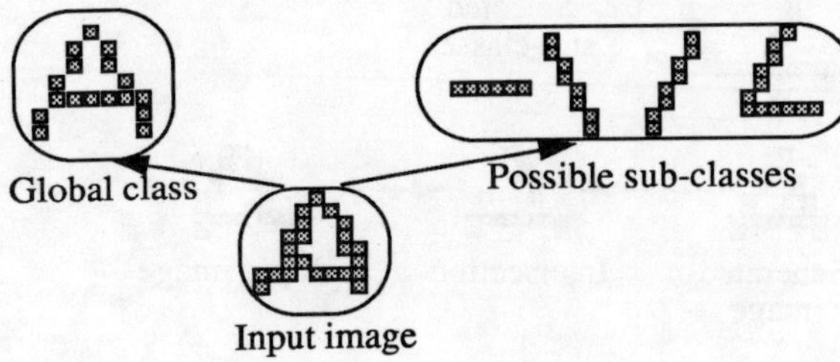

Figure 6: Class selections: At GLOBAL_MAX only one complex class is selected while at LOCAL_MAX one class in every group is selected.

The active unit GLOBAL_MAX generates a series of sub-classes associated with the complex selected class. These sub-classes in turn select the corresponding groups at LOCAL_MAX.

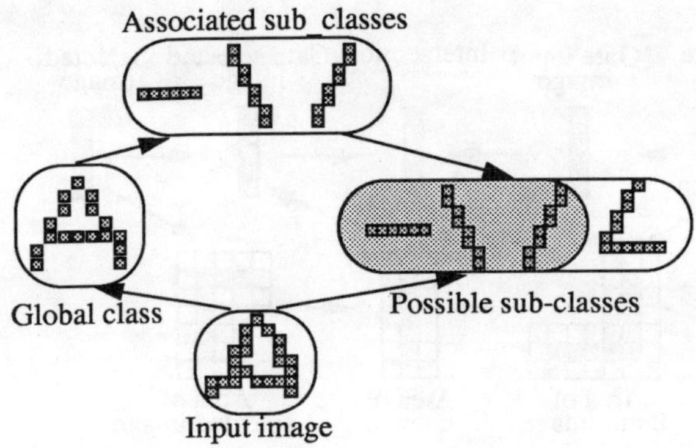

Figure 7: Generation of the associated sub-classes: The selected class at GLOBAL_MAX activates the associated sub-classes which help to select the corresponding shifted sub-images at LOCAL_MAX.

As in the ART system, an image is then generated back. This composite image is obtained by superimposing the selected primitives shifted in a way that best matches the input image. This representation enables the network to compute a better estimate of the Hamming distance between the input and the memorized image.

If the selected image is only a part of the input, a method must be used to isolate the corresponding part in the input image in order to compare the distance between

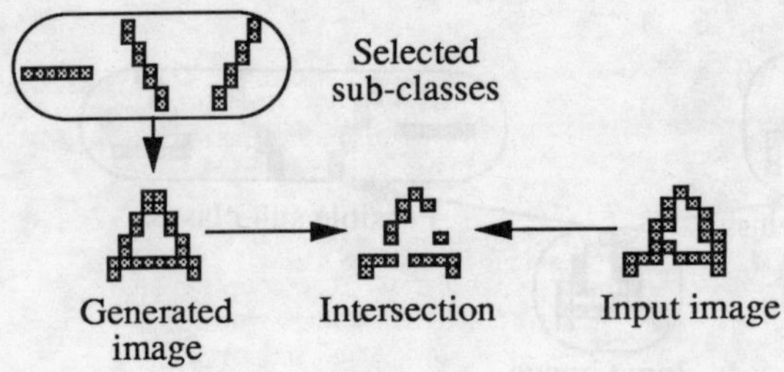

Figure 8: Generation of the selected image: The selected sub-classes from LOCAL_MAX create a distorted image that will be compare with the input for validation.

the two sub-images. This is done by a spatial delimitation process. In this process, the intersection between the areas generated around the two images to be compared is computed. These areas are generated by activating the eight neighbors around each unit in the two images. It is equivalent to a dilation process.

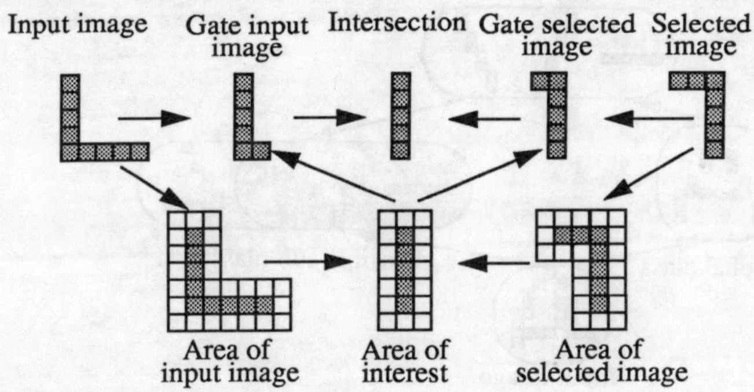

Figure 9: Area of interest: The two images to be compared generate an area around themselves. The intersection of these areas is then computed to get the area of interest. This area then gates the two images so that only the parts in common of the images are compared.

The intersection defines the area of interest. This area is then used to gate the bits of the two images which are to be compared.

A real comparison of the relative distances of these two gated images can then be computed from their intersection. If the relative distance between them is lower than a certain threshold, the selected class and sub-classes are reset, otherwise it is temporarily memorized in the upper layers, and then reset. The process is then reactivated and continues until the whole image is classified in a previously learned class or in a new one. The links of the long term memory are then modified to memorize the image in its nine shifted positions at LOCAL_MAX, the area generated by this shifting and the hierarchical grouping of the class with its selected sub-classes. A fast learning algorithm like in the ART system was used to modified the weights of the long term memory. Unfortunately, this kind of learning rule does not guarantee that the input and the memorized pattern are identical because a small distortion is tolerated.

3.1.2 Rectification and filling-in

In order to ensure a constant and undeformed representation of the memorized patterns, the input image must be rectified in the same way as the memorized one. Although the input is only a part of the selected image the input must be modified so that it matches the corresponding part of the complex image. This rectification is done by superimposing the memorized image on the input by making sure not to keep the active bits from the distorted image.

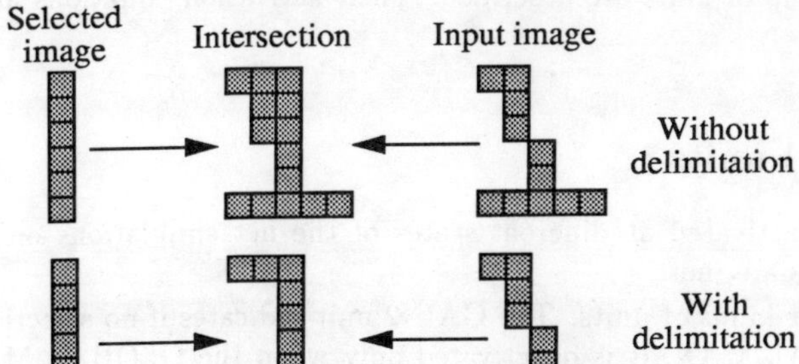

Figure 10: Effect of delimitation in the rectification process: The rectification is done by superimposing the generated image to the input while making sure that the distorted parts of the input are eliminated.

This is done by superimposing the memorized image generated from the memorizing units, while disabling the input bits that are in the area surrounding it. This method sometimes leaves gaps between the rectified part of the image and the rest of it. Therefore a simple filling-in algorithm is used in order to fill these one bit gaps: filling occurs if two of their nearest neighbors on each side are active.

Because of the symmetry of the spatial delimitation net, a complex image can be learned before its primitives and still make a hierarchical grouping. When a primitive or a small pattern is shown, the net is capable of selecting the corresponding complex images and is able to tell whether it includes the input or not. The complex class is memorized temporarily and then reset. Once the presented image is entirely classified, links between the complex classes associated with the input image are made with the constituent classes of the input.

4 Description of the units

As mentioned earlier, there exist six sections of units:

- The control units which permit the asynchronous execution of the net.
- The spatial segmentation units which delimit the area of interest between the input and the selected memorized image.
-The classification units .
- The memorizing units which temporarily keep the validated classes in memory.
- The validation units which compute the distance between the two compared images.
- The rectification and filling-in units.

Each section contains many layers of units as shown in figure 11. In the following sections each group of units are described. Their activation equations are described in the appendix.

4.1 Control units

These units are activated at different states of the net simulations only if certain output conditions are met.

There are four kinds of units. The GAIN2 unit indicates if no selection has been made by GLOBAL_MAX. It is deactivated only when the GLOBAL_MAX and SC are cleared. It essentially plays the same role as the gain control unit in an ART system.

The units IDEM and STABLE are used to check if the net is stable i.e. if the activity of the units do not change at each iteration. IDEM is activated only if no selection has been made in opposition with STABLE which is activated after a selection. These units disable the validation units in order not to evaluate the Hamming distance during the transitory state of the net.

The last unit call MEMOIRE, is activated when the system has completely classified the input image.

It is also activated when the input image is classified in a new class. In order to verify if there is a new classification, the network always keeps in memory the rank of the unit that will be used to memorize a new image. If this unit is selected, MEMOIRE is activated.

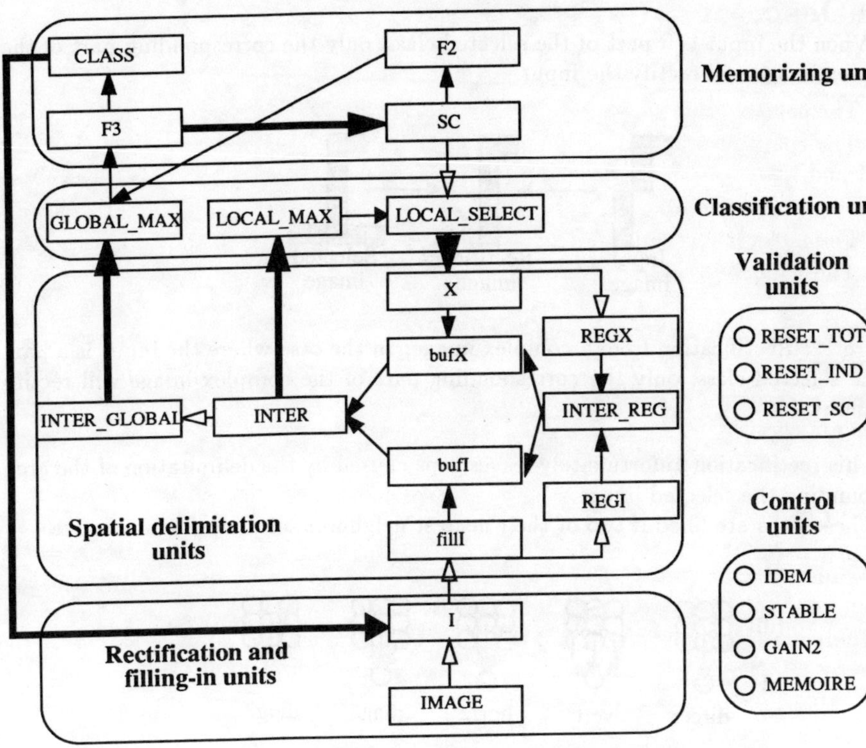

Figure 11: The ARisTotle network: The network consist of six sections: the control units, the validation units, the spatial delimitation units, the classification units, the memorizing units, the rectification and filling-in units. Large arrows represent distributed links. Open arrows represent local links and close arrows, one to one links.

4.2 Rectification and filling-in

At first, when an image is presented to the network, it passes through the input layers I and IMAGE without modification. But every time a selected class is validated, its corresponding image is fed back to the input layer so that the recognized parts of the memorized model match. If this was not done after the modification of the long term memory, the model would not be stable because of the small changes tolerated in the input.

When the input is a part of the selected class, only the corresponding part of the complex image will rectify the input.

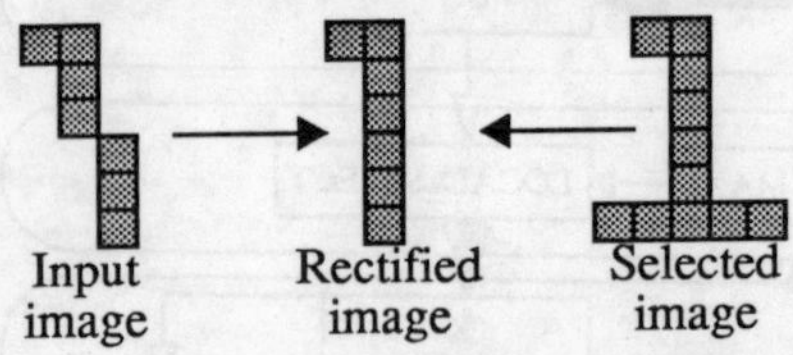

Figure 12: Rectification from a complex image: In the case where the input is a part of the selected class, only the corresponding part of the complex image will rectify the input.

This rectification unfortunately leaves gaps caused by the delimitation of the area surrounding the selected image.

These gaps are filled if two of their nearest neighbors on each side are active.

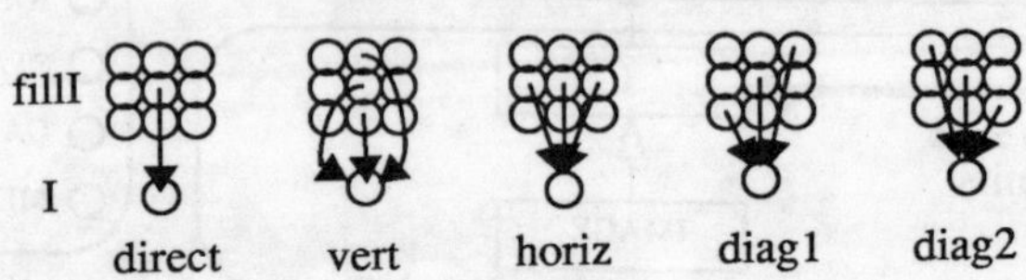

Figure 13: Weight distributions between I and fillI : In order to fill-in the gaps a simple method is used. The one bit gaps are filled if two of their nearest neighbors passing on each side of it are active.

4.3 Segmentation and focussing layers

After selection of a class, the generated image and the input image have to be compared with each other. In order to have a good evaluation of the distance between the two images, even though the selected image is a part or all of the input image, a spatial delimitation of the two images has to be performed by the network.

The input image at fillI, after rectification and filling-in, and the generated image at X produce an area around themselves at REGI and REGX. In order to generate these areas, links were established between every bit from fillI and X to their nine counterparts at REGI and REGX. All the active bits from the images generate an area around themselves and their intersection gives a surrounding area around each pattern. This process is equivalent to a dilation operation.

Figure 14: Links between X,I and REGX,REGI: The areas are generated by activating their neighbors of every active bit in the images.

The intersection of these two areas is then computed at layer INTER_REG. This area of interest helps focus the network's attention to the common bits from both images making up this area. The layers bufX and bufI are used to gate the images to be compared with the area at INTER_REG.

Finally the intersection of these two gated images is computed at INTER so the validation units can find the Hamming distance between the two images.

4.4 Classification units

There are two types of classification units: GLOBAL_MAX units which are used to memorize the area surrounding the input image and LOCAL_MAX units which memorize the input image at different shifted positions.

Before there is any class selection, the input image is transmitted through the network to INTER and INTER_GLOBAL. This last layer computes the area around the intersection image after a selection has been made. The links between INTER and INTER_GLOBAL are the same kind found between X and REG_X or I and REG_I.

The long term memory links from INTER_GLOBAL to GLOBAL_MAX generate a series of excitations. After competition between the classification units, only the one corresponding to the maximum of the excitations is activated.

Because of the dilation operation done on the image at INTER_GLOBAL, a distorted image will produce an equivalent excitation to a complex class to an input without deformations. As depicted in the next figure, the product with the shown image and the memorized region gives an equivalent excitation from a distorted than from a correct pattern.

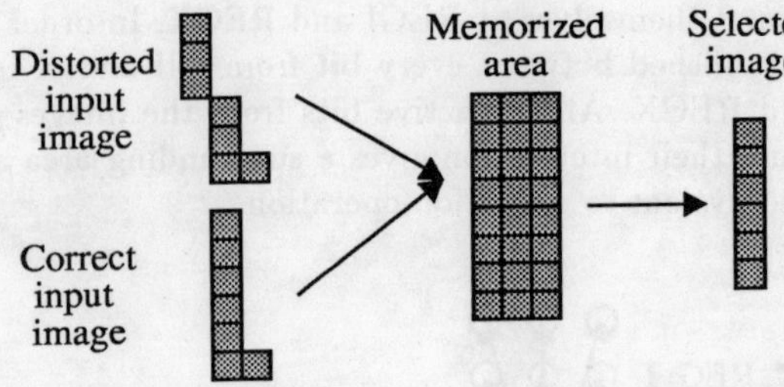

Figure 15: Selection of a complex image from a distorted one: Even though the input is distorted, the multiplication with the area memorized at GLOBAL_MAX gives the same excitation as if it was not deformed. This permits the selection of the complex image.

After the validation of the selected class the activated unit is reset if the classification of the whole image is incomplete (RESET_TOT is then active). If a class or sub-classes have already been validated and memorized, it will not be reactivated.

Contrary to GLOBAL_MAX, only one unit in every group of nine is selected at LOCAL_MAX. The selected unit corresponds to a memorized part of the image shifted so as to best match the input.

The global selection made by GLOBAL_MAX is then transmitted through F3 and a series of links activating all the associated sub-classes at SC. If the selected class has no sub-classes only the corresponding unit will be activated at SC.

These generated sub-classes gate the corresponding groups from LOCAL_MAX at LOCAL_SELECT. An image is then generated with the different shifted sub-parts selected and then compared to the input.

The F3 layer which copies the selected class from GLOBAL_MAX, is also used to memorize this class when the input is considered to be only a part of the selected class. But when a class is memorized at F3, the permanent excitation to SC has to be cancel with an adaptive threshold so that the associated sub-classes can be reactivated by another class selection.

4.5 Adaptative threshold

In order to accelerate the search through the excited units of GLOBAL_MAX and LOCAL_MAX an adaptive threshold is used [12]. This threshold enables the network to activate a classification unit only if it receives a certain percentage of the maximum possible excitation. When a distorted image is presented to the classification units, only the corresponding complex class at GLOBAL_MAX and shifted sub-parts at LOCAL_MAX will be activated.

The input excitation must also be transmitted if the input image is a part of a complex class in order to detect the complex patterns which are included in the input. The excitation is transmitted if:

$$\delta_{min}\ |fillI|\ b_{max} \geq \sum b^k_{i,j}\ INTER_{i,j} \geq \delta_{max}\ |fillI|\ b_{max}$$

where $b_{i,j}$ are weight links between $INTER_{i,j}$ and $LOCAL\_MAX_k$, $|fillI|$ is the total number of active units at fillI and b_{max} is the maximum weight value of the considered class. δ_{min} and δ_{max} are threshold where $\delta_{min} < 1$ and $\delta_{max} > 1$.

4.6 Memorization units

After a selected class has been validated the network must memorize this class with all its associated sub-classes at SC. The classes of SC are memorized at F2 if they are validated (RESET_IND is then active) but not if the image is only a part of the selected one (RESET_SC is active).

The layer CLASSE, only memorizes the selected classes from GLOBAL_MAX without their associated sub-classes. These units, connected with the rectification units I, hold the activation of the validated classes used for the rectification of the input image.

4.7 Validation units

As mentioned earlier, the validation units compute the distance between the input and the generated images.

The common parts of the two images are said to be equivalent if:

$$\beta_{min} \leq \frac{|bufX| - |INTER|}{|bufX|} \leq \beta_{max}$$

or

$$\beta_{min} \leq \frac{|bufI| - |INTER|}{|bufI|} \leq \beta_{max}$$

and $\beta_{min} < 1$ and $\beta_{max} > 1$.$|layer\_name|$ is the the total number of activated units in the layer.

The RESET_IND is activated if the the common parts of the two images to be compare are equivalent and if the input image is not a part of the selected one. Thus, it is activated if:

$$\frac{|X| - |INTER|}{|X|} \leq \rho_X$$

and if:

$$\frac{|fillI| - |INTER|}{|fillI|} > \rho_I$$

where $\rho_I < 1$ and $\rho_X < 1$

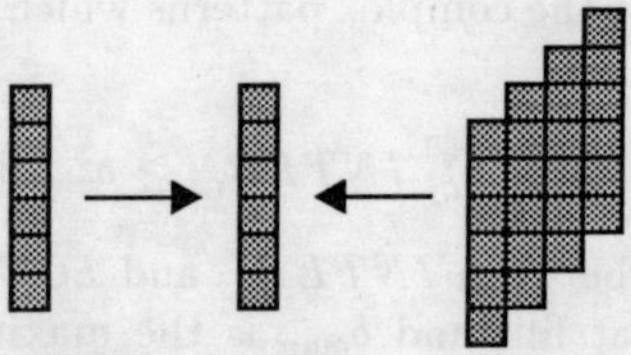

Figure 16: Case where RESET_IND is deactivated: If the generated image included the input without being of the same shape the unit RESET_IND is thus deactivated.

The unit RESET_NOT is the complement of RESET_IND. It is activated or deactivated if the conditions stated before are not met.

The RESET_SC unit is activated if the image shown is equivalent to a part of the selected class. It is activated if:

$$\frac{|X| - |INTER|}{|X|} > \rho_X$$

and if:

$$\frac{|fillI| - |INTER|}{|fillI|} \leq \rho_I$$

where $\rho_I < 1$ and $\rho_X < 1$

The three units like all the validation units are evaluated only if the net is stable after a class selection has been made (STABLE and GAIN2 are then deactivated. Also in order to prevent any spurious activations, each unit is linked to the two others in order to inhibit them after one has been activated. If the selected class is a new one (MEMOIRE is active), only the unit RESET_IND is activated.

4.8 Adaptation of the long term memory

A fast learning method like the one described by Carpenter and Grossberg in an ART1 system [2] was used to modify the weights of the long term memory.

The bottom-up weights between INTER and LOCAL_MAX and their top-down counterparts from LOCAL_SELECT and X are computed by shifting by one bit the image given by fillI around its actual position.

The weights from $INTER_{i,j}$ to $LOCAL\_MAX_k$ are given by:

$$b_{i,j}^k = \begin{cases} \frac{L}{L-1+|fillI|} & \text{if } fillI_{i,j} \text{ is active} \\ 0 & \text{otherwise} \end{cases}$$

where L = 3.

The weights from $LOCAL\_SELECT_k$ to $X_{i,j}$ are:

$$t_k^{i,j} = \begin{cases} 1 & \text{if } fillI_{i,j} \text{ is active} \\ 0 & \text{otherwise} \end{cases}$$

For every new class, a single class is created by making links with all the units from $INTER_{i,j}$ to $LOCAL\_MAX_{new}$ reducing the simulation times. The weights of this class are:

$$b_{i,j}^{new} < \frac{2}{NC * NC}$$

where NC and NL are the number of lines and column of the input image.

This value ensures that the network will select all the possible classification units before selecting a new class.

The links between $INTER\_GLOBAL_{i,j}$ and $GLOBAL\_MAX_k$ have the same values as the ones from INTER to LOCAL_MAX but are created between the active bits of the area generated at INTER_GLOBAL and the selected classification unit.

$$b_{i,j}^k = \begin{cases} \frac{L}{L-1+|fillI|} & \text{if } INTER\_GLOBAL_{i,j} \text{ is active} \\ 0 & \text{otherwise} \end{cases}$$

where L = 3.

The links between $F3_i$ and SC_k are established from the memorized sub-classes at F2. Their weight values are:

$$b_i^k = \begin{cases} \frac{1}{NF2} & \text{if } F2_i \text{ is active} \\ 0 & \text{otherwise} \end{cases}$$

where NF2 is the number of units in the F2 layer.

As mentioned earlier the direct links between $CLASSE_k$ and $I_{i,j}$ have the weight of +1 for every active bit at fillI and of -2 for the surrounding area of the image. Otherwise it is null .

$$b_k^{i,j} = \begin{cases} 1 & \text{if } fillI_{i,j} \text{ is active and } REG_{i,j} \text{ is active} \\ -2 & \text{if } fillI_{i,j} \text{ is not active and } REG_{i,j} \text{ is active} \\ 0 & \text{otherwise} \end{cases}$$

During the learning phase, the adaptive thresholds are computed from the received excitation. They are then stored in the structure of the selected LOCAL_MAX and GLOBAL_MAX units.

4.9 Resonance detection

The weights of the long term memory links are modified if the MEMOIRE unit is active and the net is stable (STABLE is inactive). However, because of the transmission delay between layers, the network must ensure that the classes are transmitted and memorized at F2 and CLASSE before changing weights.

5 Simulations

The network was simulated on a simulator developed by the university of Rochester called the "Rochester Connectionist Simulator" (RCS) [7].

In the RCS, every network is defined by three different types of C structures: units, sites and links. A network is built by creating the units to which the sites are attached. Finally, links are created between the output of some units to the sites of others. A transfer function defined by the user is associated with each structure.

During an asynchronous simulation of the network, the simulator begins by computing each site function followed by the unit and then the link functions. The output of all the units are computed in a pseudo-random order and are fed back to the link sites.

5.1 Handwritten number recognition

For illustrative purposes, the network was trained to recognize handwritten numbers from 0 to 9.

Because of the rectification process, the order of presentation of the patterns is important. In the case where the primitives are presented before the complex pattern, the integrity of the primitives are preserved but not that of the complex image (figure 17).

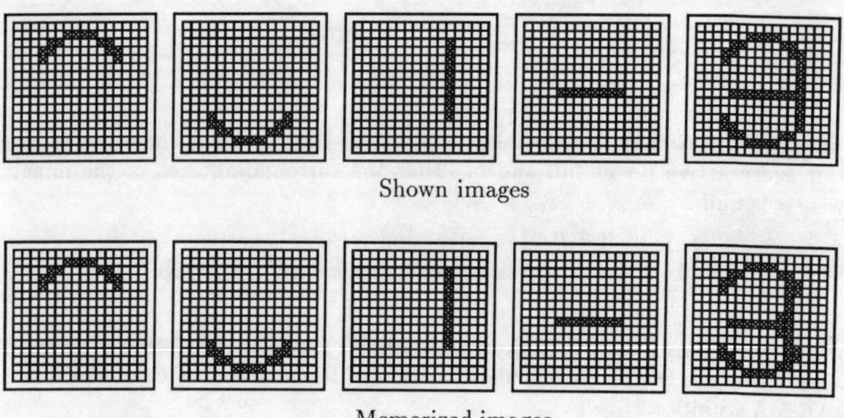

Figure 17: Effect of learning primitives before complex patterns: The integrity of primitives are kept but not that of the complex image.

However, if the complex image is preserved before the primitives, the integrity of the complex image is presented, but not that of the primitives (figure 18).

Shown images

Memorized images

Figure 18: Effect of learning complex patterns before primitives: The integrity of complex patterns are kept but not that of the primitives.

The network was first presented a series of primitives followed by one prototype. The network automatically decomposed the numbers as shown in figure 19.

The network was then shown a few partially distorted numbers, some of which were corrupted with noise. The network then execute the recognition and classification process of the shown patterns as shown in figure 20.

The simulation parameters were:

$$\begin{array}{|l} \delta_{min} = 0.85 \\ \delta_{max} = 1.5 \\ \beta_{min} = 0.2 \\ \beta_{max} = 2.0 \\ \rho_I = 0.25 \\ \rho_X = 0.2 \end{array}$$

Most of the time, the first activated class matched the number shown. However, similar numbers sometimes excited the wrong class at first, but were invalidated by the network. The proper class was then selected and validated.

Depending on the types of primitives used the recognition rate is different. In general a segmentation with more independent primitives provides a better recognition rate than one with fewer primitives.

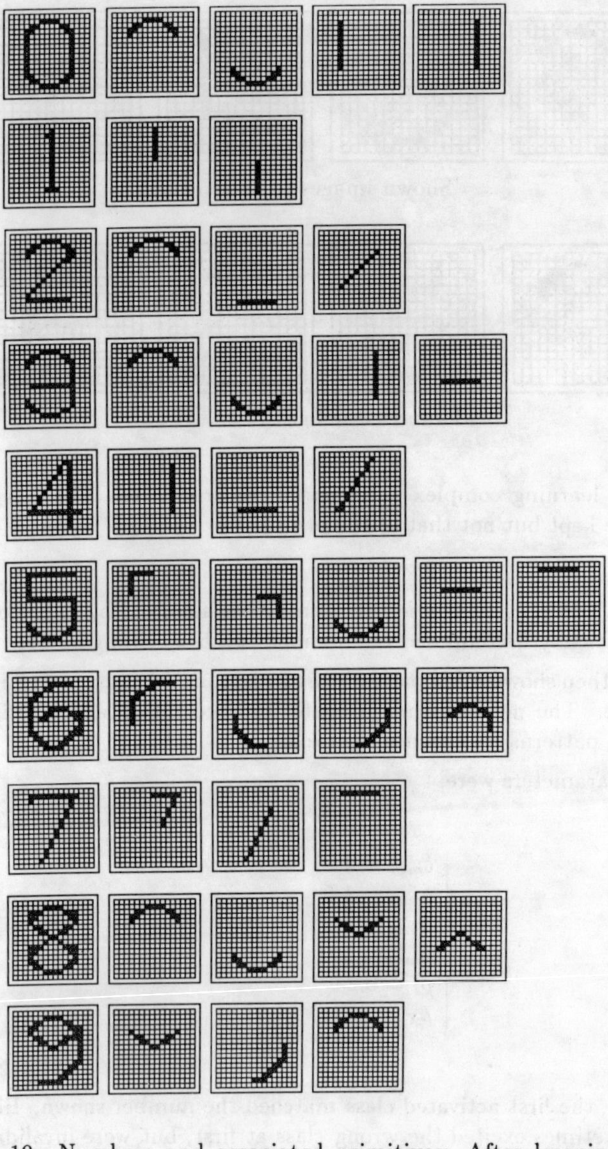

Figure 19: Numbers and associated primitives: After learning a few primitives a prototype of each number was presented. The network then decomposed the numbers into the primitives.

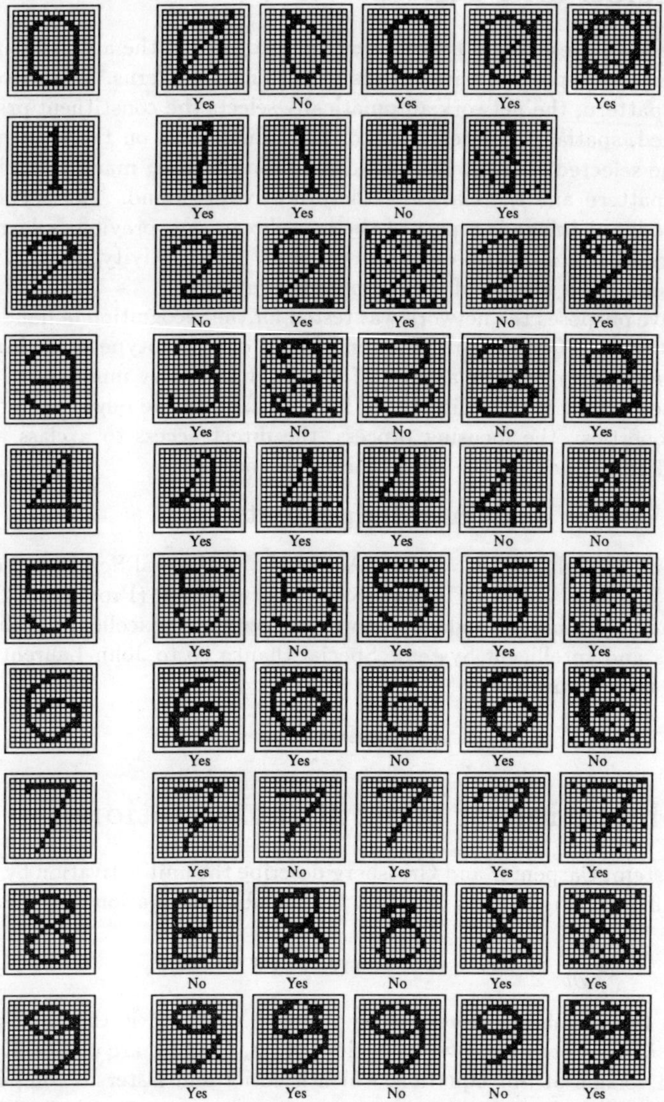

Figure 20: Recognition of distorted numbers: After the learning of one prototype the network was able to recognize distorted and noisy images.

6 Discussion and conclusion

In this paper we have presented a network for the recognition, the auto-classification and the hierarchical grouping, of partially distorted binary patterns. When presented with a complex pattern, the network automatically selects the constituent primitives previously learned, spatially isolates their corresponding parts on the input pattern and validates the selected class. A hierarchical grouping is then made with the class of the complex pattern and the classes of the primitives is found. The network can also detect if the input image is a part of the complex images previously learned. It is also able to recognize partially distorted patterns. The activity of every unit is described by a simplified version of the membrane equation.

For illustrative purposes the network was tested for the recognition of handwritten numbers from 0 to 9. After the presentation of only one prototype of each number the network was able to recognize a series of distorted and noisy numbers.

Although the present work is limited in its applicability, we ought to remember its principal properties: the focusing process, the direct access to a class and the rectification and adjustment of the input during its analysis.

Acknowledgements

This work was supported in part by grant A5274 of the Natural Sciences and Engineering Council of Canada (NSERC), grant 89EQ2830 of FCAR (Province of Québec) and funding made available through the network of centers of excellence of the Institute of Robotics and Intelligent System. Special thanks go to John Laurent for his help in revising the manuscript.

Appendix

A Simplification of the output equations

For the ART system, Carpenter and Grossberg describe the unit activation by a series of non-linear differential equations based on the membrane equation of the form:

$$\varepsilon \frac{\partial x_k}{\partial t} = -x_k + (1 - Ax_k)J_k^+ - (B + Cx_k)J_k^-$$

where x_k is the membrane potential on the k-th unit, J_k^+ is the total positive excitation and J_k^+ the total negative excitation. A, B, C and ε are constant.

A simplified version of this equation, which is easier and faster to compute, was used instead.

$$x_k = f(J_k^+ - J_k^-)$$

where

$$f(x) = \begin{cases} 1 & \text{if } x > 0 \\ 0 & \text{if } x \leq 0 \end{cases}$$

This simplified representation considerably reduces the simulation times.

Sometimes the activity of some units must vary more slowly than others. This is achieved by adjusting the time constants of the membrane equation at different levels to synchronize the order of execution in an asynchronous computation of the net. To simplify further control units were used instead of adjusting these time constants. Control units are activated only by specific states in the simulations and they are used to synchronize the activity of certain layers at specific times in the simulations.

B Activation equations

In the following sections each group of units are described by their excitation (J_k^+) and inhibition values (J_k^+).

B.1 Rectification and filling-in

In order to rectify the input during learning, the links between CLASSE and I are set at +1 for every active bit and at -2 for the surrounding area. The links between the input IMAGE and I, the rectification layer, are all set at +2 for direct units and at 0.1 for each eight surrounding bits.

The activation value for $I_{x,y}$ is:

$$J_{x,y}^+ = D \sum b_{i,j}^{x,y} IMAGE_{i,j} + E\ g(\sum t_k^{x,y} CLASSE_k)$$
$$J_{x,y}^- = E$$

where $E > D$, g() is the nominal value of excitation, $b_{i,j}^{x,y}$ are the weights of the links between each unit $IMAGE_{i,j}$ and the unit $I_{x,y}$, $t_k^{x,y}$ those between each selected unit of $CLASSE_k$ and the unit $I_{x,y}$.

The gaps leaved by the rectification process are filled with the links distribution between I and fillI as shown in figure 13.

Thus for fillI we have:

$$J_{x,y}^+ = D\ (2 * direct + horiz + vert + diag1 + diag2)$$
$$J_{x,y}^- = E$$

where $E > D$. The values *direct*, *horiz*, *vert*, *diag*1, *diag*2 refer to the multiplication of each unit $fillI_{x,y}$ with the links distribution as shown in figure 13.

B.2 Segmentation and focussing layers

The activation value of units $X_{x,y}$ is given by:

$$J_{x,y}^+ = D \sum t_k^{x,y} LOCAL\_SELECT_k$$
$$J_{x,y}^- = E + F\ GAIN2$$

where $F > D > E$.

$t_k^{x,y}$ are the links weight between the selected units of $LOCAL\_SELECT_k$ and each unit $X_{x,y}$.

From the images generated at X and fillI, the surrounding area of those images is compute at REGX and REGI. This dilation operation is done using the links distribution $t_{i,j}^{x,y}$ of figure 14. The activation values of each unit of REGX and REGI are given by:

$$J_{x,y}^{+} = D \sum t_{i,j}^{x,y} X_{i,j} \text{ or } D \sum t_{i,j}^{x,y} fillI_{i,j}$$
$$J_{x,y}^{-} = E$$

where $D > E$.

The intersection of those two areas is then computed at layer INTER_REG.

$$J_{x,y}^{+} = D\ REGX_{x,y} + E\ REGI_{x,y}$$
$$J_{x,y}^{-} = F$$

where $\max\{D, E\} < F < D + E$.

This area of interest gates the images of X and fillI at bufX and bufI. For bufX we have:

$$J_{x,y}^{+} = D\ X_{x,y} + E\ INTER\_REG_{x,y}$$
$$J_{x,y}^{-} = F$$

where $\max\{D, E\} < F < D + E$ and for bufI we have:

$$J_{x,y}^{+} = D\ fillI_{x,y} + E\ INTER\_REG_{x,y} + F\ GAIN2$$
$$J_{x,y}^{-} = G$$

where $\max\{D, E, F\} < G < \max\{D + E, D + F\}$.

Finally the intersection of those two gated images is computed at INTER where:

$$J_{x,y}^{+} = D\ bufX_{x,y} + E\ bufI_{x,y} + F\ GAIN2$$
$$J_{x,y}^{-} = G$$

where $\max\{D, E, F\} < G < \max\{D + E, D + F\}$.

Before any selection has been done, the input image is transmitted through the network to INTER and INTER_GLOBAL. After selection of a class (GAIN2 = 1), an area is computed from the intersection image like in REGI and REGX in order to memorize it.

$$J_{x,y}^{+} = D \sum b_{i,j}^{x,y} INTER_{i,j}$$
$$J_{x,y}^{-} = E\ GAIN2$$

where $D > E > 8 * 100$ and $b_{i,j}^{x,y}$ are the local links as shown in figure 14.

In layers I, fillI, bufI, INTER and INTER_GLOBAL the term GAIN2 from the control unit, which is high when a class has been selected and low otherwise, allows the network to transmit the input image all the way through the input layers, thus activating the classification layers.

B.3 Classification units

The input picture is presented from layers INTER and INTER_GLOBAL to the classification units LOCAL_MAX and GLOBAL_MAX. For GLOBAL_MAX, the units receiving the maximum excitation from the multiplication of the output values of INTER_GLOBAL by the weights of the links made during previous memorization, are activate. They are inhibited though if the unit has been previously reset by RESET_TOT or if it has already been validated and memorize at F2.

The activation values are then given by:

$$J_k^+ = D \sum b_{i,j}^k INTER\_GLOBAL_{i,j} + E\, x_k$$
$$J_k^- = \bigvee_{i \neq k} (J_i^+ - J_i^-) + F\, RESET\_TOT + G \sum_{i \neq k} x_i + H\, STATE_k + I\, F2_k$$

where

$$STATE_k = \begin{cases} 1 & \text{if } RESET\_TOT \text{ is active and } x_k \text{ is active} \\ 0 & \text{otherwise} \end{cases}$$

and $\bigvee_{i \neq k}$ is the maximum value of excitation from all the other units.

Since the maximum possible excitation that can receive a unit from the long term memory is $D_{max} = D\, L$ we then have

$$\max\{I, H\} < D_{max} + E < \min\{H + I, G\}$$

For LOCAL_MAX only one unit from each group of nine units corresponding to the nine shifted memorized image as shown in figure 4, is activated.

The activation values is equivalent to GLOBAL_MAX:

$$J_k^+ = D \sum b_{i,j}^k INTER_{i,j} + E\, x_k$$
$$J_k^- = \bigvee_{i \neq k} (J_i^+ - J_i^-) + F\, RESET\_TOT + G \sum_{i \neq k} x_i$$

where

$$F < D_{max} \text{ and } E + D_{max} < \min\{F, G\}$$

The F3 layer which copies the selected class from GLOBAL_MAX, is also used to memorize this class when the input is considered to be only a part of the selected classes (RESET_SC is then active). For F3 we then have:

$$J_k^+ = E\, GLOBAL\_MAX_k + F\, STATE_k$$
$$J_k^- = G$$

where

$$STATE_k = \begin{cases} 1 & \text{if } RESET\_SC \text{ is active and } x_k \text{ is active} \\ 0 & \text{otherwise} \end{cases}$$

and

$$E > H \text{ and } G > H$$

Every unit at F3 is linked with its counterpart to ensure that there will always be at least one associated unit at SC for each active unit at F3. The other links are determined during learning. So for SC the equations are:

$$J_k^+ = E \sum b_i^k \, F3_i$$
$$J_k^- = E \; STATE_k$$

where

$$STATE_k = \begin{cases} \sum b_i^k \, F3_i & \text{if } RESET\_SC \text{ is active and } \sum b_i^k \, F3_i > 0 \text{ is active} \\ 0 & \text{otherwise} \end{cases}$$

The units from LOCAL_SELECT select the active groups from LOCAL_MAX using the generated sub-classes at SC. We then have:

$$J_k^+ = D \; LOCAL\_MAX_k + D \; SC_k$$
$$J_k^- = E$$

where

$$2 * D > E$$

The units from F2 are activated when the active classes and sub-classes from SC are validated by RESET_IND but not when the input image is a sub-class of the selected class (RESET_SC is then active). After activation, they keep their state until the network is reset. The equation is given by:

$$J_k^+ = E \; SC_k + F \; RESET\_IND + D \; x_k$$
$$J_k^- = H + G \; RESET\_SC$$

where

$$\max\{E, F\} < H < E + F$$

$$D > I$$

$$I < E + F - G$$

Finally, the layer CLASSE memorize the active units from GLOBAL_MAX if RESET_IND or RESET_SC is active. We thus have:

$$J_k^+ = D \; F3_k + E \; RESET\_SC + F \; RESET\_IND + G \; x_k$$
$$J_k^- = H$$

where

$$D + E > H > \max\{D, E\} \text{ and } G > H$$

References

[1] Adler M.J., "Aristotle for Everybody", *MacMillan publishing Co inc.*,New-York, 1978, pp 4–167.

[2] Carpenter G.A., Grossberg S., "A Massively Parallel Architecture for a Self-organizing Neural Pattern Recognition Machine", *Computer Vision Graphics and Image Processing* , 1987, pp 54–115.

[3] Carpenter G.A., Grossberg S., "The ART of Adaptative Pattern Recognition by a Self Organizing Neural Network", *Computer*, mars 1988, pp 77–89.

[4] Cohen M.A., Grossberg S., "Some Global Properties of Binocular Resonances: Disparity Matching, Filling-in and Figure-ground Synthesis",*Mathematical Department, Boston University*, Mass., 1983, pp 1–70.

[5] Fukushima K., "Neural Network Model for Selective Attention in Visual Pattern Recognition and Associative Recall", *Applied Optics*, Dec 87, pp 4985–4991.

[6] Fukushima K.,"A Neural Network for Visual Pattern Recognition", *Computer*, Mars 1988, pp 65–75.

[7] Goddard N.H., Lynne K.J., Mintz T., "Rochester Connectionist Simulator", *Rochester University*, New-York, 1988, pp 1–181.

[8] Grossberg S., "Adaptative Pattern Classification, an Universal Recoding: I. Parallel Development and Coding of Neural Feature Detectors", *Biological cybernetics*, Springer-Verlag, 1976, pp 121–134.

[9] Grossberg S., "Adaptative Pattern Classification, an Universal Recoding: II. Feedback, Expectation, Olfaction, Illusions", *Biological cybernetics*, Springer-Verlag, 1976, pp 187–202.

[10] Marr D., Nishihara H.K.,"Visual Information Processing: Artificial Intelligence and the Sensorium of Sight",*Reading in computer vision*, edited by Fischler M.A. Firschein U., pp 616–637.

[11] Ryan T.W., Winter C.L. ,"Variations on Adaptative Resonance", *Proceedings of IEEE First International Conference on Neural Networks*, San Diego, Californie, vol II, pp 767–775.

[12] Ryan T.W., Winter C.L., Turner, C.J., "Dynamic Control of an Artificial Neural System: the Property Inheritance Network", *Applied Optics*, Dec 87, pp 4961–4971.

[13] Ullmann S., "Visual Routines",*Reading in Computer Vision*, edited by Fischler M.A. Firschein U., pp 189–328.

Visuo-Tactual Coarseness Estimation Tasks: Experiments and Modeling

Isidore Rigoutsos
Robotics and Manufacturing Laboratory,
Courant Institute of Mathematical Sciences,
New York University,
715 Broadway,
New York, NY 10003

Abstract

Visual perception has little to do with the retinal projections of the viewed scenes. Objects are perceived as 3-dimensional whereas their retinal images are 2-dimensional. The properties of perspective projection result in an object at different distances projecting to different retinal images; however, the perceived size persists independent of the variations of the viewing distance. Also, perceived objects tend to maintain their shape independent of changes in orientation. It has been suggested that humans as well as other primates must learn to "see." Touch has been considered as the most likely source for educating the visual system, presumably by adding meaning to the initially meaningless plethora of retinal images. In this work, we report on a series of experiments that focus on the task of visuo-tactual *coarseness* estimation: we examine the relation between the two senses, and give evidence for the occurrence of tactual recalibration in the presence of conflicting stimuli that extend over a relatively short period of time. The experiments are particularly interesting in that they demonstrate the occurence of visual dominance and recalibration in the case of *passive tactile* perception. A black box model, in agreement with the observed behavior, is also presented.

1 Introduction

Visual perception has little to do with the retinal projections of the viewed scenes. These retinal projections are 2-dimensional; however, a rich 3-dimensional world is being perceived, in spite of the occurring reduction in dimensionality. The retinal images of the viewed objects are sampled by distinct light-sensitive receptors positioned at the vertices of an almost hexagonal lattice [17], but one perceives single undivided objects with a concrete structure. A retinal image of a given size can be generated by either small nearby objects or larger distant ones; however, the perceived spatial relationships are quite accurate [15].

The attempts to explain the observed behavior gave rise to two schools of thought. The supporters of the first school advocate that humans (as well as other primates) are born with the ability to perceive the surrounding world. The second school takes the stance that the ability to perceive is acquired empirically rather than being innate. In the 18th century, Bishop Berkeley, along with other philosophers and psychologists, proposed that *tactual perception* provides the basis for *visual learning*: one empirically discovers the rigid characteristics of 3-dimensional objects by resorting to tactual input. Eventually, the initial plethora of meaningless images becomes a meaningful set of world objects.

Before we proceed any further, it is necessary to make more precise the meaning of the term *touch* as this is used in our work. By *touch* we mean ([14]):

> "...the active tactual or haptic perception [as that would be derived from] the position of the fingers relative to one another, [the sense of position of our body parts] and so forth ... [as well as] the passive tactile perception as when a form is impressed against the skin..."

Berkeley's approach seems plausible and is appealing at first; but, as Rock points out [15], it is non-defensible. First of all, the empirical approach advocates for the existence of an innate touch sense on which visual perception is progressively built; however, by the same logic one could equally successfully rule out any innate touch sense. Furthermore, one should not forget the drastic changes in body size (i.e. arm length, relative distances of fingers, etc.) that take place during development. It would be a bad selection, from an evolutionary point of view, if we came to life with a tactual mechanism which was already adjusted to a certain body size. Also, experimental evidence points to the innate basis of form discrimination [18]. Second, experiments with nonhuman infants which were tested on the *visual cliff* suggest that the visual sense is well developed soon after birth, or even is innate for some species. In particular, motion parallax appears to be an innate cue for depth perception for chicks, goats and lambs and is exhibited even at the age of one day [2]. Third, the sense of touch is not precise enough to account for the experienced perceptual accuracy of spatial relations. Last, as experiments on size and shape estimation [16] revealed, when vision and touch are in conflict, not only is vision dominant over touch but it also leads to "...the misperception of the 'feel' of the object..." [14, 15, 16]. When the exposure to conflicting information extended over a longer interval, some form of *recalibration* of tactual perception during size/shape estimation as well as positioning experiments was observed [15]. Visual dominance was also reported in the context of the visual cliff experiments: despite the tactually obtained information on the solidity of the glass, the animals would not function without support from their visual system [2].

The human ability to *visually* estimate roughness with great accuracy has been discussed in recent work [13]. However, to the best of our knowledge, no studies have examined the occurrence of tactual recalibration during *coarseness* estimations tasks; this is precisely what we will be discussing in the sequel. It should be stressed here that the size/shape and positioning experiments concentrated on the effect that conflicting input has on the *active tactual* perception; indeed in all of those experiments, it was the relative position of the fingers or the position of body parts that provided the sensory input. On the other hand, our experiments concentrate on *passive tactile* perception which is admittedly poorer than the active tactual perception.

The remainder of this paper is structured as follows: in Section 2 we elaborate on the experimental arrangements and present the designed procedures together with the underlying rationale. Section 3 contains the experimental results and their analysis. In Section 4 we present a model in agreement with the observed experimental behavior. Section 5 contains a discussion.

2 Experimental Details

2.1 The Templates

The objects used in our experiments were thin rectangular plates of industrial wax having a certain pattern carved on one of their two large facets (Figure 1). There was no particular reason for having chosen this texture pattern other than it was very simple to generate. This particular choice has no shortcomings, because as Gibson points out [3], in both visual and tactual perception it is the spatial information that is of importance, and not the specific *sensory pattern* generating the stimulation.

In order to machine the wax plate patterns, we made use of the N.Y.U. built CNC [1] milling machine [4]. This is a programmable drilling/carving tool that allows continuous control of the 3-dimensional position of the object that is being carved with an accuracy as fine as 0.001 inch. The control is achieved by a program generated by the ANVIL-5000 CAD/CAM package [11]. The machine also provides for interchangeable bits of different sizes and shapes. To carve our surfaces we used standard steel ball-end-mills; under program control, the table on which the wax plates were placed was moved along equally spaced line segments, and parallel to the short side of the plates.

Six distinct patterns were generated in the above manner: their perceptual differences originated from the fact that the values of the pitch parameter, i.e. the densities of the "ridges" per unit length, were *pair-wise* distinct. We used a total of six different ball-end-mill bits, with diameters: 1/16″, 1/8″, 3/16″, 1/4″, 5/16″, and 3/8″ respectively. To generate the grooves with the semi-circular shape that is shown in Figure 1, the pitch value was set to one diameter, whereas the tip of the

[1] CNC: Computer Numerical Control.

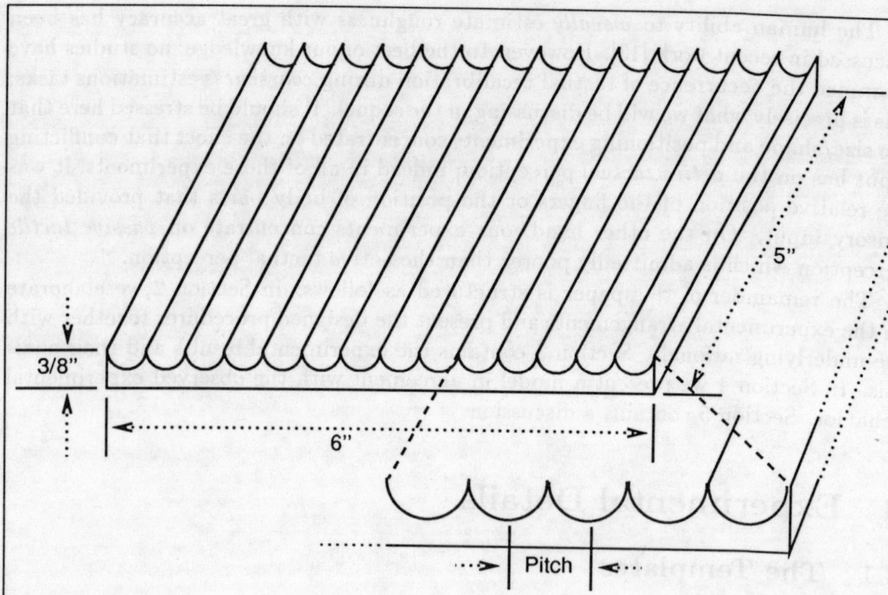

Figure 1. The plates used in our experiments: there was a total of six different values for the pitch.

bit was at a depth equal to one half of the corresponding pitch. The patterns were distinguishable both visually and tactually and one pair of plates was generated for each pattern. In what follows, we will be referring to those plates by their rank in increasing pitch order, i.e. the plate with a pitch value of 1/16″ will be plate #1, the one with a pitch value of 1/8″ will be plate #2 etc. Plates #1 through #6 comprised our "reference set."

2.2 Description of the Experiments

Our experimental procedures followed closely those observed by Rock and his colleagues during experiments on size and shape estimation [15]. A total of 37 subjects participated in our experiments: 36 of those subjects were right-handed, whereas the 37th subject was ambidextrous. The experiments were divided into two large classes.

Our goal was to determine whether persistence of conflicting visual and tactile input over an extended time interval would induce a change in tactile perception. With this in mind, the first class of experiments consisted of the following four categories:

(a) The subject was *shown* different plates in turn and was asked to *tactually* match each one of them, from our reference set. The subject was subsequently exposed to conflicting sensory input (using appropriately chosen pairs of plates) over the course of one half hour. Then, the same task was repeated.

(b) The subject was allowed to only *touch* different plates, in turn, and was asked to *visually* identify each one of the plates, in our reference set. The subject was then exposed to conflicting sensory input (using appropriately chosen pairs of plates) for one half hour. At the end of the 30 minute period, the task of visual identification of different plates was repeated.

(c) The subject was initially *shown* a standard plate[2] and *practiced* to *visually* locate it, from immediate recollection, in the reference set. The subject was subsequently exposed to conflicting sensory input (using appropriately chosen pairs of plates) over the course of one half hour. As soon as the 30 minute period was over, the subject was asked to *visually* identify the plate that was of the same coarseness as he remembered the standard plate to be.

(d) The subject *touched* the standard plate and practiced to *tactually* locate it, from immediate recollection, in our reference set. The subject was subsequently exposed to conflicting information (using appropriately chosen pairs of plates) for one half hour. At the end of this period the subject was asked to *tactually* identify the plate that was of the same coarseness as he remembered the standard plate to be.

The conflicting information was generated by using plate pairs whose pitch values were one step apart: the subject was shown plates with pitch value always one larger than the pitch value of the plate the subject touched. A typical sequence for those pairs was: (Shown, Touched)≡(6, 5) (5, 4) (4, 3) (3, 2) (2, 1) (3, 2) (4, 3) (5, 4) and (6, 5). In all of the above experiments, the subjects used their right hand. A black cloth was covering the subjects' right hand and the touched plate, throughout the exposure to the conflicting sensory input. During the match trials, the subjects were allowed a 5 second interval for each trial.

Let us now examine in more detail the purpose of these experiments. The first two categories aimed at determining whether the subject's perception changed after exposure to conflicting information for an extended time interval; the outcome of these experiments would determine the occurrence of any induced change in the involved modalities. However, and assuming that such a change would indeed be observed, we would still need to determine which of the two modalities changed. Would the visual system adapt to match the conflicting "tactual" input, or would the tactual system change to match the conflicting "visual" input? The experiments of the last two categories were designed to answer precisely this: if *vision* changed, then the subject would not be able to correctly identify *visually* the standard plate after the 30 minute interval; however, the subject would still be able to perform the identification *tactually.* Analogous reasoning would apply to the case where the sense of *touch* was affected instead.

---

[2]This was the one created with the 3/16″ bit.

In the second class of experiments, we distinguish the following two categories:

(a) The subject was presented with a pair of plates, let us call them A and B respectively, and was allowed to see (but not touch) plate A while touching (without seeing) plate B, for a period of 5 seconds. We then assessed the subject's perception by asking the subject to determine the relative coarseness (pitch value) of the two plates A, B. Each subject would typically be presented with a few pairs of plates.

(b) The same task as in (a) was repeated, but this time *after* the subject has been exposed to conflicting sensory input for an extended period.

In the first category, the subjects always employed their right hand: the experiments of that category aimed at determining whether a 5 second interval is sufficient time for the subjects to establish a unitary experience and correctly resolve the relative coarseness of the two plates. At the same time, it provided a means for calculating the probabilities that a subject would under/over-estimate the coarseness of plate B (touched plate), under normal conditions (i.e. without adaptation). In the second category, roughly half of the subjects used their left hand (b1); the remaining subjects used their right hand (b2).[3] These experiments would further corroborate conclusions from the experiments of the first class: if vision were the modality affected by the exposure to the conflicting stimuli, then the choice of the hand would be immaterial; this would no longer hold true if touch were affected instead. Throughout the experiments of this class, the template pairs that we used differed by *at most* one pitch value. The subjects were allowed 5 seconds for each of the matching trials in the experiments (b1) and (b2).

3 Experimental Results & Their Analysis

In this section, we present and analyze the results of the experiments described above. The results for the experiments of class 1 were as follows:

Class: 1 Category: a

Number of subjects:	9
Total number of matching trials:	27
Subjects *tactually* recovered plate of next *smaller* pitch:	18 times

Class: 1 Category: b

Number of subjects:	9
Total number of matching trials:	27
Subjects *visually* recovered plate of next *larger* pitch:	17 times

[3] Recall that this is the hand they employed during the exposure to the conflicting sensory input.

Class: 1 **Category:** c

Number of subjects:	10
Standard plate was *visually* recovered:	10 times

Class: 1 **Category:** d

Number of subjects:	9
Standard plate was *tactually* recovered:	2 times

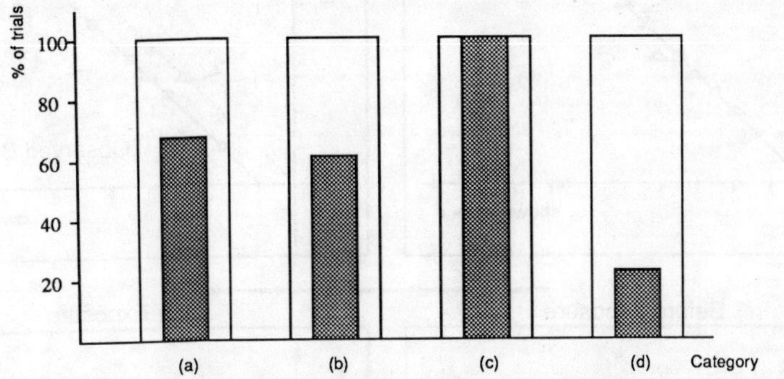

Figure 2. Class 1: graphical representation of the experimental results. The shaded regions indicate the following: (a) times plate of next smaller pitch was tactually recovered; (b) times plate of next larger pitch was visually recovered; (c) times standard plate was visually recovered; (d) times standard plate was tactually recovered.

In Figure 2, the above numerical results are graphically represented in bar form. Figure 3 shows some results for two pairs of subjects: in order to be able to represent the results on the same graph, the experimental points of the second subject in each pair were shifted to the right and upwards by one half unit.

We next present the results for the second class of experiments.

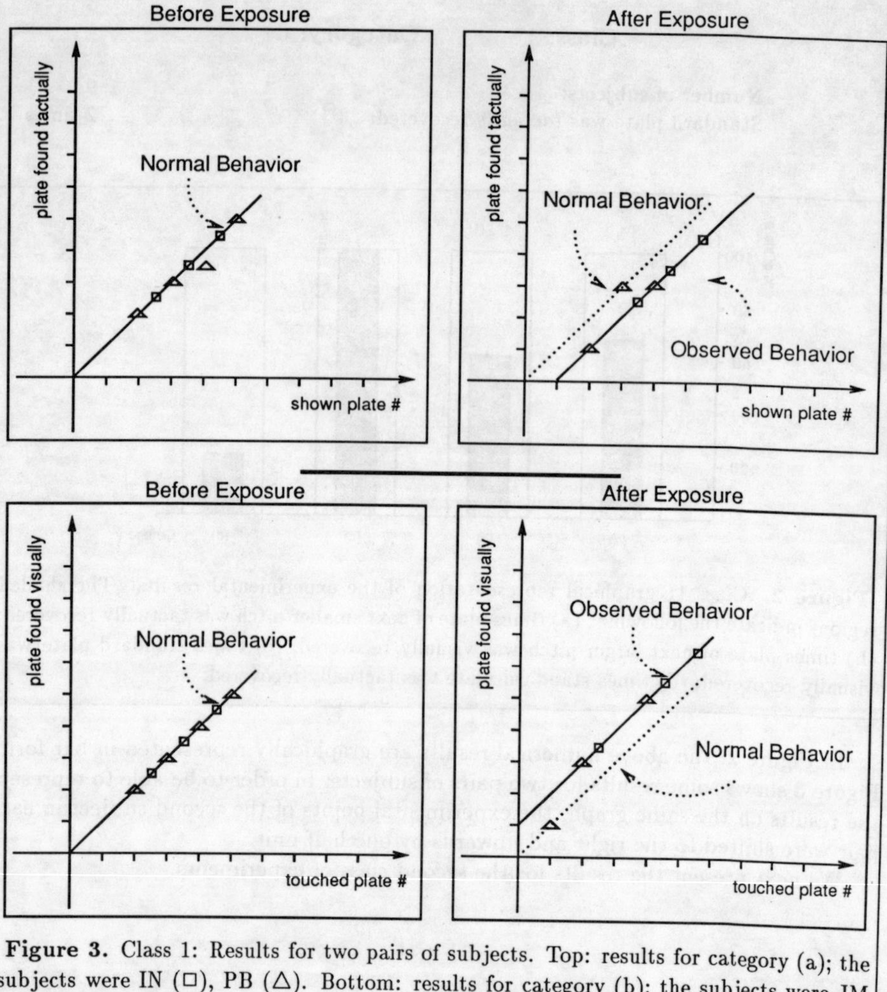

Figure 3. Class 1: Results for two pairs of subjects. Top: results for category (a); the subjects were IN (□), PB (△). Bottom: results for category (b); the subjects were JM (□), SF (△). For presentation purposes, the experimental points corresponding to the second subject were in each case shifted to the right and upwards by one half unit.

Class: 2 Category: a

Number of subjects:	19
Total number of trials:	76
Relative coarseness was correctly estimated:	70 times
Coarseness of touched plate was **underestimated**:	1 time
Coarseness of touched plate was **overestimated**:	5 times

Class: 2 Category: b1

Number of subjects:	7
Total number of trials:	28
Relative coarseness was correctly estimated:	26 times

Class: 2 Category: b2

Number of subjects:	8
Total number of trials:	32
Relative coarseness was correctly estimated:	9 times

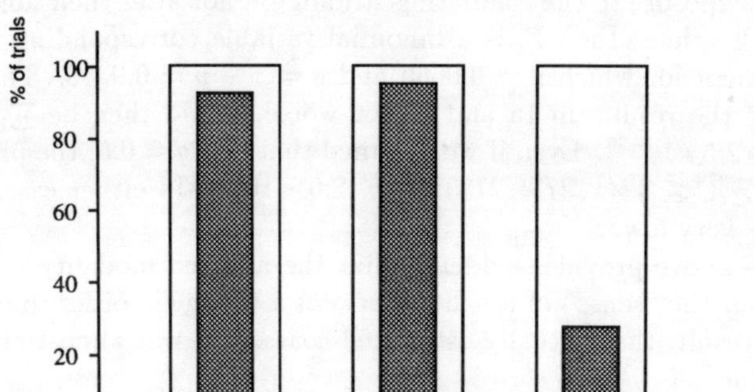

Figure 4. Class 2: graphical representation of the experimental results. Shaded regions indicate percentage of trials during which the relative coarseness was correctly estimated. The light gray (resp. black) region in (a) corresponds to the percentage of trials the relative coarseness was overestimated (resp. underestimated).

Figure 4 graphically represents the above numerical results in bar form.

The implications of the experimental results are clear. First, it is obvious that 5 seconds suffice for the subjects to develop a clear understanding and establish the actual relation between the physical characteristics of the objects they handled. Indeed, in 92% of the trials (class 2, category a), the subjects had a definite unitary experience of the templates' coarseness. In what regards the results from categories (a) and (b) of the first class, we can see that extended exposure to conflicting sensory input leads to changes in the subject's perception; the results from categories (c) and (d) of the same class resolve the issue of which of the two modalities undergoes a change. Vision did not change after the exposure to conflicting information: the subjects were able in all the cases to visually identify the standard template from memory. On the other hand, only 2 out of 9 times were subjects able to tactually recover the standard plate from memory: the templates now felt coarser than they really were. These conclusions are corroborated by the results of categories (b1) and (b2) of the second class.

In order to further appreciate the results of categories (a) and (b) of the first class, we proceed with a standard sign test analysis [10]. Let X_i (resp. Y_i) be the template number corresponding to the *perceived* tactual (resp. visual) stimulus during the i-th matching trial. Also let

$$Z_i = \begin{cases} 1, & \text{if } Y_i - X_i \leq 0 \\ 0, & \text{if } Y_i - X_i > 0 \end{cases}$$

If $p \equiv Pr(Z_i = 1)$ then from 2a we can see that the probability of overestimating the coarseness of the touched plate is $p \approx 5/76 = 0.0660$. Let us now hypothesize that the subjects' exposure to the conflicting stimuli did not alter their ability to tactually estimate pitch values; then Z_i is a binomial variable corresponding to a single trial of an experiment for which $p = 0.0660$ and $q = 1 - p = 0.9340$. The probability of occurrence of the results in 1a and 1b, or worse, would then be $\sum_{k=18+17}^{27+27} b(k, 27 + 27, 0.0660) = 2.5 * 10^{-28}$. Even if we assumed that $p = q = 0.5$, the probability would be equal to $\sum_{k=18+17}^{27+27} b(k, 27 + 27, 0.5) = 2.0 * 10^{-2}$. In either case, the calculated probability is very low.

All of the above provides evidence that the affected modality was indeed touch and not vision: the "sense" of touch underwent a change in order to match the visual input. As a result, the tactually estimated coarseness was almost always shifted by one pitch value.

In previous size and shape estimation experiments [16], the conflicting stimuli were created by a distorting optical lens of whose existence the subjects were not aware. This guaranteed that they would not resort to a conscious decision of which sense to rely on. An interesting observation was made during our experiments: in a number of cases the subjects reported they suspected they were being exposed to conflicting sensory input. However, this did *not* prevent their sense of touch from adapting and eventually matching the visual input. During the matching trials these subjects behaved similarly to those that had not suspected anything. Apparently, adaptation occurred despite their suspicion of the existence of a conflict. It appears that *nature* has chosen to give more credit to the sense of vision and has consequently *hard-wired* the sensory pathway in such a way that adaptation will be induced even if other sources appreciate it as unnecessary. As we already mentioned in Section 1 similar behavior was observed during experiments on the visual cliff.

4 The Model

We next describe a model that is in agreement with the results presented in the previous section.

With regard to the sensory pathway pertaining to the senses of vision and touch, the observed experimental behavior suggests an organization like the one shown in Figure 5. This model is general enough to account for the observed behavior during size and shape estimation experiments as well.

The environment provides the input to the touch[4] and vision sensors, whose appropriately thresholded outputs enter the stimulus associator (an instance of a pattern associator). Normally, the associator's output is a perception of the coarseness of the object causing the given activation pattern; the associator is also able to determine the occurrence of conflicts between the perceptions induced by the visual and tactual stimuli. When such a case arises, both "consciousness" and the "teacher" are made aware of it. The latter subsequently generates a correction signal and induces an adjustment of the connection strengths in the tactual input association matrix only (visual capture). Such an adjustment signal can be determined using any variation of basic Hebbian learning [8, 12].

This closed-loop system is another instance of the basic (unsupervised) self-regulating architecture used to model the actions of adaptive biological organisms within an everchanging environment [1, 5, 6, 7, 9]. Notice the dashed line between the "consciousness" box and the "teacher." As the experimental results suggest, this link is either weak or non-existent. It should also be evident why the induced changes should occur in the pattern associator rather than be located at the corresponding sensors: after exposure to the conflicting stimuli, the physical signal itself (i.e. the response of a sensor) still remains in agreement with the stimulus which produced it; what changes is the *perception* that the subject now associates with the given stimulus. The suggested model, although simple, accounts well for the observed behavior.

5 Discussion

In the previous sections we presented and studied the issue of visual capture during coarseness estimation tasks. Through a series of experiments we were able to provide evidence that during coarseness estimation tasks the following occur: (a) vision dominates in the presence of conflicting sensory input, and (b) if the sensory conflict

[4]Recall our definition of "touch" from Section 1.

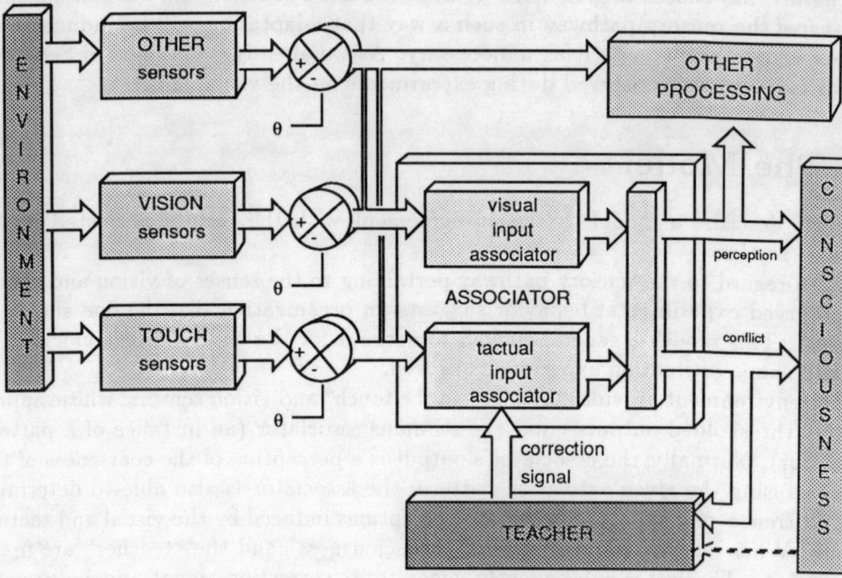

Figure 5. A model in agreement with the observed experimental behavior.

persists over an extended period, the sense of touch undergoes a change in order to match the visual input.

Our experiments are particularly interesting in that they demonstrate the occurrence of visual dominance and recalibration in the case of *passive tactile* perception. All of the previous experiments, namely the size/shape estimation and positioning ones described in [15], had dealt with the aspect of *active tactual* perception. A model in agreement with the accumulated experimental evidence was also suggested.

However, a number of questions still remain unanswered. Assume for instance that the subjects were exposed to conflicting stimuli for time periods much longer than one half hour, and that upon termination they were tested to verify that adaptation has occurred. Subsequently, they were not allowed to use their right (adapted) hand for yet another extended time interval, after which the coarseness estimation ability of that hand was again tested. Would adaptation still be observed? Or, would the tactual pattern associator have "unlearned" the associations induced by the exposure to the conflicting stimuli? The first alternative is not an interesting one. On the other hand, if the associator had unlearned, this would have implied the existence of a *quiescence state* (i.e. a "generic" association matrix) in which the associator was prior to the induced adaptation. But, should there exist such a "preferred" set of associations in the first place? And, since this associator is *programmable* in the broad sense, how was this particular set of associations established in the first place? A possible explanation that would be favored by nativists could of course be based on the assumption of innate hardwiring. But as we already mentioned, an empiricist could counter that attempt by pointing out that such an explanation would have to ignore the drastic change in size of the human body during development. Last, it would also be interesting to examine the specifics of the dynamic evolution of the system towards the alleged equilibrium state. Clearly, all of the above questions and their ramifications would extend to the case of size/shape estimation and positioning experiments as well. Research work in this direction is expected to shed some light on these topics.

Acknowledgements

The author would like to thank Fred Hansen and Elias Pavlakos of the Robotics and Manufacturing Lab. for carving the templates used in the experiments. The author's thanks also go to all the volunteers that generously offered some of their valuable time and helped with the experiments.

References

[1] Bullock, D. and S. Grossberg. *Neural and Synergetic Computers*, chapter title: Self-Organizing Neural Architectures for Eye Movements, Arm Movements, and Eye-Arm Coordination. Springer-Verlag, 1988.

[2] Gibson, E. and R. Walk. "The Visual Cliff." In *Readings from the Scientific American*, chapter 33. Freeeman and Company, 1973.

[3] Gibson, J. "Observations on Active Touch." *Psychological Review*, 69:477–491, 1962.

[4] Greenfeld, I., F. Hansen, J. Fehlinger and E. Pavlakos. "MOSAIC: System Description, Specification and Planning." Technical report, Robotics Lab., New York University, June 1989.

[5] Grossberg, S., editor. *The Adaptive Brain.* North-Holland, 1987.

[6] Grossberg, S., editor. *Neural Nets and Natural Intelligence.* MIT Press, 1988.

[7] Grossberg, S. and M. Kuperstein. *Neural Dynamics of Adaptive Sensory-Motor Control: expanded edition.* Pergamon Press, 1989.

[8] Hebb, D. *The Organization of Behavior: a neurophysiological theory.* Wiley, 1949.

[9] Held, R. "Plasticity in Sensory-Motor Systems." In *Readings from the Scientific American*, chapter 38. Freeeman and Company, 1973.

[10] Hoel, P. *Introduction to Mathematical Statistics.* Wiley and Sons, 1984.

[11] Manufacturing and Consulting Services Inc. *ANVIL-5000 Sun Site Support*, 1987.

[12] McClelland, J. and D. Rumelhart. *Parallel Distributed Processing*, volumes 1 & 2. MIT Press, 1986.

[13] Pentland, A. "Shading into Texture." *Artificial Intelligence*, 29:147–170, 1986.

[14] Rock, I. *The Logic of Perception.* MIT Press, 1983.

[15] Rock, I. and C. Harris. "Vision and Touch." In *Readings from the Scientific American*, chapter 26. Freeeman and Company, 1973.

[16] Rock, I. and J. Victor. "Vision and Touch: An Experimentally Created Conflict between the Two Senses." *Science*, 143:594–596, 1964.

[17] Watson, A. "An Orthogonal Oriented Quadrature Hexagonal Image Pyramid", December 1987. NASA Technical Memorandum 100054.

[18] Zimmerman, R. and C. Torey. *Behavior of Nonhuman Primates*, volume 2, chapter 11. Academic Press, 1965.

3-D Object Model Synthesis in a Monocular Imaging System

A. Mark Earnshaw
Andrew K.C. Wong
Pattern Analysis & Machine Intelligence Lab
Department of Systems Design Engineering
University of Waterloo
Waterloo, Ontario, Canada, N2L 3G1

Abstract

One common method of 3-D computer vision is stereopsis. A single camera system which has been implemented using this approach is presented here. The *MAP* (Matching Algorithm for Points) program is able to synthesize relatively accurate models of unknown polyhedral objects (*i.e.* straight edges and planar faces) of simple to medium complexity. Edge data is first extracted from sequences of digitized images and then processed to extract corner points and remove background noise. Corresponding points are matched between successive images and the resulting information is used to triangulate the (x, y, z) coordinates of the original object vertices. This then permits the synthesis of a wire-frame model depicting the item under consideration. Sample results are presented to demonstrate the capabilities of the system. Possible directions for future development are also briefly discussed.

1 Introduction

A common method for equipping computers with the ability to acquire visual data in three dimensions is the technique of stereopsis. This approach uses the offsets between corresponding points in paired images captured from different locations to recover the original coordinates of observed features via a triangulation process. If the camera positions are known, this ranging operation may be performed relatively easily and accurately.

This form of data acquisition has many potential applications, both in academic research and the industrial world. Coupled with an object recognition program, for example, such a system would be capable of constructing models of, and then identifying unknown items in its environment. Other examples might include tasks such as automated parts inspection.

The Pattern Analysis & Machine Intelligence Lab at the University of Waterloo has been studying the field of 3-D computer vision for some time. Previous work has concentrated on matching simple constellations of points between stereo images [9, 17] and on the recognition of objects in real-world scenes using predefined models [14]. Based on these investigations, research emphasis was then placed on developing the ability to generate object models from visual data without the need for any a priori knowledge about the items.

Current research work concentrating on the area of stereoscopic vision has resulted in a system, entitled *MAP* (Matching Algorithm for Points), which is capable of synthesizing models of objects of medium complexity with reasonable accuracy. The immediate application of the program is the ongoing development of an autonomous workcell by providing an Intelledex 705 robot with a visual interface through which it can interact with the real world.

2 Background & Motivation

Over the years, a number of vision systems using stereopsis have been developed and implemented. Typically, the primary area of concern with this approach appears to be the identification of corresponding features between paired images. Once these correspondences have been determined, the triangulation process to recover three-dimensional profiles is relatively simple. For example, Alvertos *et al* [1, 2] performed detailed analyses for different possible stereo camera configurations. The actual features selected for matching tend to vary, but are generally individual pixels, lines, or corner points (*i.e.* line intersections). The latter two represent high level features and their use generally reduces the necessary processing time significantly due to their smaller numbers.

If a camera system has been accurately calibrated, it is possible to pair individual pixels between images to recover a range map. Kim and Aggarwal [10] matched edge points using local gradient patterns as a guide. A three-view approach implemented by Ito and Ishii [8] paired points between the traditional two views, but also used a third image to verify all hypothesized matches. Both Hoff and Ahuja [6] and Olsen [13] integrated the processes of feature matching and surface interpolation by using a surface smoothness constraint in an iterative fashion to aid in identifying correct and incorrect pairs.

Lines are a higher level feature which may also be used to recover three-dimensional structure. McIntosh and Mutch [12] presented one algorithm for accomplishing this task, although their range calculation method was not accurate for partially occluded lines. A graph theory approach was used by Horaud and Skordas [7] to derive a correspondence graph which represented all feasible line mappings. The optimal set of feature mappings was obtained by identifying the largest clique in this graph. Even curved edge contours can be utilized as was illustrated by Sherman and Peleg [15] who used attributes such as shape similarity, equal contrast gradients, and relative contour ordering to facilitate the correspondence process.

The intersection points of linear edges or corner points represent yet another feature that may be matched for stereo purposes. Jenness [9] began some preliminary research into this area using constellation arrangements as an additional guiding attribute. However, the resulting system was not capable of constructing a complete object model from the source data.

There are a few problems with these existing approaches to stereo vision. Triangulation error is one primary area of concern. It is usually caused by the image discretization process and/or imprecise feature localization. Ideally, the baseline between camera positions should be as large as possible to minimize this error. Further improvements result if the cameras are oriented so that their optical axes are at right angles [5]. However, this often causes the two views to appear radically different, thus complicating the feature correspondence process. In addition, only a portion of the object being studied will be visible from any single location. A number of features will be occluded and therefore only a partial reconstruction of the item may be performed.

A possible solution to both of these problems is the integration of information from multiple (*i.e.* 20-30) images. For instance, the technique used here was to move a camera around an object, capturing images at successive 10° rotation increments (this value was experimentally found to yield data whose similarities facilitated the feature matching process). Corresponding points could then be tracked relatively easily through the sequence of images. This approach provided more information for the triangulation procedure and was thus expected to improve measurement accuracy. In addition, a full circumnavigation of the object implied that all of its features would be visible in at least several of the images (assuming a reasonably convex item) so the occlusion problem was also addressed.

3 System Overview

There are four major components in the *MAP* system as illustrated by the flowchart of Figure 1. Visual data enters the system through a CCD video camera. The resulting digitized images are then passed through an edge detector which extracts line features and identifies corner points by searching for intersecting lines. Some heuristic techniques may also be used to remove background noise from the edge data. Matching corner points are then tracked through successive images so that they can be correlated for triangulation purposes, this being the next stage of the program. Finally, the individual images are integrated to synthesize a wire-frame model of the original object.

It should also be mentioned that the system is programmable using a simple interpreted language developed specifically for this purpose. This allows the separate

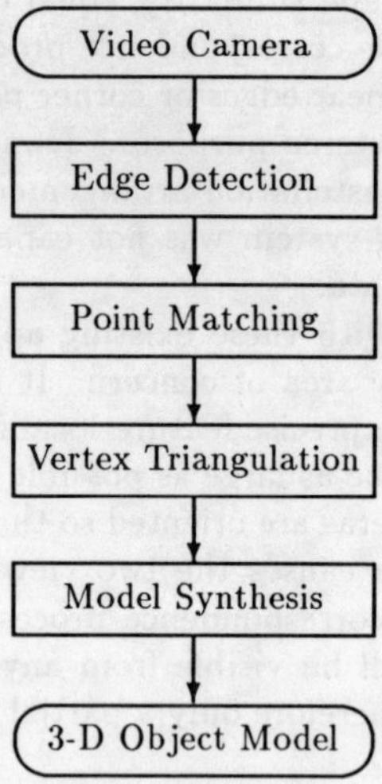

Figure 1: Flowchart of System Components

functions to be performed in an order deemed to be most appropriate for a specific application. For instance, it may be necessary to acquire and process an increased number of images for a more complicated object. This programmability results in significant flexibility and allows the system to be reconfigured quickly and easily.

4 Model Synthesis Algorithm

4.1 Edge Detection

There are four basic steps in the edge detection process. Gradient values representing possible edges are first extracted from a digitized gray level image. These are linked together to form lines which are then extended to identify corner points. Finally, an optional set of routines recover the camera's position and orientation from a visual reference pattern.

The direction of a gradient is defined as being perpendicular to the edge to which it belongs. Many edge detection schemes calculate gradients for several different directions, but the *MAP* system only computes horizontal and vertical values as follows:

$$H_{x,y} = [p_{x+2,y} + 2p_{x+1,y}] - [2p_{x-1,y} + p_{x-2,y}]$$

$$V_{x,y} = [p_{x,y+2} + 2p_{x,y+1}] - [2p_{x,y-1} + p_{x,y-2}]$$

where $H_{x,y}$ is the horizontal gradient at (x, y), $V_{x,y}$ is the corresponding vertical gradient, and $p_{i,j}$ is the gray level intensity at the specified coordinates. The 5×1 weighted mask was selected to identify edges primarily from the immediately adjacent pixels, although the outlying values are also used to smooth the gradient and reduce the effects of noise. To position the edge point, successive gradients that are higher than a specified magnitude are searched for local maxima or minima. These typically represent boundaries of areas with high contrast and thus the best locations for situating edge crossings.

It is then necessary to group collinear points together to form lines. A nucleus is first identified by locating two adjacent gradient values and extrapolating to predict the position of a third point which is then verified. Once such a line has been successfully initialized, it is possible to track the entire edge by continuing to project the line at both ends and searching for further gradient points in the vicinity of the predicted locations.

The slopes and intercepts of these lines are determined using standard linear regression formulas which minimize the sum of squared error terms. Although this method can cause problems when outlying points are included in the data set, the line tracking algorithm excludes such points, so excessively misplaced edges are not an area of concern. Since it is only necessary to maintain a few summation values for each edge, it is relatively simple to add new points to the line by modifying the appropriate tabulations and then updating the slope and intercept parameters. This capability is important for the edge tracking process which functions by incrementally adding points to an existing line.

The identified lines are then extended at each end to produce corner points. By first fitting the edges to the observed data and then locating the intersections, it is theoretically possible to position these points with subpixel accuracy, although this level of precision is unlikely to be achieved in the real world. The task is accomplished by considering each line in turn and searching in the local vicinity of both ends for any other nearby lines which may form potential intersections. Since multiple edges may not intersect exactly at a common location, the actual point coordinates are calculated so as to minimize the sum of squared perpendicular distances from each line. Before finalizing a corner point, various tests are made such as checking if two lines are collinear and should therefore be combined, or if the intersection falls in the middle of one line which should then be split in half.

In order for the object reconstruction process to function accurately, it is essential to know the camera's position and orientation for each image. This may be accomplished through the use of an easily identifiable reference pattern (as shown in Figure 2) which is included in the image. Since the pattern's dimensions and construction are known, it is not difficult to use this information, together with its observed shape, to derive a 3-D transformation (consisting of a rotation and a translation) giving the position of the symbol relative to the camera [5, 11]. This may

Figure 2: Visual Reference Pattern

then be inverted to obtain the camera position and orientation in a common global coordinate system.

Figure 3 shows the horizontal and vertical gradient values for a sample image. As can be seen, these correspond primarily to vertical and horizontal lines respectively. The edges which were extracted before and after corner point identification are shown in Figure 4. Clearly, the latter information is potentially more useful than the former.

4.2 Point Matching

In order to recover the coordinates of the original object vertices, it is first necessary to match corresponding points between different views. Since successive images are normally captured at rotation intervals ranging from 10° to 20°, they are typically similar and the problem becomes one of feature tracking which can be performed quickly and accurately. In addition, the use of a monocular system implies that only 36 images need be matched over a full revolution. Therefore, the processing time is much less than for a stereo two-camera system performing the same task (*i.e.* matching images captured from the same number of observation points). Requiring fewer images also reduces the probability of matching errors that might adversely affect the final model.

The entire matching process is centred around a cost function which evaluates a hypothesized point mapping on the basis of such factors as relative (x, y) coordinates, connections to already matched points, and lengths, angles and numbers of attached lines. For each specified pair, a cost figure is returned, with lower values indicating better matches than higher ones. If a match is considered to be completely unsuitable, it is assigned a very large penalty cost which effectively removes it from further consideration. Since this routine is typically accessed thousands of times during the synthesis of a single model, it has been highly optimized for processing efficiency through the use of techniques such as look-up tables for frequently calculated values.

When a mapping of features from one image to the next in a sequence is performed,

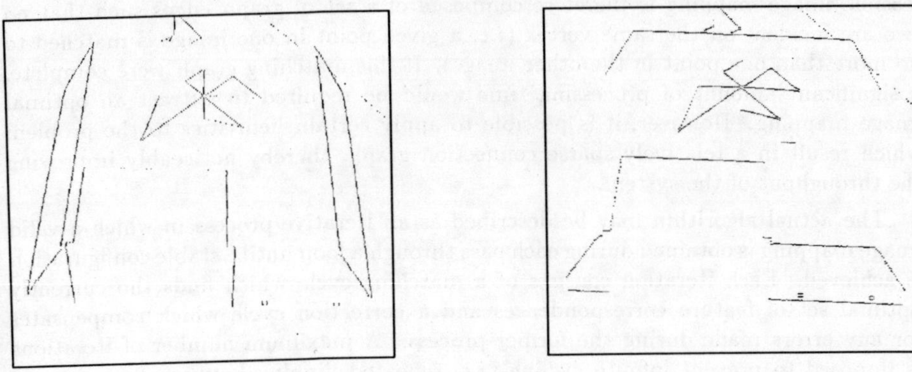

Figure 3: Sample Horizontal and Vertical Edge Gradients

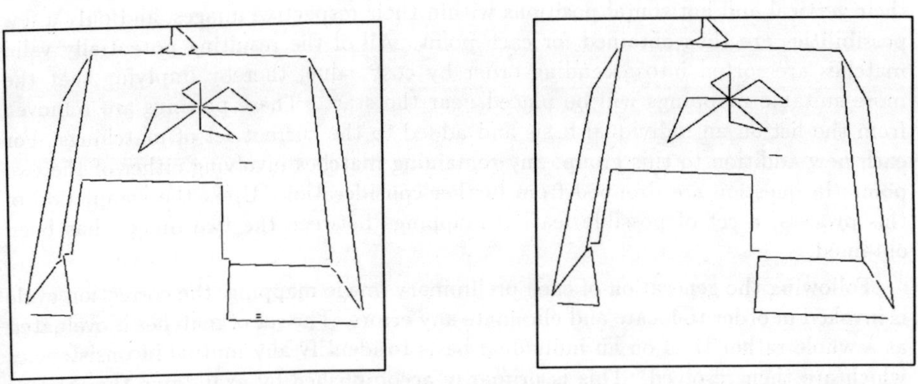

Figure 4: Initial and Final Edge Identification

the corner points are represented as vertices in a bipartite graph. The points from each image form each of the two vertex groups. Individual matchings are specified using the connecting arcs which are weighted with the appropriate cost values. A feasible image mapping is therefore composed of a set of graph edges such that no two are incident on the same vertex (*i.e.* a given point in one image is matched to no more than one point in the other image). If this matching graph were complete, a significant amount of processing time would be required to extract an optimal image mapping. However, it is possible to apply certain heuristics to the problem which result in a relatively sparse connection graph, thereby noticeably improving the throughput of the system.

The actual algorithm may be described as an iterative process in which a valid image mapping is obtained during each pass through a loop until a stable configuration is achieved. Each iteration consists of a matching cycle which finds the currently optimal set of feature correspondences and a correction cycle which compensates for any errors made during the former process. A maximum number of iterations is imposed to prevent infinite cycling (*i.e.* repeated flipping between two or more different image mappings), although this has not yet manifested itself as a problem during testing. Typically, only two or three iterations are required to obtain an optimal image mapping.

During the matching cycle, the cost of pairing each point in the first image with each point in the second is nominally evaluated. Assuming about n points per image (n is typically between 15 and 20 at the present time), this would be an $O(n^2)$ process and therefore computationally expensive and inefficient. However, due to image similarities, potential matching candidates can be constrained according to their vertical and horizontal positions within their respective images, and only a few possibilities are thus obtained for each point. All of the resulting potentially valid matches are sorted into ascending order by cost value, thereby implying that the more suitable mappings will be placed near the start. These pairings are removed from the list on an individual basis and added to the current set of matchings. For each new addition to this group, any remaining matches involving either of the two points in question are dropped from further consideration. Upon the completion of this process, a set of possible feature mappings between the two images has been obtained.

Following the generation of each preliminary image mapping, the correction cycle is invoked in order to locate and eliminate any errors. The set of matches is evaluated as a whole rather than on an individual basis to identify any mutual inconsistencies which are then resolved. This is primarily accomplished by examining the relative positionings and connectivities of paired points between the two images. The cost of each matching is recomputed, taking into consideration any information known about the local neighbours, and any infeasible correspondences are removed. The program then attempts to complete the total set of matches by searching for valid pairings between the remaining leftover points.

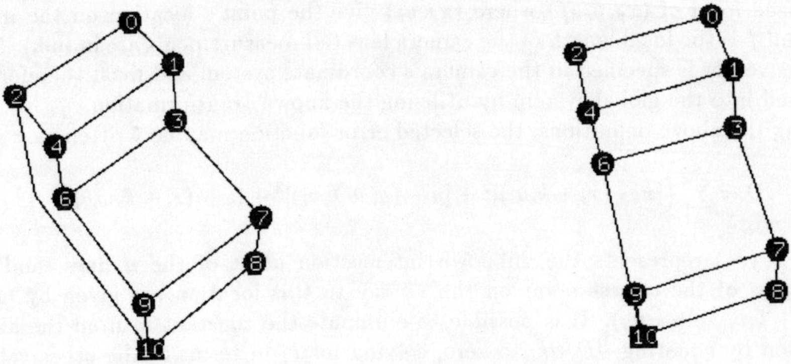

Figure 5: Sample Point Matching Between Two Images

A sample point matching produced by the *MAP* system may be seen in Figure 5. The numeric labels represent corresponding pairs of points which have been matched to each other.

4.3 Vertex Triangulation

If the camera position and orientation are known for a given image and a pinhole camera model is assumed, an object point can be constrained to lie on a ray extending from the appropriate position on the CCD array and through the centre of the camera lens. If multiple rays can be determined for the same point (as matched between successive images), it is theoretically possible to compute their intersection point which will represent the location of the original object vertex.

In reality, it is desirable to have at least several rays for a given object point to compensate for the inaccuracies of the camera positioning, image digitization, and edge detection processes. In addition, it is improbable that all of these lines will intersect precisely at a common location. Thus, it is necessary to determine the best "point of intersection" as evaluated according to some error function. The *MAP* system uses the sum of squared perpendicular distances between the point and the lines as its error function and positions the object vertex so that this value is minimized.

For ease of calculation, rays are expressed in a parametric form such as:

$$\mathbf{r_i}: \quad (x_i, y_i, z_i) + k_i(u_i, v_i, w_i)$$

where (x_i, y_i, z_i) are the coordinates of a point known to be on the ith ray (*i.e.* the lens centre), k_i is an unknown constant to be determined, and (u_i, v_i, w_i) is a normalized direction vector. An actual point on the line may be specified by calculating the appropriate value of k_i. If the x and y axes define the plane of the CCD array and the z axis represents the camera axis, then (u_i, v_i, w_i) may be expressed as the normalized form of (x_p, y_p, f) where (x_p, y_p) give the point's location on the image plane and f is the focal length of the camera lens (all measurements are in *mm*). Note that this vector is specified in the camera's coordinate system and must therefore be translated into the global system by utilizing the known transformation.

Using the above definitions, the selected error function may be written as:

$$D = \sum_{i=1}^{n} \left\{ [x - (x_i + k_i u_i)]^2 + [y - (y_i + k_i v_i)]^2 + [z - (z_i + k_i w_i)]^2 \right\}$$

where (x, y, z) represents the unknown intersection point of the n lines, and the coordinates of the closest point on the ith ray to this location are given by $(x_i + k_i u_i, y_i + k_i v_i, z_i + k_i w_i)$. It is possible to eliminate the constant k_i from the above expression by equating $\partial D / \partial k_i$ to zero, solving for k_i in terms of the other values, and then substituting the result back into the equation for D. After setting the partial derivatives of D with respect to x, y and z equal to zero and performing some manipulation, the following system of three equations may be obtained.

$$\left(n - \sum_{i=1}^{n} u_i^2\right) x + \left(-\sum_{i=1}^{n} u_i v_i\right) y + \left(-\sum_{i=1}^{n} u_i w_i\right) z = \sum_{i=1}^{n} \left[(1 - u_i^2) x_i - u_i v_i y_i - u_i w_i z_i\right]$$

$$\left(-\sum_{i=1}^{n} v_i u_i\right) x + \left(n - \sum_{i=1}^{n} v_i^2\right) y + \left(-\sum_{i=1}^{n} v_i w_i\right) z = \sum_{i=1}^{n} \left[-v_i u_i x_i + (1 - v_i^2) y_i - v_i w_i z_i\right]$$

$$\left(-\sum_{i=1}^{n} w_i u_i\right) x + \left(-\sum_{i=1}^{n} w_i v_i\right) y + \left(n - \sum_{i=1}^{n} w_i^2\right) z = \sum_{i=1}^{n} \left[-w_i u_i x_i - w_i v_i y_i + (1 - w_i^2) z_i\right]$$

Assuming these to have the form $a_i x + b_i y + c_i z = d_i$ for ease of notation, the final solution for the vertex coordinates can be written as:

$$y = \frac{f_2 g_3 - f_3 g_2}{e_3 f_2 - e_2 f_3} \qquad e_i = b_i - \frac{a_i b_1}{a_1}$$

$$z = \frac{e_2 g_3 - e_3 g_2}{e_2 f_3 - e_3 f_2} \qquad f_i = c_i - \frac{a_i c_1}{a_1}$$

$$x = \frac{d_1 - b_1 y - c_1 z}{a_1} \qquad g_i = d_i - \frac{a_i d_1}{a_1}$$

4.4 Model Synthesis

The final stage of the *MAP* system consists of the actual synthesis of the object model. After the points have been triangulated, the model is defined in terms of its 3-D vertices, but it is not known which points are connected to each other with edges. However, this information is present in the original source images and the mappings from image points to model points are known. When these vertex links are copied to the model, a count is kept of the number of source images in which a given edge appears. If this total is lower than a specified threshold value, the corresponding edge is assumed to be due to noise and/or errors in the point matching process and is deleted.

Point matches involving sections of partially occluded lines are also identified and eliminated at this time since these unduly affect the positioning accuracy of the model vertices. Typically, these points are in the form of a T with the crossbar representing the edge of the occluding face and the remaining stroke the portion of the line which is still visible. Thus, occluded features may be identified by examining each sequence of successively matched points. If a fraction of these are of the T format while the remainder are not, the former points probably represent views in which the line was partially occluded and are therefore removed from further consideration. In such an instance, the coordinates of the affected model vertex are also recalculated from the updated information.

5 Results

The vision system has been tested with both real and simulated data using objects such as those shown in Figures 6 and 7. The former demonstrated that the program could function effectively in real-world situations, and the latter were used to evaluate the theoretical accuracy of the reconstruction process when supplied with ideal input data. In both cases, camera positions and orientations were derived by processing images of the visual reference pattern. The real images were obtained from a camera mounted on a robot arm and represented a 220° revolution of the item, whereas the simulated data covered a full 360° rotation. These latter images were generated by shading faces with a common gray level intensity and then smoothing the data by replacing each pixel with its 3×3 local average. The maximum dimensions in the (x, y, z) directions are also given for each item in *mm*.

The observed average measurement errors and standard deviations are listed in Tables 1 and 2. These errors represent the distances between each triangulated vertex and its expected position. As can be seen for the real images, average triangulation discrepancies ranged from about 3 to 6 *mm*. These values are for a camera to object distance of approximately 650 *mm* so the relative measurement error is less than 1% in each case. When evaluating the simulated data, results were calculated for both 18 (20° rotation increment) and 36 (10° rotation increment) source images. The average

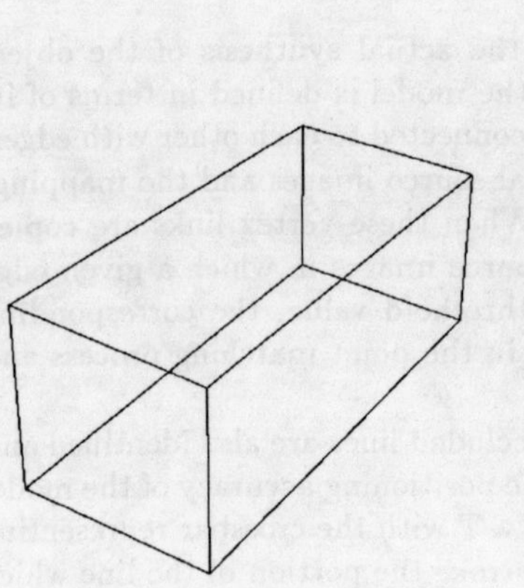

Box

(200 × 60 × 90)

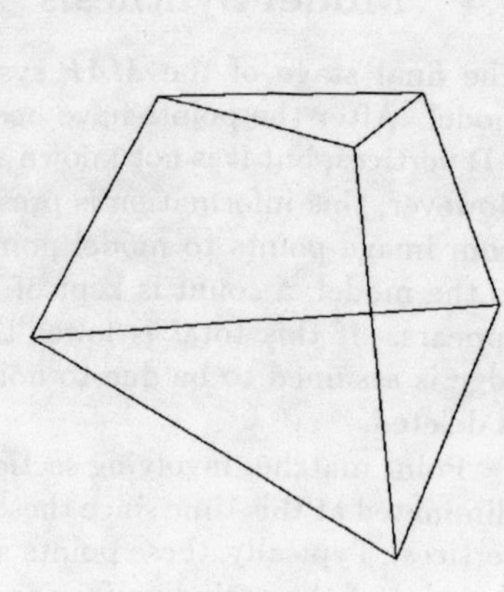

Truncated Tetrahedron

(240 × 100 × 210)

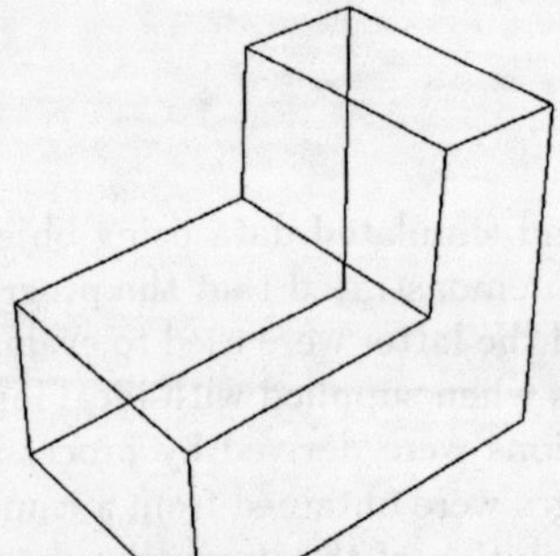

Widget

(100 × 100 × 150)

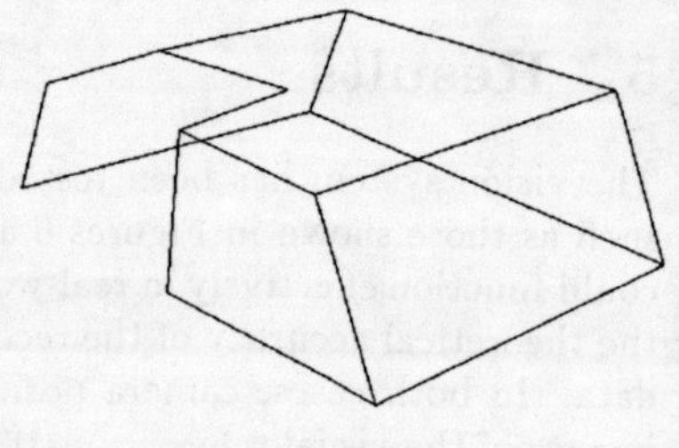

Robot Cover Plate

(320 × 70 × 240)

Figure 6: Real World Test Objects

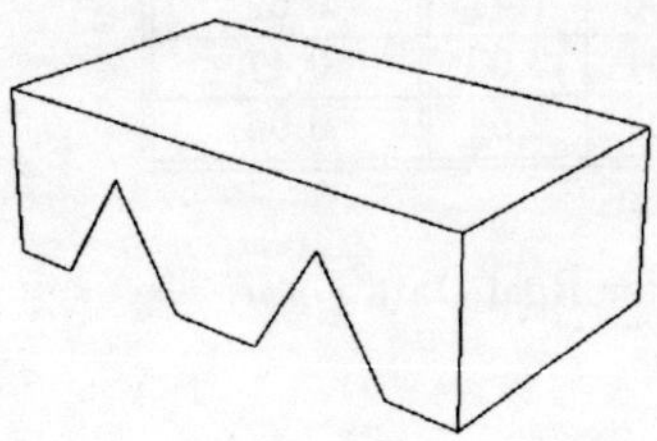

Slide Case

(110 × 30 × 60)

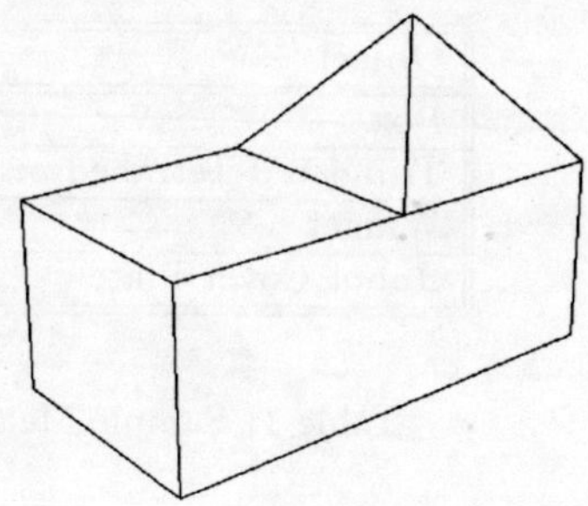

Pyramid Box

(300 × 175 × 150)

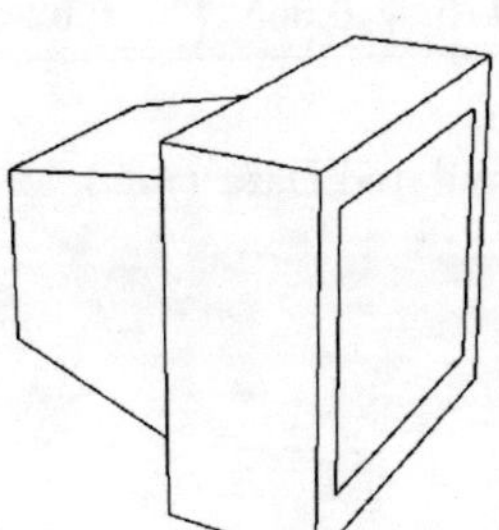

Monitor

(500 × 400 × 500)

Rubik's Cube

(120 × 120 × 120)

Figure 7: Simulated Test Objects

Object	# Points	$\overline{x}$	s	Rel Err %
Box	8	3.84	2.25	0.59
Truncated Tetrahedron	6	4.00	1.41	0.62
Widget	12	3.04	2.03	0.47
Robot Cover Plate	12	6.27	2.54	0.96

Table 1: Sample Measurement Errors for Real Data (*mm*)

	18 Images			36 Images		
Object	$\overline{x}$	s	Rel Err %	$\overline{x}$	s	Rel Err %
Slide Case	0.111	0.058	0.04	0.091	0.047	0.03
Pyramid Box	0.159	0.060	0.02	0.215	0.115	0.03
Monitor	1.127	1.237	0.06	0.643	0.705	0.03
Rubik's Cube	0.184	0.096	0.03	0.115	0.065	0.02

Table 2: Sample Measurement Errors for Simulated Data (*mm*)

	Execution Times				
Object	Edge	Synth (18)	Synth (36)	Total (18)	Total (36)
Slide Case	1.41	0.89	1.64	26.27	52.40
Pyramid Box	1.42	0.80	1.54	26.36	52.66
Monitor	1.45	1.16	2.23	27.26	54.43
Rubik's Cube	1.65	5.11	8.28	34.81	67.68

Table 3: CPU Execution Times for Simulated Data (*sec*)

error was typically less than 1 *mm* and an improvement in accuracy was usual for an increased number of source images.

The simulated data demonstrates the capabilities of the system when "perfect" input data is used. Clearly, this will usually not be the case as illustrated by the larger measurement errors obtained for real-world data. It is likely that these discrepancies can be attributed to such factors as slight errors in the camera positioning, inexact edge localization, and the use of imprecise physical camera attributes. In particular, no calibration procedure to measure the intrinsic camera parameters such as exact focal length and lens distortion has been integrated into the program as of yet, so this represents one area in which improvements may be realized.

Sample CPU execution times for the simulated images are shown in Table 3. These were obtained by averaging the results for a large number of program iterations on a Sun SPARC 330. The edge detection times are for a single image, whereas the model synthesis values are for matching the corresponding number of images and producing a final model. The total expected times for the entire process have also been listed. As would be expected, increased object complexity tended to result in increased execution times, especially during the point matching phase. Although these values are for simulated data, running times for real data should be similar.

6 Future Work

A significant amount of research remains to be carried out in order to improve the performance and enhance the capabilities of the *MAP* system. Some possibilities currently being considered for development are discussed briefly below.

At the present time, no explicit camera calibration has been performed and the relevant parameters are assumed to conform to the manufacturer's specifications. This is not necessarily a realistic assumption and may contribute towards the observed measurement errors. Consequently, it would be useful to incorporate some method for estimating these parameters into the existing system [4, 16]).

The primary problem at present is the edge detector component. It is not always possible for the computer to fit accurate edges to the digitized images and this affects the accuracy of the final model. Incorrectly placed and/or missing lines may contribute towards possible point matching errors which would also degrade the structural integrity of the synthesized model. The image formatting heuristics used when identifying corner points partially address this problem, but it would definitely be advantageous to increase the robustness of this system component.

Another possibility for reducing or eliminating matching errors would involve generating a preliminary model of the object and then using it to verify all of the hypothesized point matches. This could easily be accomplished by projecting it back onto the original images. Any identified errors would be corrected and the model revised to reflect these changes. This approach would also aid in ensuring that edge features, particularly in areas of low contrast, had been extracted properly.

7 Summary

The *MAP* system has demonstrated an ability to generate reasonably accurate models of unknown polyhedral objects using video image data as its only input. These images are currently gathered using a camera mounted on an industrial robot arm. Execution times appear to be in the range required for real-time operation. At the present time, some supervision of the process is necessary to ensure that accurate and valid results are obtained. Work is now underway on improving the vision interface to develop a workcell environment where a robot can function autonomously without the need for operator intervention.

References

[1] N. Alvertos, D. Brzakovic, and R. C. Gonzalez. "Correspondence in pairs of images acquired by camera displacement in depth". In *SPIE Conf. on Intel. Robots and Comp. Vision*, pages 131–136, 1986.

[2] N. Alvertos, D. Brzakovic, and R. C. Gonzalez. "Camera geometries for image matching in 3-D machine vision". *IEEE Trans. on Patt. Analysis and Mach. Intel.*, 11(9):897–915, 1989.

[3] J. A. Bondy and U. S. R. Murty. *Graph Theory with Applications.* Elsevier Science Publishing Co., Inc., 1976.

[4] B. Caprile and V. Torre. "Using vanishing points for camera calibration". *Int. J. of Comp. Vision*, 4(2):127–140, 1990.

[5] A. M. Earnshaw. *3-D Object Synthesis in a Robotic Workcell.* Master's thesis, Dept. of Systems Design, Univ. of Waterloo, 1991.

[6] W. Hoff and N. Ahuja. "Surfaces from stereo: Integrating feature matching, disparity estimation, and contour detection". *IEEE Trans. on Patt. Analysis and Mach. Intel.*, 11(2):121–136, 1989.

[7] R. Horaud and T. Skordas. "Stereo correspondence through feature grouping and maximal cliques". *IEEE Trans. on Patt. Analysis and Mach. Intel.*, 11(11):1168–1180, 1989.

[8] M. Ito and A. Ishii. "Three-view stereo analysis". *IEEE Trans. on Patt. Analysis and Mach. Intel.*, PAMI-8(4):524–532, 1986.

[9] J. D. Jenness. *A Constellation Matching Algorithm with Applications to Machine Vision.* PhD thesis, Dept. of Systems Design, Univ. of Waterloo, 1987.

[10] Y. C. Kim and J. K. Aggarwal. "Positioning three-dimensional objects using stereo images". *IEEE J. of Rob. and Auto.*, RA-3(4):361–373, 1987.

[11] Y. Liu, T. S. Huang, and O. D. Faugeras. "Determination of camera location from 2-D to 3-D line and point correspondences". *IEEE Trans. on Patt. Analysis and Mach. Intel.*, 12(1):28–37, 1990.

[12] J. H. McIntosh and K. M. Mutch. "Matching straight lines". *Comp. Vision, Graphics, and Image Proc.*, 43:386–408, 1988.

[13] S. I. Olsen. "Stereo correspondence by surface reconstruction". *IEEE Trans. on Patt. Analysis and Mach. Intel.*, 12(3):309–315, 1990.

[14] K. D. Rueb and A. K. C. Wong. "Knowledge-based visual part identification and location in a robot workcell". *Int. J. of Mach. Tools and Manufacture*, 28(3):235–249, 1988.

[15] D. Sherman and S. Peleg. "Stereo by incremental matching of contours". *IEEE Trans. on Patt. Analysis and Mach. Intel.*, 12(11):1102–1106, 1990.

[16] R. Y. Tsai. "A versatile camera calibration technique for high-accuracy 3D machine vision metrology using off-the-shelf TV cameras and lenses". *IEEE J. of Rob. and Auto.*, RA-3(4):323–344, 1987.

[17] A. K. C. Wong and R. Salay. "An algorithm for constellation matching". In *Proc. of the Eighth Int. Conf. on Patt. Rec.*, pages 546–554, 1986.

From Slice Data to Suggestive Parts

Emanuele Trucco
Department of Artificial Intelligence
University of Edinburgh
Edinburgh, EH1 2QL
Scotland

Abstract

This paper describes a simple but powerful two-stage technique for inferring volumetric descriptions of articulated objects from parallel planar cross-sections or *slices*. The number of slices can vary and their density is, in general, unknown. The algorithm identifies the relevant components or *parts* of an object and fits volumetric primitives to them. The data are first segmented into parts, then *sticks* and *blobs* are fitted to the parts. The segmentation is completely model-independent. It exploits regularities like *transversality* to generate a set of perceptually plausible object subparts which, in principle, could be described by any volumetric primitives. Our volumetric primitives are convex. They are meant to be suggestive more than literal: they describe the object in terms of type, connectivity, position and orientation of its subparts. Attractive features of the approach include its simplicity, its stability with respect to variations of imaging conditions, and the model-independent nature of volumetric segmentation.

1 Introduction

Automatic acquisition of model-based object descriptions from planar slices is relevant in several domains, including analysis of medical tomographic data (e.g. Soroka [1], inspection of industrial parts (White [2]), and object recognition (Trucco [3]). This paper describes a technique for inferring volumetric descriptions of articulated objects from a variable number of cross-sectional slices. We assume the slices are parallel, but no assumption is made on their spatial resolution: the slice density per centimeter could be irregular and is, in general, unknown.

The hypothesis of non-uniform slicing makes it difficult to adopt curvature-based surface segmentation techniques, because wide gaps can lead to considerable errors in the curvature estimates. We keep therefore our conjectures about the nature of

the surfaces to a minimum, just expecting them to satisfy the general convexity and transversality regularities explained in the next section. Using regularity, we build hypotheses of plausible cross-sections of volumes going through each slice and coalesce such section hypotheses to form consistent 3-D volumes.
The models used to describe the segmented volumes are a subset of the volumetric primitives of Fisher's Suggestive Modelling System (SMS), the *stick* and the *blob* [4]. The stick represents straight or bent elongated volumes, e.g. a pen or a finger. The blob represents compact structures with no privileged elongation, e.g. a human head.

An example of industrial application of slice-based methods is given by White [2], who presents a two-step system for reconstructing accurate models of cylindrical pipes in different orientation from a sequence of slices. Grimson and Lozano-Pérez [11] discuss a technique for identifying simple cylinders in range data by fitting ellipses to arbitrary cross-sections. Although robust, such techniques seem confined to simple cylinders. An example of medical application is Soroka's system [1], which infers linear straight generalized cylinders from tomographic slices. The technique cannot cope with bent subparts and seems prone to overfragmentation. A certain interest has recently been directed to the problem of inferring volumetric models from full range images, in particular fitting models to segmented parts (Pentland [12]), Ferrie *et al.* [7], Solina and Bajcsy [13]). The methods proposed are usually mathematically elegant and well-founded but computationally expensive, involving optimization and variational techniques.

The rest of this paper is organized as follows. First the assumptions made in our work are presented and motivated. Then our algorithm and its implementation for recovering SMS descriptions from slices are introduced. Some experimental results are reported and discussed briefly. The last section mentions some related work.

2 Assumptions

The following assumptions are made in order to detect boundaries between object subparts:

- *Hoffman and Richards's transversality regularity* [5]. This states basically that interpenetrating parts, arbitrarily shaped, form contours of concave discontinuities of their tangent planes (do Carmo [6]). Transversality suggests to segment parts along the loci of negative minima of each principal curvature, which could be tackled satisfactorily by curvature-based segmentation techniques if the data were dense (e.g. Ferrie *et al.* [7]). We'll sketch in the next section how the subparts recovered by our technique are separated by contours which approximate the loci of minima in negative curvature.

- *Shape preservation in the data.* This means that the data must be informative enough. The slice resolution must be appropriate to represent the object, so that discontinuities detected across slices are due to discontinuities in the scene and not to insufficient sampling frequency.

- *Convexity.* Convexity is used in an intuitive more than mathematically precise sense. It means that the surfaces of the parts forming the objects must be formed by elliptic convex or parabolic convex points. From the mathematical point of view, this hypothesis could be better renamed *non-concavity.*

3 Outline of the algorithm

The input to the algorithm is a sequence of parallel planar slices of an articulated object. In our experiments, the slices were obtained from dense range images, either sysnthetic or acquired by a laser striper available in our laboratory. The output is a file describing the structure of the object, i.e. type, parameters, position, orientation and connectivity of its subparts. The descriptive file is written in SMS (Fisher [4]), a language which allows to express models in terms of line, surface and volumetric features. Only volumetric features are inferred here.
The algorithm is divided in two sequential stages: *model-independent segmentation* and *model fitting.*

3.1 Stage one: Segmentation

The *segmentation* is meant to recover the object subparts exploiting the transversality regularity. This is carried out in three steps: *cross-section detection, local section coalescing* and *overfragmentation analysis.*

3.1.1 Cross-section detection

First the contour on each slice is filtered with a median filter to eliminate noisy spikes, which would be likely to upset the curve segmentation. Then the filtered curve is segmented into convex sections at points near local minima of negative curvature. Such sections are taken as plausible cross-sections of object subparts going through the slice. The local minima of negative curvature are estimated by means of a scale-independent curve segmentation algorithm. This algorithm is a fast split-and-merge technique which splits the curve recursively at maximum deviation points until their length becomes less than a user-defined threshold or they approximate sufficiently well a straight line. The linearity check is made scale-independent by using the ratio of length to maximum deviation for each curve arc. The points taken as border between sections are a subset of the points identified by the algorithm.
The reason for accepting the approximation introduced by such a deviation-based segmentation is twofold. First, the volumetric descriptions to be obtained are meant to be suggestive more than literal: we are interested in discovering the type and

position in space of the object subparts as significant distributions of mass, not their exact contours. Second, although less accurate than those involving explicit curvature estimates, deviation-based algorithms for curve segmentation are fast, simple and efficient, and give results in good accord with human judgement for a number of tasks (Fischler and Bolles [8]).
The result of this step is a collection of hypotheses of cross-sections of parts going through the slice. The contour of each cross-sectional hypothesis is represented by a linear piecewise approximation enclosed between two concavities of the cross-sectional curve. This representation constitutes a significant reduction of the original volume of data.

3.1.2 Local section coalescing

This step checks local shape continuity across slices to form preliminary volumes. Cross-section hypotheses lying on adjacent slices are projected on the same plane and their degree of overlap is computed by a *O(n log n)* polygon intersection algorithm (Kundu [9]). Two sections are coalesced if their overlap exceeds a user-defined threshold. A similar technique was devised by Boissonat for surface reconstruction from slice data [10]. Low overlap figures indicate sharp discontinuities and therefore contours of subparts. An additional exploration detects "smooth gradient" contours which could not be identified by the crude overlap analysis, e.g. the border between head and trunk of the mouse in figure 5, or between body and neck for the bottle in figure 4.

3.1.3 Overfragmentation analysis

Bent subparts can be erroneously segmented by the overlap analysis owing to their orientation (see figure 3). This step identifies and solves such situations (called *snake cuts*). The axes of the two volumes involved are estimated locally around the joint by means of a simplified medial-axis-transform technique. Two border sections (one per volume) are then computed by slicing the two subvolumes involved with slices perpendicular to the estimated axes. Shape continuity is tested on the resulting sections and the two subparts merged if sufficient continuity is found.
The overall result of the segmentation stage is a final orientation-independent set of volumes or *subparts* and a *subpart connectivity graph*, to be fed into the model fitting procedure.

3.2 Stage two: Model Fitting

The second stage, *model fitting*, generates the SMS file describing the subparts identified by the segmentation. Ten parameters are recovered for each part: type, position, orientation, structural parameters (length, thickness and bend radius for a stick, three main radii for a blob).

A subpart is first classified as stick or blob by means of heuristic estimates of its maximal and minimal lengths. Then the axis is computed. Blobs pose no serious problems for axis computation, since there is no privileged elongation direction. Blob axes are assumed to be straight. On the contrary, the axis of a stick can be bent. If the stick was assembled by the overfragmentation analysis (step 1c) its axis is obviously bent and a suboptimal circle fitting algorithm is used to fit a circular arc to the stick's axis. The current version of SMS allows only circular axes for bent sticks. If the stick was not overfragmented then a linearity test is run on an initial estimate of the axis. Least squares fit is performed on the centroids of the sections to recover straight axes. Suboptimal circle-fitting is adopted again for bent axes.

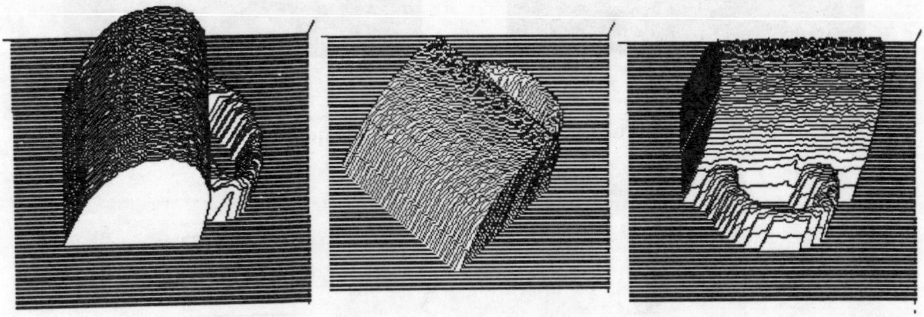

Figure 1: Mug data in different orientations.

4 Some experimental results

We have implemented the above algorithm in C on a SPARC workstation. We report in this section the results of a few experiments. The input slices were obtained from 128x128 full range images, sliced by hand. This implies that only one side of the object was visible; the full cross-sections were inferred by symmetry. This would not be necessary with real tomographic data.

A mug - Figure 1 shows several slice images of a mug in different orientations. An example of the SMS description files generated by the program is given in the appendix. Figure 2 shows a 3-D pictorial rendering of the reconstructed model. The length threshold in pixel and the overlap threshold were set to 15 and 0.5 respectively. Parameter variations in the generated models (position, orientation, shape of primitives) indicated good stability against rotation. The handle was always overfragmented (figure 3) and successfully recovered by the overfragmentation analysis (step 1c). The circular shape of the handle in figure 2 is due to the current SMS limitation

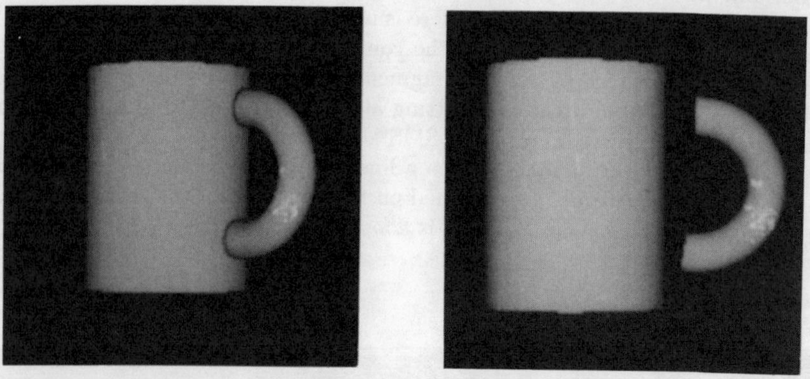

Figure 2: Reconstructed mug model.

Figure 3: Overfragmentation after coalescing local section.

for the axis of a bent stick (circular arc only). Notice however that the data generated by stage 1 allows fitting with more complex curves. The correct connectivity is preserved, although a certain amount of shrinking is introduced.

A bottle - Figure 4 shows a slice image of a bottle and the volumetric model generated by the program. The length and overlap threshold were the same as for the mug. This example shows a case where the shape of the object changes smoothly (slicing from the body into the neck) and the overlap between adjacent sections is always high. Body and neck are therefore segmented by the analysis of "smooth gradient" contours of step 1b.

A plasticine toy mouse - Figure 5 shows a plasticine mouse and its sliced version. The corresponding range image was acquired using a laser striper developed in our department and analyzed in different orientations. The plasticine mouse was 12 cm wide and 16 cm long. The 256 height values covered a range of 8 cm from the

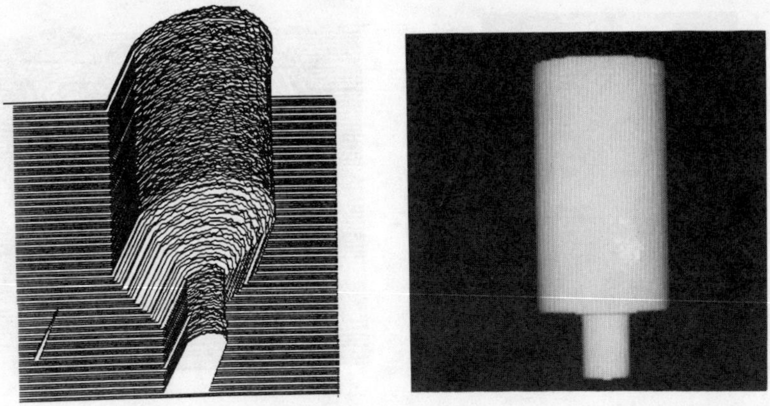

Figure 4: Bottle data and reconstructed bottle model.

background plane. The virtual camera was placed above the object at a distance of 50 cm. Figure 6 is a 3-D rendering of the recovered subpart model. Some shrinking has been introduced again for some subparts, owing to a few peripheral slices being dismissed as noise. The image pixel defined in the striper was not square. This distortion was not corrected and resulted in the image being stretched. Consequently, the ears and the left hand were classified as sticks and the left leg as blob. A correction is obviously trivial to introduce. Notice that the result still makes good sense in spite of the distortion.

Limitations: a complex mechanical part - Figure 7 (left) shows the sliced data of a mechanical part from a Renault car. Notice that this object is not articulated and its "parts" not immediately obvious. Figure 7 (right) shows that the segmentation has recovered succesfully a plausible number of parts in plausible positions (apart from the noisy protruding stick). However, the model fitting generates a visually poor reconstruction. This is due to the combined effect of the oversimplification of the reconstruction and the model fitting being too simplistic for this case.

5 Discussion

We have presented a two-stage algorithm for recovering a SMS volumetric description from tomographic slices of objects. Attractive features of the algorithm include model-independent segmentation of the data into subparts, based solely on the transversality regularity; rotational stability; the ability of correctly describing arbitrarily curved objects (for the segmentation phase); simplicity and computational inexpensiveness.

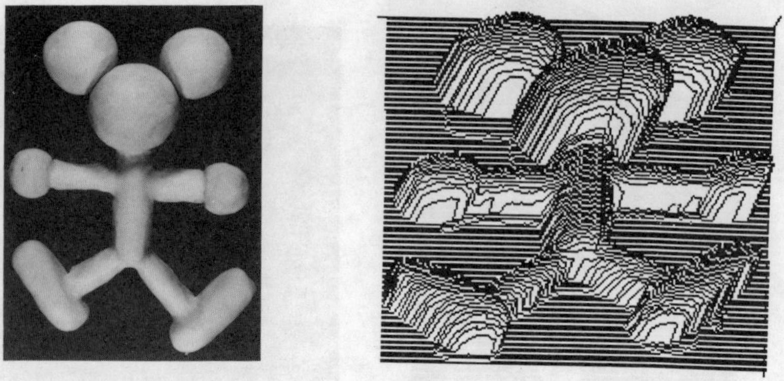

Figure 5: Plasticine mouse and mouse data.

Each step of the segmentation algorithm is local in nature and therefore suitable for parallel implementation. We observed that decreasing the number and the resolution of the slices degrades smoothly the accuracy of the reconstruction.

In practice, the algorithm estimates roughly the minima of negative curvatures (concavities) along two perpendicular directions, along and across the planar slices. Apart from sampling considerations, sources of imprecision include the fact that the minima detected along one direction can become weak for particular object orientations. The combined action of the two analyses in step 1b, however, made up for it in most of our experiments.

Another error is introduced by imposing that parts are sequences of parallel cross-sections, thus neglecting in some cases the first and last sections encountered. This may results in underestimated subpart diameters, but a correction would not be difficult to introduce.

Hyperbolic surface patches (saddles) lead to different segmentations in different orientations with our algorithm. However, such patches arise solely as borders between interesting convex parts, which are always segmented correctly. Hyperbolic patches are therefore no trouble as long as only convex subparts are interesting.

Acknowledgements

Thanks to Bob Fisher for his constant advice and to KP Naidu for his help with the laser striper. This work has been supported by a "Stimulation Action" Grant (contract no. 030640/3.10.89) of the Commission of the European Communities.

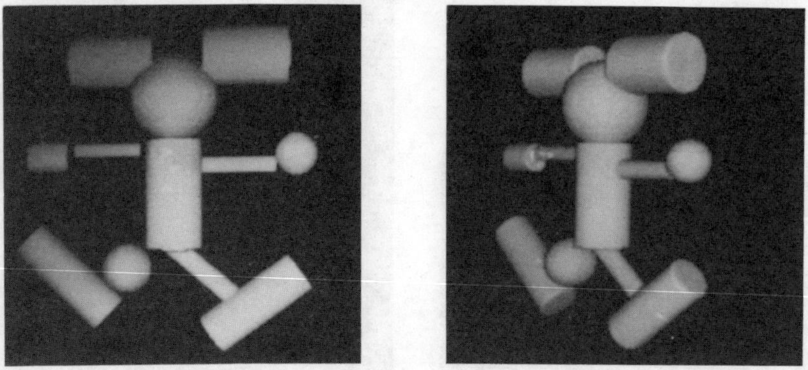

Figure 6: Reconstructed mouse model.

References

[1] B. Soroka, "Generalized cones from serial sections", *Computer Graphics and Image Processing*, vol. 15, pp. 154–166, 1981.

[2] S. White, "Technology innovations and product design issues in machine vision", in *Proc. British Machine Vision Conference*, (Oxford, England), pp. xi–xvi, 1990.

[3] E. Trucco, "Inferring convex subparts from slice data", *Pattern Recognition Letters*, vol. 12, no. 11, pp. 707–715, November 1991.

[4] R.B. Fisher, "SMS: a suggestive modelling system for object recognition", *Image and Vision Computing*, vol. 5, pp. 98–104, 1987.

[5] D. Hoffman and W. Richards, "Parts of recognition", *Cognition*, vol. 28, pp. 65–96, 1985.

[6] M. do Carmo, *"Differential geometry of curves and surfaces"*, Prentice-Hall, 1976.

[7] F.P. Ferrie, J. Lagarde and P. Whaite, "Recovery of volumetric object descriptions from laser rangefinder images", in *Proc. First European Conference on Computer Vision*, pp. 387–396, 1990.

[8] M. Fischler and R. Bolles, "Perceptual organization and curve partitioning", *IEEE Transactions on Pattern Analysis and Machine Intelligence*, vol. 8, pp. 100–105, 1986.

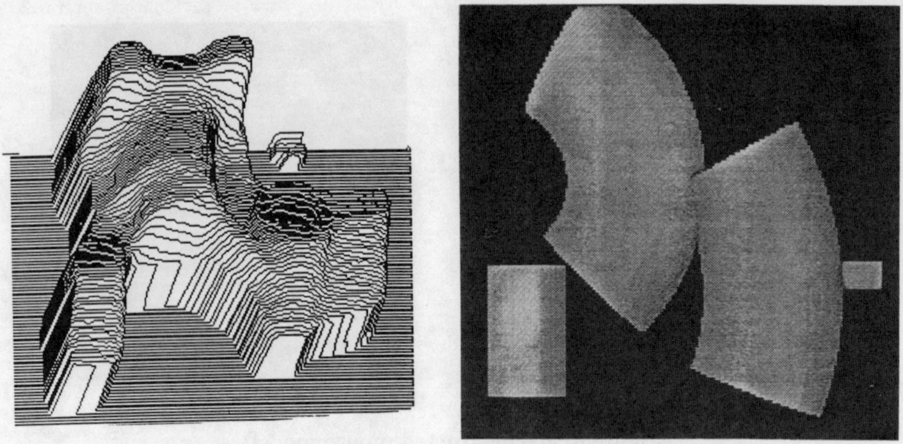

Figure 7: Renault part data and reconstructed Renault part model.

[9] S. Kundu, "A new *O(n log n)* algorithm for Computing the intersection of convex polygons", *Pattern Recognition*, vol. 20, pp. 419–424, 1987.

[10] J.-D. Boissonat, "Surface reconstruction from planar cross-sections", in *Proc. IEEE Conference on Computer Vision and Pattern Recognition*, pp. 393–397, 1985.

[11] W.E.L. Grimson and T. Lozano-Pérez, "Finding cylinders in range data", in *Proc. IEEE Conference on Computer Vision and Pattern Recognition*, pp. 202–207, 1985.

[12] A. Pentland, "Automatic extraction of deformable part models", *International Journal of Computer Vision*, vol. 4, pp. 107–126, 1990.

[13] F. Solina and R. Bajcsy, "Recovery of parametric models from range images", *IEEE Transactions on Pattern Analysis and Machine Intelligence*, vol. 12, pp. 131–147, 1990.

A APPENDIX

The following is the SMS file generated by the program from the slices of the mug. C-style comments were inserted by hand.

```
(STICK stick0                       /* body */
        LENGTH           83.75
        CROSS_RADIUS     30.81
        BEND_RADIUS      0
        (NONE))

(STICK sn_stick321                  /* handle */
        LENGTH           79.68
        CROSS_RADIUS     6.32
        BEND_RADIUS      24.19
        (NONE))

(PRIMARY_ASSEMBLY mug

        /* positions & connectivity */

        (DEFAULT_POSITION AT
                TRANSLATION     (40, -80, 500)
                ROTATION RST    (HALFPI, 0, 0))

        (VARS (NONE))

        ((ASM_ALT (
                        (PLACED_FEATURE stick0 AT
                                TRANSLATION     (76, 24, 0)
                                ROTATION RST    (1.57, 0, 0)
                                SCALE           1.0)

                        (PLACED_FEATURE sn_stick321 AT
                                TRANSLATION     (34, 40, 0)
                                ROTATION RST    (3.14, 0, 0)
                                SCALE           1.0))))

        ( (CONNECTED  stick0 sn_stick321) ))

(VDFG mug (             /* visibility information */
                        /* front view */
```

```
    (VIS_GROUP (stick0 sn_stick321)
     TAN_GROUP (NONE)
     CONNECT_CONSTRAINTS (NONE)
     NEW_FEAT_CONSTRAINTS (NONE)
     POSITION_CONSTRAINTS (NONE))))
STOP
```

A Microcomputer-Based Supervised System for Automatic Scoring of Mitotic Index in Cytotoxicity Studies

M. Garza-Jinich, C. Rodriguez,
IIMAS, UNAM
Apdo. 20-726
Mexico D.F. 01000
G. Corkidi,
CIUNAM
R. Montero, E. Rojas and P. Ostrosky-Wegman
IIB, UNAM

Abstract

The supervised classifier contained in an automatic PC based mitotic index scoring system is presented. An interactive statistical training based on size and circularity is combined with fast cluster identification techniques to produce an accurate count of cells and mitotic clusters. The system is validated comparing the results obtained automatically with those obtained visually in a set of 1000 images.

1 Introduction

Mitotic index is the percentage of cells which are in the process of division. It has proven to be a widely used marker to evaluate and compare cell proliferation rate [11, 18]. Normal cell proliferation can be affected by exposure to physical, chemical, and biological agents. This can lead to cytotoxicity, which is expressed as a decrease in cell proliferation, lowering the mitotic rate among other parameters, like DNA synthesis and cell viability. Cell toxicity analysis is used in chemical safety evaluation of new substances, mainly those involved in food and drug products. It is included as an in vitro test for every batch of vaccines. Cell proliferation evaluation is also used in the screening of new anticancer drugs, due to the fact that their action is to stop cell growth [19, 14]. Conversely, in clinic, an increase in mitotic index in certain tissues is used as a diagnostic marker for some types of cancer.

The determination of mitotic index in cell cultures is traditionally done by light microscope analysis of slide preparations. The analyst identifies and counts thousands of cells, and reports the percentage of mitotic counts found among the cells. It is a long and tedious process, and the problem of subjective and biased identification is unavoidable. The automation of such a process would undoubtedly benefit research in this area. Much work has been done in the development of digital computer image analysis of chromosomes, cells, and tissue preparations. Most systems, the so called metaphase finders, have been implemented for the quick detection of good metaphase chromosome spreads, specifically applied to kariotyping cells. Some of them were developed for fully automatic performance [8, 9], and even commercial machines are available for the purpose of finding potential metaphase cells automatically and turn the analysis to the cytotechnologist [12]. These systems are oriented to the identification of mitotic spreads and ignore all the rest of the objects in the slide, even the 'not so good' mitotic clusters. Work has also been done using sophisticated hardware [2], which based on morphologic, photometric, and textural parameters allows recognition of cells in different stages of the cell cycle, but requires scanners, photomultipliers, and even microphotometers besides the computer and the microscope. More recent solutions involving densitometric approaches have been developed, which distinguish cells in different phases of the cell cycle on the basis of their DNA content, aided with the information provided from the cell histogram without recognizing mitotic figures and without giving the mitotic index as result [1]. There is work specifically oriented to pathologists in the area of tissue analysis [10], but mitotic counts per field are introduced manually. Another work for histological analysis [6] allows the measure of histometric features by examining detailed nuclear images, but the mitotic index is estimated as a function of depth in the epithelium. None of these systems perform simultaneous recognition of cells and mitotic spreads, necessary for mitotic index estimation in cell cultures.

In the current study we introduce a new methodology oriented to cytologists. It includes a supervised personal computer based classifier, initially reported elsewhere [7], contained in a fully automated mitotic index scoring system for cell cultures. The classifier is first trained to identify objects using a 20X objective. It classifies them based on cell size, cell shape, chromosomes size and shape. Due to the nature of the objects to be classified, the parallel computer techniques of propagation and shrinking in pyramids of images were chosen as the most adequate and adapted to PC capabilities for cluster identification. The validation of the system in a set of more than 1000 images containing lymphocyte cultures is discussed.

2 Material and Methods

2.1 Culture preparation

$0.5ml$ of heparinized blood are added to $6.0ml$ RPMI-1640 culture medium, supplemented with glutamine ($2mM$) and non essential aminoacids ($0.1mM$), in the presence of bromodeoxiuridine $32\mu M$. 2.8% phytohemagglutinin is used to stimulate lymphocytes growth.

Cultures are incubated at $37°C$ for 72 hrs. Two hours before fixation of cells, Colcemid ($2\mu g$ per culture) is added in order to stop cycling cells in mitosis. Fixation is done with Carnoy solution (methanol-acetic acid 3 : 1) after incubating cells in KCI $0.075M$ for 30 minutes.

2.2 Slide preparation

Quality of slides is a critical point for a good automated analysis of mitotic index. Therefore, fresh solutions ought to be used in fixation: cell pellets must be thoroughly washed as many times as needed, in order to eliminate cell debris. Pellets are resuspended in $0.5ml$ fresh Carnoy and dropped onto cold alcohol moistened slides. Staining is done as reported previously [13].

2.3 Hardware and software specifications

The program was written in TURBO PASCAL 4.0 using an IBM 80386 PC computer operating with MS-DOS version 3.30, with 640 Kbytes of memory, and an additional 2 Mbytes RAM disk. The PC was employed as the core of the image acquisition and processing system. A frame grabber (Matrox PIP 1024B) plug-in card was used, together with a dedicated analog RGB monitor (NEC Mutisync) and a CCD RS-170 monochrome TV camera (Cohu), attached to an Olympus BH-2 microscope. 512 x 480 pixel resolution was used, with 256 grey levels. The acquisition of the images was done using IMAL System (Biocom, France). The automatic scanning of the slide was done with MarsHauser hardware.

2.4 Digitization

The objective of the microscope setup, the initial field selection and image digitization through a standard CCD RS-170 B/W camera are performed interactively. Before the first frame of the desired field is digitized, adjustment of the input video signal's offset and gain is performed in a continuous frame grabbing mode. This is done by manipulating the digitizer's input Look-up Table (LUT) together with the input A/D converter offset and gain, and the microscope's illumination adjustments. In this way, saturation is avoided and the full dynamic range of the digitizer's A/D converter is employed on the image to be analyzed. Once these parameters are established, a whole set of images containing 512×480 pixels is digitized under similar illumination conditions. A motorized stage was used where slides were placed for observation at the microscope under a 20X objective. The analysis consists on the digitization of contiguos fields in the inner area of the slides, where most of the material is randomly distributed.

2.5 Training process

The system requires an initial training process in order to obtain the set of statistics used for identification and classification, which might vary from slide to slide. The expert chooses a set of images (around six in the current study) containing representative objects of the whole slide, and predefines the n classes of cells that will be used during classification. This number n is greater than or equal to one, and allows the user to obtain better results when there is great variability in cell size. He then displays each image, and points to those cells representing each class. In addition, the expert should point to a set of representative chromosomes, which are assumed to belong to class $n+1$. The system generates a disk file, accesible from any spreadsheet type software containing class number, area, and circularity of every identified object. At the end of the training process, another file is generated automatically containing the statistics of each class: mean area, standard deviation of area, and minimum circularity. The training procedure is shown in Fig. 1: small dark cells are assigned to class one, larger and lighter cells to class two, and chromosomes to class three. The data are assumed to be normally distributed and therefore, can be analyzed by the usual discrimant analysis software. Once this is done, the system is ready to automatically quantify the whole slide.

2.6 Automatic segmentation

During analysis, images are segmented automatically and independently of each other, using the BCV Method [15]. This method was chosen due to its nonparametric and unsupervised characteristics, and its capability of selecting an optimal threshold in histograms with flat and broad valleys, and extremely unequal peaks in height. Besides, the images used are a typical example of a problem where there is no a priori knowledge except for the gray level histogram. After digitization, the histogram of the image is obtained, and the optimal threshold that maximizes separability between the two classes representing objects and background is selected and used to produce a binary image.

2.7 Cell identification

After automatic segmentation is done, and the image is binarized, it is scanned and all 4-connected components are identified using classical algorithms [3]. Their area

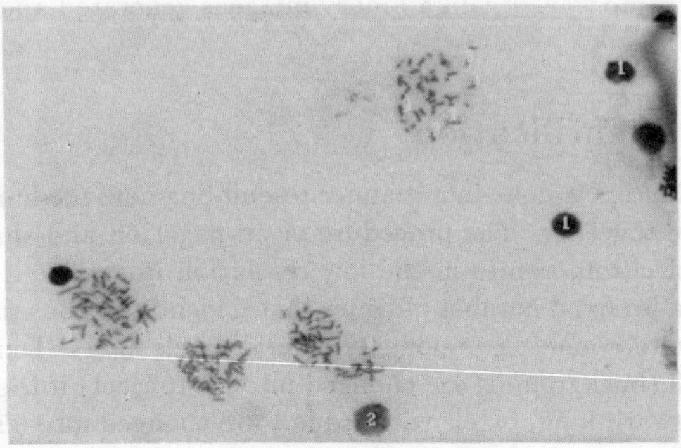

Figure 1: Monitor showing image during training process.

and perimeter are computed, and together with their extreme coordinates are kept in a table. Components in the borders of the image are eliminated because they are likely to be misclassified. 'Very large' objects (with area larger than three times the maximum area of the cell classes), regardless of their circularity, are eliminated from the image, as well as 'very small' objects (a couple of pixels in size). All the remaining objects are processed by the cell classifier. Their one-dimensional Mahalanobis distances to each cell and chromosome class are computed [4] using variable area. The object is assigned to the nearest class, provided its circularity is above the minimum circularity of the class. Otherwise, the object is assigned to the 'unknown' class. Only those objects belonging to a cell class are counted and eliminated from the image.

The rest of the objects are assumed to be either isolated chromosomes, clusters of chromosomes, or small isolated artifacts. Their coordinates are kept in a table and will be used during mitosis identification. The image is reduced in size using pyramids techniques, in order to accelerate CPU time, and afterwards is processed by the mitosis identifier.

2.8 Pyramids

The usage of pyramids technique [17] was chosen in order to accelerate processing time during mitosis identification. It is considered that cluster characteristics are preserved in a smaller resolution image. As objects corresponding to the chromosome class have already been identified, it is of no importance the loss in shape accuracy that results using less resolution. Once cells and large artifacts are identified and removed from the image (see cell identification), the 512×480 binary image assumed to contain only chromosomes, clusters of chromosomes, and small isolated artifacts is used as input at the bottom of a pyramid. The next level of the pyramid is computed averaging 2×2 windows, and a 256×240 image is obtained. A 128×120 third level is obtained in the same way. Again a binary image is generated and processed by the mitosis identifier.

2.9 Mitosis identification

Classification of clusters is done in a manner resembling how the human eye identifies small objects close together. The procedure of propagation and shrinking is used to identify clusters of chromosomes in the low resolution image [16]. Propagation and shrinking is done a prefixed number of times that depends on the amplification objective used, in order to connect components several pixels apart. During propagation, pixels with value 0 (background) are changed into 1's (object) if they have any 1's as neighbors. During shrinking, pixels with value 1 are changed into 0's if they have any 0's as neighbors. As a result, large connected components appear over the clusters of chromosomes in the original image.

As propagation and shrinking techniques are specifically defined for parallel processing, it was necessary to develop a serial algorithm that could make feasable the use of these techniques. A published method [5] was modified in order to analyze in each iteration only those pixels of interest. During propagation, pixels with value 1 lying in the border are searched, and their neighbors with value 0 are changed into 2's (meaning it is a border pixel in propagation iteration number 1) and kept in a list used in the next iteration. This procedure is repeated the predefined number of times. During shrinking, the lists are analyzed again, but only those pixels having 0's as neighbors are changed into 0's.

The image produced by the propagation and shrinking procedure is again analyzed by the connected components identifier. In order to decide which object corresponds to a mitotic cluster, it is necessary to count how many chromosomes in the original image intersect it. The table obtained during cell identification containing chromosomes and artifact coordinates is analyzed, and only those objects intersecting a considerable amount of chromosomes (we established at least 10) are classified as mitotic clusters, which are counted, and together with the number of cells found, are stored as a final result in a disk file.

3 Results

The system we developed has been tested with preparations containing lymphocyte cultures. A typical image, as the one shown in Fig. 2, is processed and analyzed in approximately nine seconds, with two classes of cells and one class of chromosomes. The result is shown in Fig. 3, where dark objects were assigned to either class one or class two, two mitosis labelled with 'M' were found as clusters of small objects (chromosome class), and the rest of the gray objects (unknown class) were ignored.

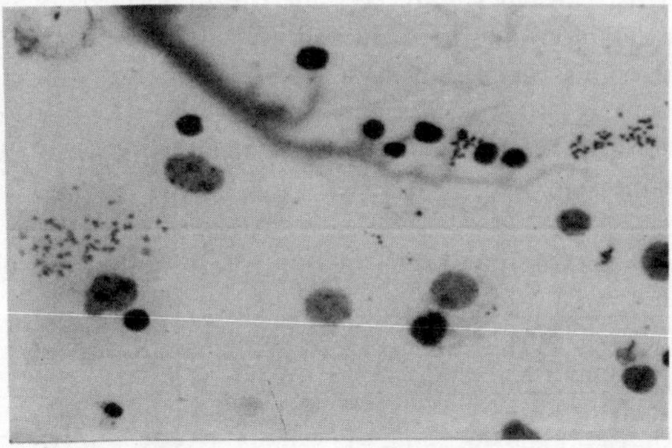

Figure 2: Monitor showing a typical image to be classified.

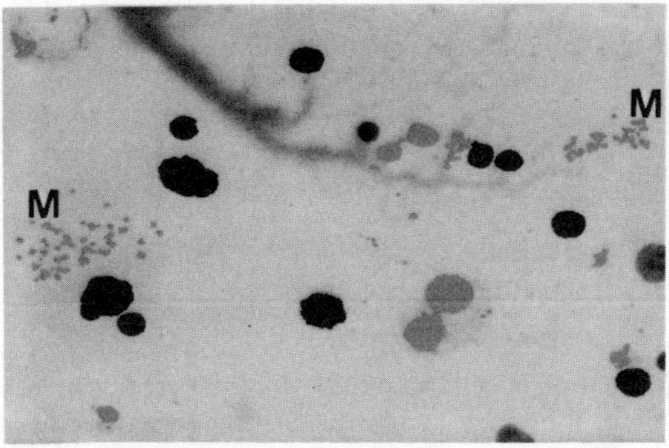

Figure 3: Monitor showing the result classification of image in Fig. 2.

A set of 1196 images covering one slide was processed automatically, using 20X objective. Every image was visually analyzed in parallel by the user. The mitotic index obtained by the system differs 0.09% from the one obtained visually (see Table 1). An analysis with the chi square test for the hypothesis of equal proportion of identified mitosis and non-identified mitosis was done, and these proportions were found to be equal. This test resulted in a non-significant difference at an α–value probability (type I error) which can be as large as 0.50; the p-value was clearly non-significant. Therefore the fact that the automatic classification is being used induces no bias on the estimation.

Table 1. Comparison of manual vs. automatic results.

	MANUAL	AUTOMATIC
Cells Identified	2809	2149
Mitosis Identified	124	97
Total Objects Identified	2933	2246
MITOTIC INDEX	0.0423	0.0432

In order to calculate the error with which our system is working, a second expert analyzed, in the traditional way, the same slide looking at different fields and obtained a mitotic index of $p = 0.038$, which was used as the 'real' proportion in the preparation. The tolerance of our result was computed as follows: The computer analyzed a sample of $N = 2246$ objects and $\hat{p} = 0.0432$ was the proportion of mitosis estimated, which has an approximate normal distribution with mean p and variance $\frac{p(1-p)}{N}$. Therefore, the tolerance at roughly a 95% confidence is:

$$Tol = 2\sqrt{\frac{p(1-p)}{N}} = 0.008,$$

which covers the 0.0009 error observed. (In these laboratory tests such a tolerance is smaller than the traditionally accepted.)

Depending on the type of cells analyzed, different errors could be introduced and it should be desirable to minimize such error. Although in this present study the result obtained is within the accepted limits, there are several possibilities of improving the classification process. Non- identification of very sparse mitosis, or confusion of compact mitosis with either cells or artifacts can be avoided by increasing the number of mitotic classes. The system allows the usage of as many classes as needed, which can help identify different types of mitosis, provided the classification time does not increase significantly.

4 Conclusions

The software we developed for obtaining the mitotic index of a cell culture has been proven to be reliable within the conditions of the present study. As in all supervised classification problems, its success rests on the quality of the training process, and the quality of the slide preparations is of great importance. The classifier is the core of an automatic mitotic index scoring system, integrated to a microscope with automatic slide device. Although several problems with misclassification exist, i.e., round-like artifacts and compact mitotic clusters are sometimes confused with cells, and very sparse clusters are likely to be ignored, results obtained so far are greatly encouraging. More studies are needed to confirm whether our results are reproducible in a larger randomized series of slides, and whether our system can, indeed, be a valuable tool in daily practice for cytotoxicity studies.

Acknowledgements

The authors wish to thank Dr. Larry S. Davis, Director of the Institute for Advanced Computer Studies, University of Maryland, for his ideas and encouragement; Dr. Federico O'Reilly, Head of the Statistics Department, and Patricia Romero, M.Sc., IIMAS, UNAM, for their advice in statistics.

References

[1] T. Aboussaouira, Y. Moustafa, and S. Idelman, "Image analysis of cell proliferation in rat thymus throughout development," *Thymus*, vol. 12, pp. 167–186, 1988.

[2] G. Brugal, C. Garbay, F. Giroud, and D. Adelh, "A double scanning microphotometer for image analysis: hardware, software and biomedical applications," *The Journal of Histochemistry and Cytochemistry*, vol. 27, no. 1, pp. 144–152, 1979.

[3] K. R. Castleman, "Digital image processing," *Prentice-Hall, Inc.*, New Jersey, p. 317, 1979.

[4] C. Chatfield and A. J. Collins, "Introduction to multivariate analysis," *Chapman and Hall*, London, p. 135, 1980.

[5] S. J. Crabtree, Jr., L. Yuan, and R. Ehrlich, "A fast and accurate erosion-dilation method suitable for microcomputers," *CVGIP: Graphical Models and Image Processing*, vol. 53, pp. 283–290, 1991.

[6] H. E. Dytch, M. Bibbo, P. H. Bartels, J. H. Puls, and G. L. Wied, "An interactive microcomputer-based system for the quantitative analysis of stratified tissue

sections," *Analytical and Quantitative Cytology and Histology*, vol. 9, pp. 69–78, 1987.

[7] M. Garza-Jinich, C. Rodriguez, G. Corkidi, R. Toledo, R. Montero, and P. Ostrosky, "Supervised mitotic index scoring as a tool in cytotoxicity studies," in *Proceedings of Vision Interfase*, (Calgary, Alberta, Canada), pp. 150–153, June 1991.

[8] D. K. Green and P. W. Neurath, "The design, operation and evaluation of a high speed automatic metaphase finder," *The Journal of Histochemistry and Cytochemistry*, Vol. 22, No. 7, pp. 531–535, 1974.

[9] E. T. Johnson and L. J. Goforth, "Metaphase spread detection and focus using closed circuit television," *The Journal of Histochemistry and Cytochemistry*, vol. 22, no. 7, pp. 536–545, 1974.

[10] A. Marchevsky, J. Gil, D. Silage, "Computerized interactive morphometry as a potentially useful tool for the classification of Non-Hodgkins lymphomas," *Cancer*, vol. 57, pp. 1544–1549, 1986.

[11] A. V. Pisciotta, D. W. Westring, C. DePrey, and B. Walsh, "Mitogenic effect of PHA at different ages," *Nature*, vol. 215, p. 193, 1967.

[12] K. Preston, Jr. and P. H. Bartels, "Automated image processing for cells and tissue," in *"Progress in Medical Imaging"*, V. L. Newhouse, ed., Springer-Verlag, New York, 1989.

[13] P. Ostrosky-Wegman, G. Garcia, R. Montero, B. Perez Romero, R. Alvarez Chacon, and C. Cortinas de Nava, "Susceptibility to genotoxic effects of niclosamide in human peripheral lymphocytes exposed in vitro and in vivo," *Mutation Research*, vol. 173, pp. 81–87, 1986.

[14] P. Ostrosky, R. Montero, N. Hernandez, L. Ruiz, L. Gasque, L. Herrera, M. Ruiz, C. Cortina, R. Rodriguez, "Lymphocyte proliferation kinetics as a cytostatic screening system," *Environmental and Molecular Mutagenesis*, vol. 11, suppl. 11 p. 80, 1988.

[15] N. Otsu, "A threshold selection method from gray-level histograms," *IEEE Transactions on Systems, Man, and Cybernetics*, vol. SMC-9, pp. 62–66, 1979.

[16] A. Rosenfeld and A. C. Kak, "Digital picture processing," *Academic Press*, New York, vol. 1, p. 362, 1976.

[17] A. Rosenfeld and A. C. Sher, "Detection and delineation of compact objects using intensity pyramids," *Center for Automation Research, University of Maryland*, CAR-TR-235, 1986.

[18] D. Scott, S. Galloway, R. Marshall, M. Ishidate, D. Brusick, J. Ashby, and B. Myhr, "ICPEMC, Genotoxicity under extreme culture conditions," *Mutation Research*, vol. 257, pp.147–204, 1991.

[19] S. Shrivastav, R. Bonar, K. Stone, and D. Paulson, "An in vitro assay procedure to test chemotherapeutic drugs on cells from human solid tumors," *Cancer Research*, vol. 40, pp. 4438–4442, 1980.

Analysis of Retinal Images Using Mathematical Morphology

Piotr Jasiobedzki, Chris J. Taylor
Wolfson Image Analysis Unit,
Dept. of Medical Biophysics,
University of Manchester,
Manchester M13 9PT, UK

Abstract

An automatic analysis of angiographic retinal images is discussed. A detection method based on texture differences between perfused and non-perfused regions is described. First the image is tessellated into a large number of primary regions using the positions of vessels and other structures in the image. The image is represented as a region adjacency graph. Arcs in the graph are classified as vessels using features obtained from grey level profiles in the original image. For each region the degree of non-perfusion is estimated by measuring textural properties using mathematical morphology. Adjacent regions are subsequently merged using information on the degree of perfusion.

1 Introduction and Medical Background

The optical system of the eye permits direct observation of retinal blood vessels including some capillaries. Diabetes Mellitus, which affects 2% of the population, is one of the most important diseases which causes abnormalities in the retinal capillaries. Such Diabetic Retinopathy is the commonest cause of blindness in people in developed countries.

Changes in the retinal vascular network in diabetes include occlusion of capillaries with formation of non-perfused regions. This process is accompanied by local dilation of adjacent vessels and formation of microaneurysms, small (25-100 μm in diameter) circular blood filled sacks, which tend to surround areas of non-perfusion. Microaneurysm counting has been used as an indirect method of grading the severity of Diabetic Retinopathy [7]. After an initial increase in the microaneurysm number there is a subsequent decrease as the retinopathy progresses by extensive capillary closure. Therefore a microaneurysm count can only be regarded as a reliable measure of the severity of early retinopathy.

Fluorescein angiography is an established clinical technique for observing the retinal circulation. Fluorescent dye is injected into a vein in the arm and a sequence of photographs of the illuminated retina is taken through the pupil. Standard images taken for clinical tests cover a region about 10 mm in diameter around the fovea and the optic disc (figure 1).

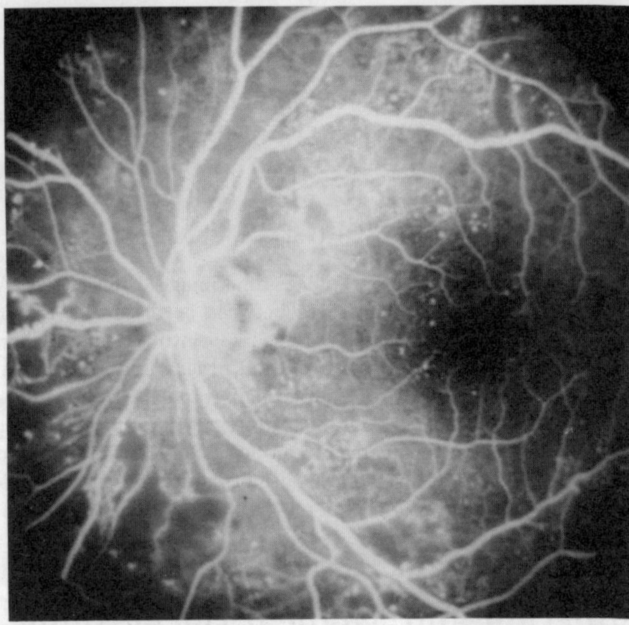

Figure 1: Angiographic image of the retina.

Vessels filled with dye are observed as bright ribbons on a non-uniform background. The vessels are roughly cylindrical and assuming uniform flow of the dye and low attenuation of light by the dye and blood, the grey level profile taken across the vessel can be approximated by an ellipse. The maximum intensity level for each vessel depends on its diameter. Major vessels can be easily detected, but contrast decreases with decreasing vessel diameter. Separate microcapillaries can be observed in the region around the fovea where there is only a thin layer of microcapillary net and the contrast is good. Further from the fovea where the microcapillary net is much thicker, there is less contrast with the background and detection of individual capillaries is not feasible. We are attempting to measure degree of perfusion (density of microcapillary net) directly in the retinal image.

2 Previous Work

Previous work on analysing retinal images has concentrated mainly on detecting vessels and microaneurysms. Detection of vessels was usually performed in two steps:

- detection of vessel segments,
- identification of vessel tree by joining segments and resolving connectivity at bifurcations and crossing points.

Vessel segments have been detected using various methods. Tanaka [15] used adaptive thresholding followed by binary thinning to detect major vessels. Katz [6] segmented and thinned output from the Sobel edge detector to outline vessels. Only major vessels were detected since smaller ones could not be distinguished from other structures. Chaudri [2] described a more robust approach based on detecting local vessel direction by convolving the image with a set of masks. The masks were obtained by rotating a linear feature detector of a gaussian cross-section which roughly matched the grey level profile across vessels. This method allowed detection of much smaller vessels than found previously [6]. The technique was only applied at a single scale but application of a range of kernel sizes would have allowed a much wider range of vessel widths to be detected. Application of a different linear feature detector ("road detector") has been also investigated [1].

The structure of the vessel tree has been obtained either directly from detected ribbons [6] or from branching and crossing points which were identified in the binary image [15]. Vessel segments joining these points were then extracted from the image. Akita [1] described the use of relaxation labelling to provide a connectivity interpretation for the vessel segments and to label vessels as arteries or veins. Caliber of vessels at arterio-venous crossings is used for ranking hypertension.

Lay detected microaneurysms using a morphological "top hat" transformation of the positive image [8]. The top hat transformation was performed by subtracting the image opened using linear structuring elements from the original. The resulting image contained objects of circular shape and of diameter less than the structuring element length.

The problem of detecting non-perfused zones has received little attention. Goldberg [4] described results of clustering regions of similar grey level. The image was smoothed and segmented, with a fixed threshold, into perfused and non-perfused regions. This approach did not take into account non-uniform background and did not try to relate detected regions to any constant objects (major vessels, optic disc, etc.) or to monitor increase in number and extent of non-perfused regions over a period of time. We are not aware of any other published work in this area.

3 Outline of the Proposed Method

The requirements for the analysis of retinal images for the purpose of detecting and monitoring the early stages of Diabetic Retinopathy can be summarised as follows:

- to detect and measure non-perfused regions (without a visible microcapillary net),
- to relate the position of non-perfused regions to landmarks in the image (optic disc, fovea, position of major vessels),
- to detect and compare the same regions in later images.

The approach we have taken follows the method used in the manual detection of non-perfused regions by an ophthalmologist. Regions which contain a fine and regular structure of capillaries are regarded as well perfused. Areas where this structure is not visible are regarded as non-perfused. Larger vessels do not directly contribute to perfusion. Boundaries of non-perfused regions are usually aligned with the position of vessels or are created by joining adjacent groups of microaneurysms. The automatic processing of each image is performed in the following steps:

- detecting the network of vessels,
- defining primary regions,
- measuring texture in primary regions,
- merging similar adjacent regions.

Detection of the network of vessels is based on detecting linear structures in the image and identifying vessels among them.

Defining primary regions is based on partitioning the image into irregularly shaped areas using the positions of vessels and chains of microaneurysms as boundaries.

As mentioned in the introduction, it is possible under good conditions to detect single capillaries in a small area around the fovea. In other regions of the observed image the background has less contrast and capillaries form a more complicated and thick layer of entangled vessels. Our approach is based on measuring texture in order to grade regions. The texture is measured for each region separately using mathematical morphology.

Adjacent regions are merged together if their textural properties are similar and their boundaries are not major vessels. It is essential for the purpose of this work to relate the position of each detected region to a constant set of local landmarks such as the fovea, optic disc and major vessels. Our approach aligns the regions to the position of vessels and allows us to map detected regions to these elements of the retina which do not change even over a long period of time. One of the problems in analysing textured images is reliable identification of the boundaries between regions of different texture. We solve this problem by using non-textural information to define the boundaries.

4 Detecting the Network of Vessels

Detection of the network of vessels is performed in two stages. First linear structures corresponding to vessels, lines of adjacent microaneurysms and other structures present in the image are detected, and represented as a network of arcs and nodes. In the next step each of the arcs is processed separately by referring to the original image and labelled as vessel or other structure.

The original image (figure 2) is smoothed by morphological opening and resam-

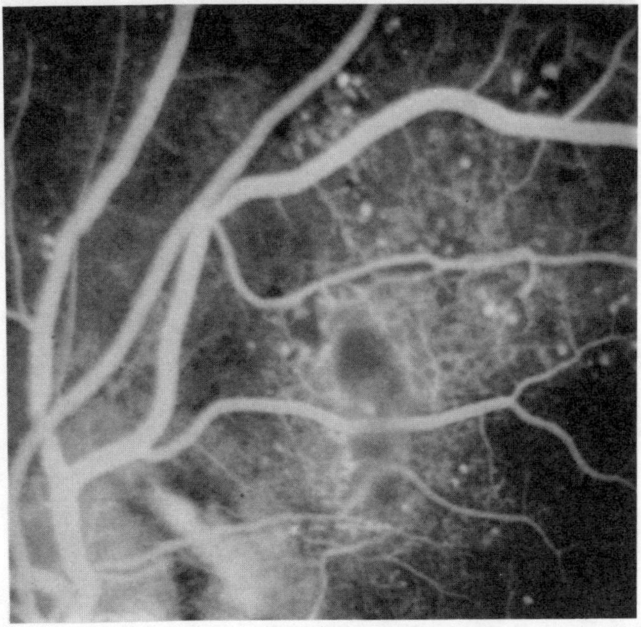

Figure 2: Fragment of the image selected from figure 1.

pled giving a low resolution version of the image. The resulting image is thinned to give one pixel wide lines. Thinning is performed by conditional erosion in a 3 x 3 neighbourhood using twelve directional masks. This is based on the method of Goetcherian [3] which uses only eight masks. Four additional masks are used for pruning the image of lines with one unconnected end.

For each mask two neighbourhoods are defined: L and H. If the value of the central pixel $g\ (m,\ n)$ is higher than all pixels in the L neighbourhood and lower or equal to pixels in the H neighbourhood this pixel is set to the maximum value of the pixel in the L neighbourhood. If not it is left unchanged:

$$\left(g(m,n) > \max_{(i,j)\in L}(g(i,j))\right) \wedge \left(g(m,n) \leq \min_{(i,j)\in H}(g(i,j))\right) \Rightarrow g(m,n) = \max_{(i,j)\in L}(g(i,j))$$

Each mask is applied in turn.The processing is continued until the image is unchanged by subsequent thinning passes by the set of masks. Thinning creates plateaus of uniform intensity which are surrounded by ridges. The ridges are one pixel wide with the exception of junctions, where ridges of different intensities meet.The connectivity of all connections of ridges is retained (figure 3).

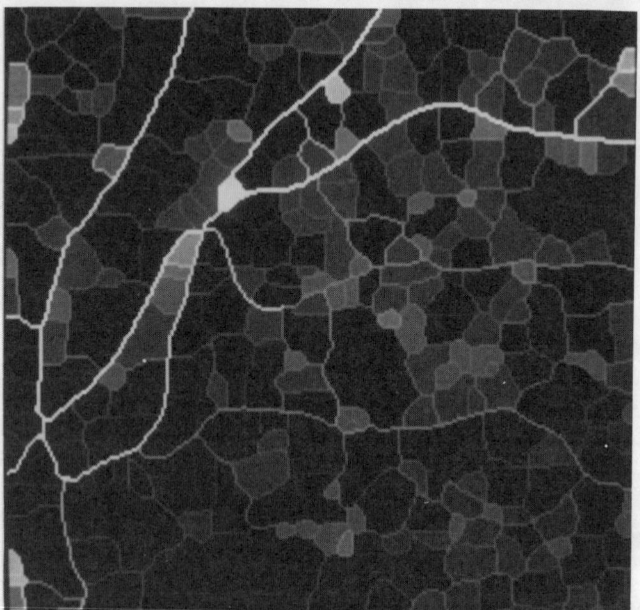

Figure 3: Smoothed, resampled and thinned version of image from figure 2.

The thinned image is segmented by detecting inner points on the plateaus i.e. points which have all 8 - connected neighbours of equal value. The remaining points are binary thinned and give truly one pixel wide arcs (figure 4). The arcs form a network with all end points connected.

The arcs in the binary image correspond to different structures in the original image. Some of them may represent vessels, lines of microaneurysms or boundaries

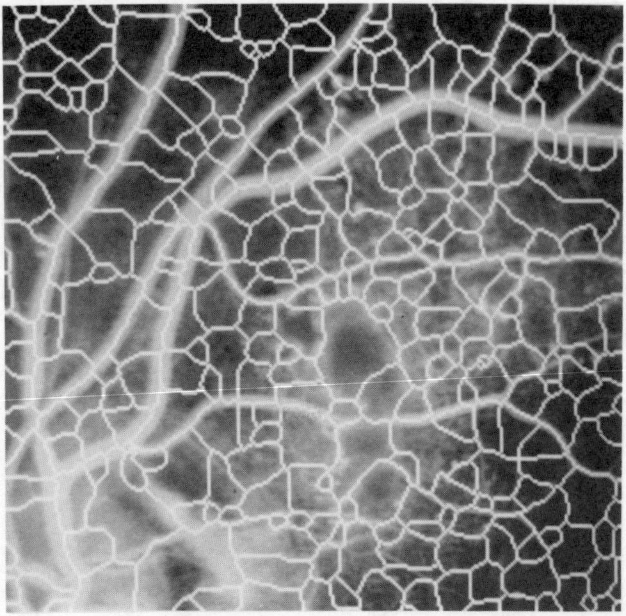

Figure 4: Network of arcs detected in the image from figure 3.

between regions of different intensity or texture. The arcs are stored in a region adjacency graph and are used for three purposes:

- to delineate primary regions,
- to merge primary regions into larger patches,
- to enable identification of vessel tree.

For the purpose of region delineation information on the type of object (vessel, line of microaneurysms, artifacts) an arc corresponds to is not essential. However as experiments on extracting textural features show it helps to reliably calculate textural features if the region is reduced so as not to include parts of larger vessels. In order to merge adjacent regions into large patches or to identify the vessel tree it is necessary to detect vessels among the arcs in the graph.

To identify the actual type of object that each arc represents, it is necessary to refer to the original image and analyse the grey level distribution around the arc. This is performed by generating a sequence of lines perpendicular to the arc and sampling the original image along the lines. The length of lines is larger than the maximum vessel width. Each line is placed in such way that it is perpendicular to the arc and the middle point on the line coincides with a point on the arc. Grey level profiles are obtained by sampling the image along these lines. Each profile is convolved with a derivative mask to give a gradient profile. The positions of two extreme values of the gradient, to the left and right of the centre are detected. The fragment of the original profile between the detected points is selected as the most likely location for a vessel. An ellipse is fitted to this section of the profile using a least square error criterion. Figure 5 shows the grey level profile taken across a vessel and it's approximation by an ellipse. Width, height and residual error of the fitted ellipse are measured.

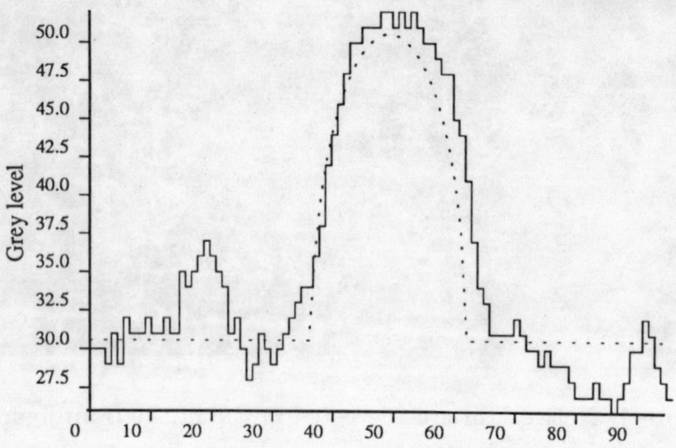

Figure 5: Profile taken across a vessel and it's approximation by ellipse.

For each arc several profiles are sampled and analysed, and following parameters are derived:

- mean and standard deviation of the width of the fitted ellipses,
- mean and standard deviation of the aspect ratio (ratio of height and width) of the fitted ellipses,
- mean and standard deviation of the residual error of the fitted ellipses.

These six features are used for classification of arcs into vessels or other structures currently using a box classifier. Limiting values for each feature have been obtained by training on images with a wide range of vessel widths.

Classification works well for isolated vessels. At some bifurcations or crossings of vessels the grey level profile taken in direction perpendicular to the arc is no longer of elliptical shape, and the arc is not properly classified. This does not affect detection of primary regions because of the method used for drawing detected arcs (see below).

5 Detecting Primary Regions

Having detected vessels in the image it is possible to delineate primary regions in images in such way that their boundaries are aligned with the edges of vessels. Boundaries of regions are obtained by drawing the arcs which have been classified as vessels and dilating them using a circular structuring element of diameter equal to their width. Other arcs are drawn as lines of unit width. Figure 6 shows the primary regions detected after labelling arcs in the network in figure 4.

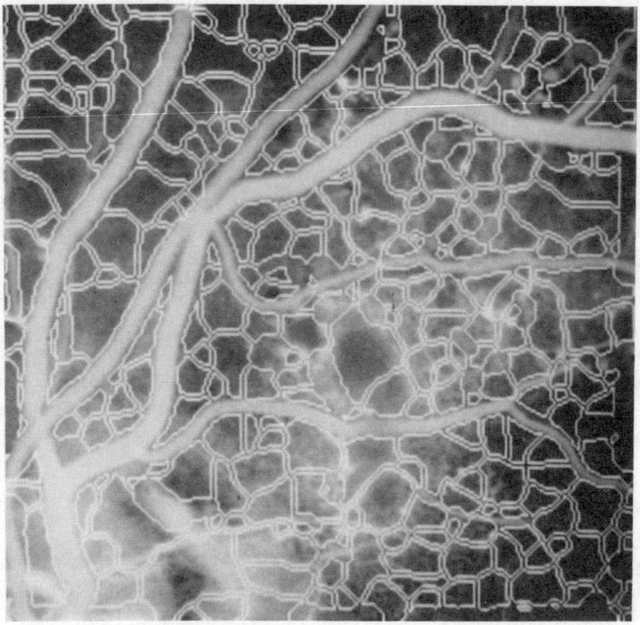

Figure 6: Primary regions detected in the image in figure 2.

The image partitioned into regions in this way, is represented as a region adjacency graph. Information regarding topological relations between regions, connectivity of arcs forming boundaries and properties of regions and boundaries is stored in the graph.

6 Measuring Texture in Primary Regions

Mathematical morphology provides a way of describing sizes and shapes of objects and their relative positions in binary or grey level images. The basic morphological operations for a function $f(y)$ using a binary structuring element B are defined as [13]:

$$\textit{erosion} : (f \ominus B)(x) = \min_{y \in B_x}(f(y))$$

$$\textit{dilation} : (f \oplus B)(x) = \max_{y \in B_x}(f(y))$$

$$\textit{opening} : f_B(x) = ((f \ominus B)(x) \oplus B)(x)$$

$$\textit{closing} : f^B(x) = ((f \oplus B)(x) \ominus B)(x)$$

Following detection, each of the primary regions is analysed separately in order to measure its texture. We are interested in differentiating between regions which contain a dense and regular structure of microcapillary net and regions where this structure is severely reduced. The number of capillaries and areas between them reflect the degree of perfusion of a region.

Werman [16] calculates morphological erosions and dilations using linear structuring elements of increasing lengths, oriented at four directions. This yields a feature matrix for each texture which is then used for classification purposes. Serra [13] uses random set theory to model texture. A texture is described by a probabilistic model. Covariance estimated for the model and measured for the image are used in classification. Maragos [10] defines pattern spectrum as a function of structuring element size. The value of the pattern spectrum for a given size corresponds to the difference between images opened (closed) with consecutive sizes of structuring elements. The entropy of the pattern spectrum is used a texture measure.

A box measure of a fractal surface as described in [9] can be implemented using morphological dilations and erosions with a three dimensional structuring element. This method was described for the first time in [11] without actually calling it morphological. An algorithm corresponding to dilation with a pyramid shaped structuring element was used. Rigaut [12] describes segmentation of biological images by measuring fractal dimension in windows. Textured regions do not need to have fractal properties over a large range of scale to be classified properly.

The main problem in using morphological operations on images to describe texture is that it is possible either to use a combination of many measures as in [16] or to try to model the process which generates the texture, estimate its parameters and limit interest to a class of applications. In our application we are interested in differentiating between regions which contain dense structure of microcapillari es and regions where this structure is severely reduced. Major vessels do not contribute to the degree of perfusion of the region. The size of small areas between microcapillaries indicates the degree of perfusion.

The one dimensional case is illustrated in figure 7. Diagrams *a* to *d* represent grey level profiles obtained by sampling the image along lines in regions with different degrees of perfusion. Profile *a* corresponds to a region with a relatively dense and regular network of capillaries. In profile *b* this density is lower, while in *c* only one large vessel is present. Profile *d* is a more realistic illustration of the expected appearance. Closing with a structuring element B_1 fills the valleys in the profile *a* increasing the average grey level but does not change the profiles *b* and *c*. Increasing the structuring element size to B_2 fills the valleys in profile *b* but *c* remains unchanged. For a periodic texture one may expect a significant increase in average grey level intensity for a precisely defined size of the structuring element. In the real case we obtain profiles more similar to *d*. Then for a region containing a microcapillary network we obtain responses for a wide range of structuring element sizes. For non-perfused regions we obtain some response due to the presence of noise and other visible structures unrelated to the capillaries.

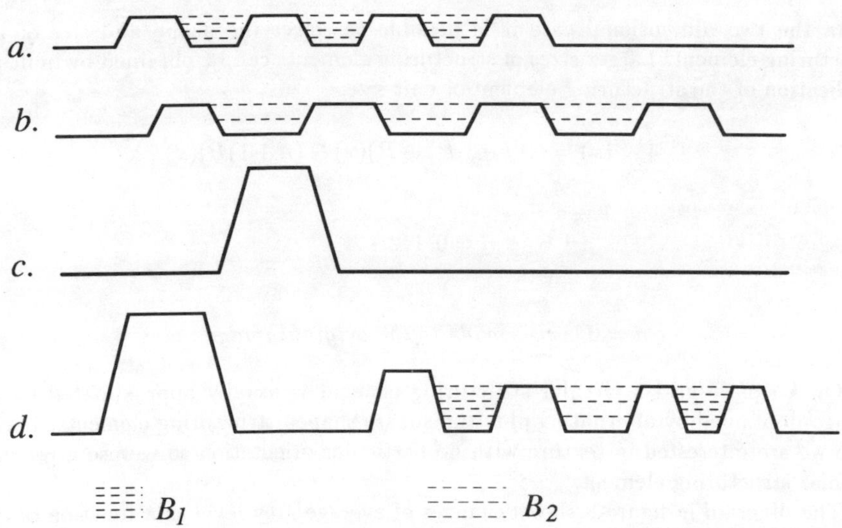

Figure 7: The effect of closing on 1D profile: (a) profile across well perfused region, (b) profile across less perfused region, (c) profile across non-perfused region, (d) profile across a "real" region, (e) two structuring elements of different sizes.

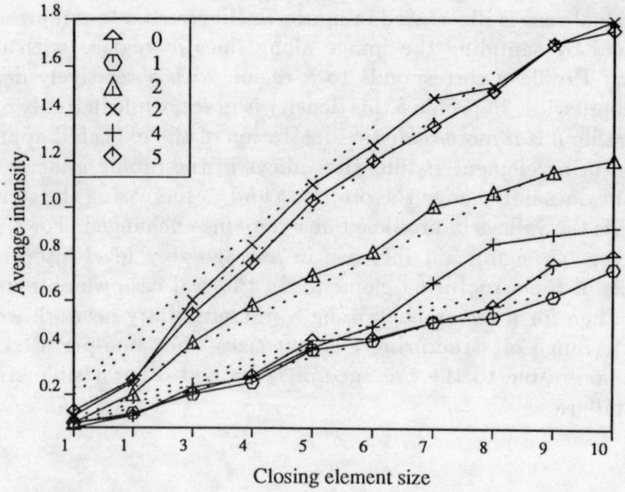

Figure 8: The effect of closing the image on average intensity for selected regions from figure 6: regions 0, 1, 4 are non-perfused, regions 2, 3, 5 are perfused.

In the two dimensional case it is possible to select the shape and size of the structuring element. Larger sizes of structuring elements can be obtained by multiple application of the structuring element of unit size:

$$f^{(n+1)B}(x) = (((f \oplus nB) \oplus B)(x) \ominus (n+1)B)(x)$$

$$n = 0, 1, 2, ...$$

$$n = 0 \textit{ corresponds to the original image}$$

On a square grid a circular structuring element is usually approximated by an octagon obtained by alternating plus and square shaped structuring elements. In our case we are interested in texture with no particular orientation so we use a pseudo-circular structuring element.

The diagram in figure 8 shows changes of average grey level as a function of the size of structuring element for number of perfused and non-perfused regions. The points on each of the curves correspond to the difference between the closed and the original image. The difference is calculated for the whole region and normalised by its size. For non-perfused regions the slope is low (curves *0, 1, 4*), for per fused ones (curves *2, 3, 5*) the slope is high. We calculate the least square error (LSE) straight line for the initial part of each curve. The number of points for each fit is selected so as to limit the error to a predefined constant value. We use the slope of this line as a feature for classification purposes.

In addition we use the opening of the image in a similar way in order to measure the size and number of the capillaries in the region. We measure the normalised difference between original and opened images as a function of the structuring element size. The slope of the LSE line fitted to the initial fragment of each of the curves is used as a second feature to describe regions.

Thus we currently use two features. One of which depends on the size and number of gaps (density of capillaries) in the region, whilst the other depends on the size and number of capillaries. In practice it is unlikely that these features are truly independent but they encode essentially different properties of the texture.

The image in figure 9 illustrates feature values obtained for closing which have been linearly mapped into the corresponding regions of the image shown in figure 6. Low intensity corresponds to regions of low perfusion, high to regions with visible capillary structure. Black gaps between regions correspond to boundaries between regions where the texture was not measured.

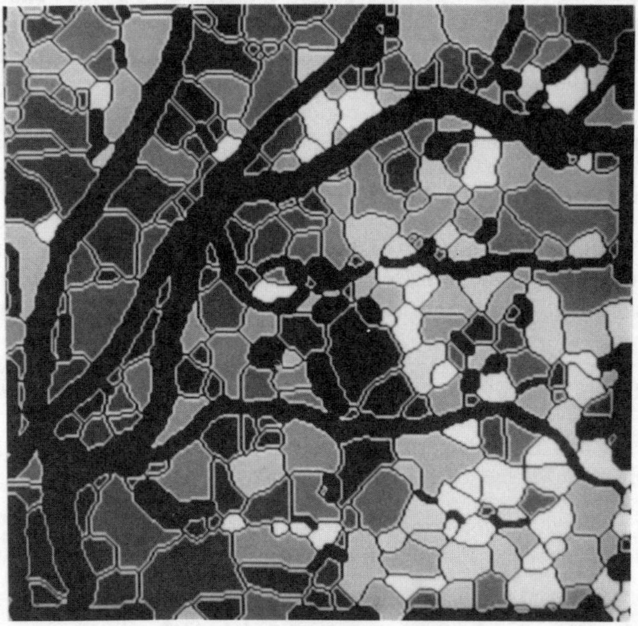

Figure 9: Feature map calculated by closing the image and linear mapping of the value.

Comparing results of manual grading of regions by an ophthalmologist with obtained feature maps we have found a good correspondence between large, non-perfused regions. The correspondence decreases for small regions due to the fact that texture is a property which can be measured more reliably in large regions. This means that textural features calculated for a small region have to be treated with lower confidence than for a large one. Small regions could be merged into bigger ones before calculating their textural features.

7 Conclusions and Further Research

There is a good conformation between the results of our method of tessellating images and the boundaries defined by ophthalmologists. There is one major parameter which controls the process of detecting regions: the size of structuring element used for opening. This parameter has a natural interpretation for it corresponds to the minimum diameter of vessel used for delineating regions and thus indirectly determines the number and size of identified regions.

The features which are used for grading the images are sensitive enough to detect non-perfused regions. The usefulness of these features for classifying regions into different grades and their correlation with manual grading needs further work.We also want to use the presence and position of microaneurysms and local dilation of vessels to modify the classification of regions derived from their texture since both these features accompany non-perfused regions.

Acknowledgements

This research has been supported by a grant from the Kershaw Trust. We are grateful to Prof D. McLeod from Dept. of Ophthalmology, University of Manchester for proving us retinal images and for helpful advice.

References

[1] Akita K., Kuga H.: "Pattern Recognition of Blood Vessel Networks in Ocular Fundus Images." *Proc. SPIE*, vol. 375, 1982, pp. 436-441.

[2] Chaudri S. et al.: "Detection of Blood Vessels in Retinal Images Using Two-Dimensional Matched Filters." *IEEE Trans on Medical Imaging*, vol. 8, no. 3, pp. 263-269.

[3] Gotecherian V.: "From Binary to Grey Tone Image Processing Using Fuzzy Logic Concepts." *Pattern Recognition*, vol. 12, 1980, pp. 7-15.

[4] Goldberg E.G. et al.: "Quantification of Progressive Diabetic Macular Non-perfusion. " *Ophthalmic Surgery*, vol. 20, no. 1, 1989, pp. 42-45.

[5] Jagoe J.R. et al: "Quantification of Retinal Damage During Cardiopulmonary Bypass: Comparison of Computer and Human Assessment." *IEE Proc.*, vol. 137, pt. I, no. 3, June 1990, pp. 170-175.

[6] Katz N., Goldbaum M., Nelson M., Chaudri S.: "An Image Processing System for Automatic Retina Diagnosis." *Proc. of SPIE* vol. 902, 1988, pp. 131-137.

[7] Kohner E.M., Sleightholm M.: "Does Microanerysm Count Reflect Severity of Early Diabetic Retinopathy." *Ophthalmology*, vol. 93, May 1986, pp. 586-589.

[8] Lay J., Badouin B.: "Computer Analysis of Angioflourograms." *Proc. of VII Int. Conf. on Pattern Recognition*, 1984, pp. 927-929.

[9] Mandelbrot B.: "The Fractal Geometry of Nature. " *Freeman*, 1982 .

[10] Maragos P.: "Pattern Spectrum and Multiscale Shape Representation." *IEEE Trans. on PAMI*, vol. 11, no. 7, July 1989, pp. 701-715.

[11] Peleg et al. : "Multiple Resolution Texture Analysis and Classification." *IEEE Trans. on PAMI*, vol. 6, no. 4, 1984, pp. 518-523.

[12] Rigaut J.P.: "Automated Image Segmentation by Mathematical Morphology and Fractal Geometry." *J. of Microscopy*, vol. 150, 1988, pp. 21-30.

[13] Serra J.: "Image Analysis and Mathematical Morphology. " *Academic Press* 1982.

[14] Takahashi N., et al.: "Distribution of Capillary Nonperfusion in Early-stage Diabetic Retinopathy." *Ophthalmology*, vol. 91, December 1984, pp. 1431-1438.

[15] Tanaka M., Tanaka K.: "An Automatic Technique for Fundus Photograph Mosaic and Vascular Net Reconstruction." *Proc. of MEDINFO 80*, Tokyo, pp. 116-120.

[16] Werman M., Peleg S.: "Min-Max Operations in Texture Analysis." *IEEE Trans. on PAMI*, vol. 7, no. 6, 1985, pp. 730-733.

Behaviour Based Robotic Assembly using Vision

Prabhas Chongstitvatana
Alistair Conkie
Department of Artificial Intelligence
University of Edinburgh
5 Forrest Hill, Edinburgh EH1 2QL
Scotland

Abstract

The design of, and experimentation with, vision sensing methods in robotic assembly is discussed in the context of a behaviour based approach. This approach leads to an elegant way of incorporating sensing into an assembly system. The method used does not need a coordinate system that is common to the subcomponents of the assembly system, and has minimal representation. It has strong coupling between sensing and action. A complete assembly task was tested to show the robustness of the system.

1 Introduction

The motivation for the work described here comes from dissatisfaction with today's assembly robot systems which cannot cope with uncertainties nor handle sensing adequately, and are in general hard to program.

The behaviour based approach to robotic assembly is one in which planning is done in terms of behavioural modules [12, 13]. Behavioural modules, amongst other things, are an abstraction mechanism allowing assembly planning without detailed consideration of the real world. Each behavioural module is designed to achieve a particular task in a robust manner, at run time, in the robot cell. Behaviours can interact with each other via the real world. For example, one behaviour could pick up a part then put it down at a predetermined location for another behaviour to reorient it.

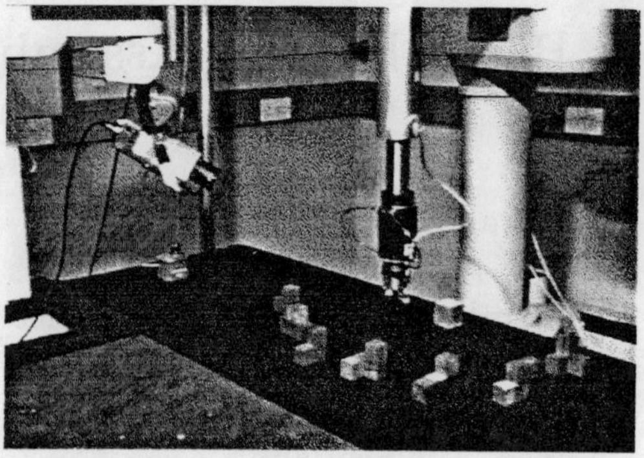

Figure 1: The Assembly Cell and Soma Parts

Malcolm[11] has demonstrated a successful automatic assembly system for the Soma[1] world using this approach. The Soma assembly planner (Somass) is an automated assembly planning and execution system. It plans and assembles the Soma parts according to a specified shape of final assembly. The plan specifies ideal motions of parts. The behavioural modules, at run time, execute the robot motions to achieve the required part motions.

In the Somass system, the implementation of behaviours uses no sensing. Instead it employs constrained motion to locate the parts. This paper describes extensions to the Somass system which use vision sensing for the part-acquisition behaviour. This extension avoids the use of a global coordinate system for using the vision data and employs a self-calibration procedure to help achieve robustness. The position of the camera does not need to be known.

As will be seen, the experiments also demonstrate that two robots, one holding a camera, and the other performing the assembly, can co-operate with each other in a very natural way without either having to know exactly where the other is.

2 Review of previous work

Visual feedback has been used to guide robots in "hand-eye" coordination tasks since the early days of robotic research. Jones [8] used a simple tracking method to control

[1]Soma is a set of parts made by considering all the different ways in which 4 or fewer cubes of the same size can be abutted face-to-face so as to form irregular solids. There are seven such parts [4], (see Fig. 1).

a camera's aim and derived useful information from the image in terms of pixel coordinates. To increase assembly robot capabilities, a geometric modelling system was introduced. This was used to reason about spatial occupancy, for collision avoidance, and to plan robot actions. The use of geometric modelling together with the control of the robot hand in a Cartesian coordinate system gives rise to the need of mapping objects into a common coordinate frame. Most robotic assembly systems build this map [7, 14, 15, 9, 6].

Vision systems have been used to build up 3D models of objects by various techniques, for example binocular stereo vision [7], or laser range finder systems with model-based object localisation [9]. These systems incorporate explicit geometric models of the environment, the robot, and the camera. The success of operation relies heavily on the accuracy of these models and calibration of the vision system with respect to robot arm and environment.

In dealing with a dynamic world, Andersson [1] combines both real-time performance of visual feedback and manipulation that demands high speed response. He uses special hardware and sophisticated calibration techniques to achieve the accuracy required. It is worth noting that the technique of using image sequences, rather than static scene analysis, helps to simplify stereo calculations and is useful for deriving depth information [2, 5].

3 Problems and Possible Solutions

One thing that is apparent from the previous work in this area is that systems that require an accurate world model and accurate calibration tend to be large and complicated possibly unweildy. This is a situation that it would be preferable to avoid. Various principles can be adopted to help make things simpler. The general underlying motivation is to reduce the cost of representation and increase robustness.

The first principle is to try and replace, as far as is possible, calculations by sensing. This equates to preferring perception over representation. Several techniques are used that contribute to this end. Frequent vision sensing, for example, allows motions in the world to be viewed as linear approximations, and only short term predictions of motion need be made for following an object. The tracking algorithm adopted is similar to that used by others [8, 2].

The system is purposely made as calibration free as possible. Relative quantities are used and self-calibration is done while carrying out tasks. This approach is possible because adequate sensing allows self-calibration. The need for calibration to establish the mapping of image data to world coordinates is absent from the system described here. The camera position need not be known.

The idea of having an explicit world model is rejected in favour of using heuristics to guide the robot hand, in contrast to much previous work. Contextual information is also exploited in the same spirit as others have used it [7, 8]. The system does not rely on explicit models of the objects nor does each assembly agent need to know

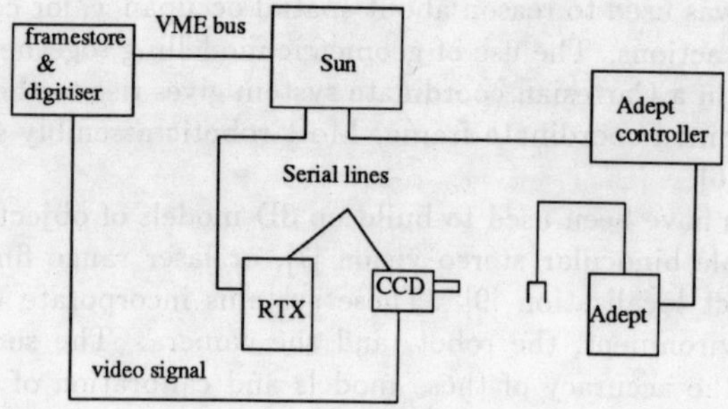

Figure 2: System Architecture

about the other(s). The system does not maintain representations of the Soma pieces, the camera, the robots nor of possible uncertainties due to inaccuracies of the model with respect to the real world, unlike other systems.

This philosophy of avoiding using a world model and instead relying directly on vision to pick up the Soma pieces has benefits as regards uncertainties. Uncertainties are dealt with directly, rather than being treated as an afterthought that would fall outwith the formalism of the system. Direct observation and feedback allow the possibility of removing many uncertainties without resorting to complex analysis.

4 The Experiments

4.1 System Architecture and Setup

The system adopted consists of two robots and a Sun workstation with some vision processing hardware (see Fig. 2).

One robot is an AdeptOne industrial robot, with 5 degrees of freedom. This robot carries out the assembly task. It has a simple pneumatic close-open parallel gripper. The fingers are painted white and the rest of the hand painted black.

A second robot arm (an RTX made by UMI Ltd.) is used to hold a small camera in the second experiment. The robot/camera combination is moved to track the movements of the Adept arm.

The Sun machine contains two MaxVideo boards: an image digitiser and a framestore. A subset of the data in the framestore is selected and analysed on the Sun. The hardware is driven in interrupt mode. Both robots are controlled via serial lines from the Sun. The software is written in C and Prolog.

The camera is a CCD video camera with automatic gain control, a Sony DXC-101P with a 16 mm lens. The work cell is lit by a white light lamp, and the surface of the table is covered by a black cloth, which gives adequate contrast.

The Soma set we used is made of wood and is not painted. It is in its natural texture and colour. The plan for the complete assembly is generated by Soma Planner [11]. Only the pick-up behavioural module is substituted into the plan.

Two experiments were carried out. In the first a behaviour was devised to pick up a Soma part from the work area. Starting with the robot hand above the nominal (or predefined) position of the part, and the part in approximately its nominal position, a limited amount of image data combined with a simple strategy to move the hand can guide the hand to pick up the part. This was done without referring to robot, part, or camera coordinates.

The second experiment supplements the first. A behaviour capable of finding the robot hand is used. The camera is then able to follow the hand to a position directly above the nominal position of the part. This tracking behaviour thus aims the camera at the correct location for the pick-up behaviour to start, replacing the manual camera movements of the first experiment.

In the context of a complete assembly, the part-acquiring behavioural module is used to locate each part. The vision system does not have to recognise the individual part. Before a part is picked up the hand hand moves to directly above the nominal position of the part. Because the fingertips can be located the position of the hand in the image is known. The area below the hand in the image is searched for the part. The camera is moved to aim at the correct part each time the new part is to be acquired.

4.2 First Experiment : Picking up a Soma part

In this experiment the camera is moved manually, for each part, so that both the robot hand and the part to be picked up are within the field of view. The part has a nominal position with the translation variance of the part being about twice its cube size and the rotation variance being plus or minus 30 degrees.

The camera looks down at an angle that can vary between approximately 30 and 60 degrees, and approximately plus or minus 15 degrees horizontally, from the ideal "head-on to the robot fingers" position. A wide range of positions are possible. The goal is to use the information derived from the image data to guide the hand to align the finger tips with one edge of the top cube of the part. The height of the part is assumed to be known, and the part lies flat on the working surface with the cube to be grasped pointing up. The technique used requires that both the fingers and the specified edge are in clear view.

The strategy is as follows (see Fig. 3): A plane is defined, parallel to the work surface at a known height and above the tops of the parts, which is called the "approach plane". There are two points of interest that are then projected onto the "approach

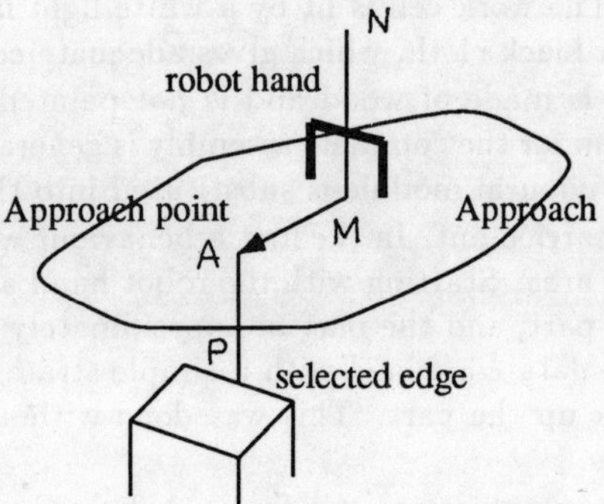

Figure 3: Strategy Diagram

plane". The first is the midpoint between the two fingers of the gripper. The second is the midpoint of the top edge (in the image) of the part. The projection of this point is called the "approach point". The robot is driven in the "approach plane" so as to minimise the distance between these two points.

There is a continuous 1-to-1 mapping from the image plane to the approach plane. This means that, in practice, we can work with the image data only. The robot hand is driven vertically a known distance to lie in the "approach plane". The resulting pixel displacement of the hand allows us to deduce, the "approach point" in the image. Successive estimates are made of how the robot should move which are refined based on the actual movements (again in the image plane), until the error is reduced to an acceptable level. This process is repeated twice. Finally the hand is rotated so that it is parallel to the edge and then it can move down to the final position.

The analysis is in two parts. The first part treats the method of visual servoing the hand and the second part treats the analysis of the perspective distortion of projecting the image into the camera.

4.3 Visual servoing

Consider the mapping of the motion in a plane in the robot coordinate frame to the image plane.

$$\mathbf{v} = \mathbf{T}\,\mathbf{m} \tag{1}$$

where $\mathbf{m}$ is a vector PQ joining the point P to the point Q in the robot coordinate frame which lies on the approach plane. $\mathbf{v}$ is the corresponding vector in the image plane formed by projecting P, Q into p, q. $\mathbf{v}$ = pq. $\mathbf{T}$ is the transformation of the points in the approach plane into the image plane.

The robot hand has two translational and one rotational degrees of freedom. Only the translational components are considered here. The transformation $\mathbf{T}$ is estimated iteratively. The first estimate of $\mathbf{T}$, $\mathbf{T}_1$, comes from observing the vector between the fingertips, with *a priori* knowledge about their position in the robot coordinate frame. In general, $\mathbf{T}_i$ can be used to estimate $\mathbf{m}_i$.

$$\mathbf{m}_i = \mathbf{T}_i^{-1}\,\mathbf{v}_i \tag{2}$$

The robot is then moved by $\mathbf{m}_i$. The observed movement is $\mathbf{v}'_i$

$$\mathbf{v}'_i = \mathbf{T}\,\mathbf{m}_i \tag{3}$$

and

$$\mathbf{v}_i - \mathbf{v}'_i = (\mathbf{T} - \mathbf{T}_i)\mathbf{m}_i \tag{4}$$

From this a new estimate of $\mathbf{T}$, $\mathbf{T}_{i+1}$ can be calculated. The next move $\mathbf{v}_{i+1}$ is given by

$$\mathbf{v}_{i+1} = \mathbf{v}_i - \mathbf{v}'_i \tag{5}$$

The iteration is stopped when the hand is close to the goal, i.e., when the term $\mathbf{v}_i - \mathbf{v}'_i$ is smaller than a preset value.

4.4 Analysis of the perspective distortion

The approach point is found by moving the hand downward by half of the vertical distance from the top of the object thus making the line PA parallel to MN and of equal length (see Fig. 3). Because of the perspective distortion, there will be a discrepancy QA, (see Fig. 4) which will create an error AA' in determining the approach point.

QA is determined as follows: d_1, d_2 are measured along the optical axis of the camera and f is the focal length of the camera.

From the triangle OMN,

$$\frac{ON}{\sin\beta_1} = \frac{MN}{\sin\gamma} \tag{6}$$

$$\frac{OC}{\sin\alpha_1} = \frac{BC}{\sin\gamma} \tag{7}$$

$$BC = h\,\frac{OC}{ON}\,\frac{\sin\beta_1}{\sin\alpha_1} \tag{8}$$

and similary the triangle OPQ, where $BC = DE$

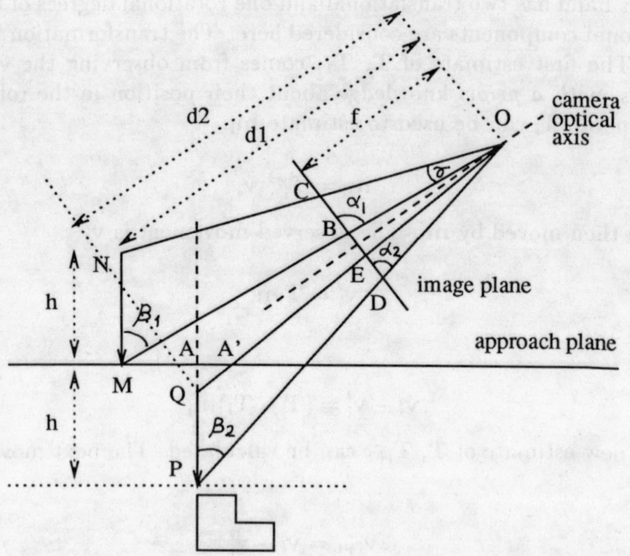

Figure 4: Determination of the perspective distortion

$$DE = PQ \frac{OE}{OQ} \frac{\sin \beta_2}{\sin \alpha_2} \tag{9}$$

from $\frac{OC}{ON} = \frac{f}{d_1}$; $\frac{OE}{OQ} = \frac{f}{d_2}$

$$PQ = h \frac{d_2}{d_1} \frac{\sin \beta_1}{\sin \beta_2} \frac{\sin \alpha_2}{\sin \alpha_1} \tag{10}$$

$$QA = h - PQ \tag{11}$$

$$QA = h \left(1 - \frac{d_2}{d_1} \frac{\sin \beta_1}{\sin \beta_2} \frac{\sin \alpha_2}{\sin \alpha_1} \right) \tag{12}$$

The camera position can be arranged such that QA is small. This does not mean that the exact position of the camera must be known. The permissible range is large enough such that in practice the camera can be positioned without any measurement tools.

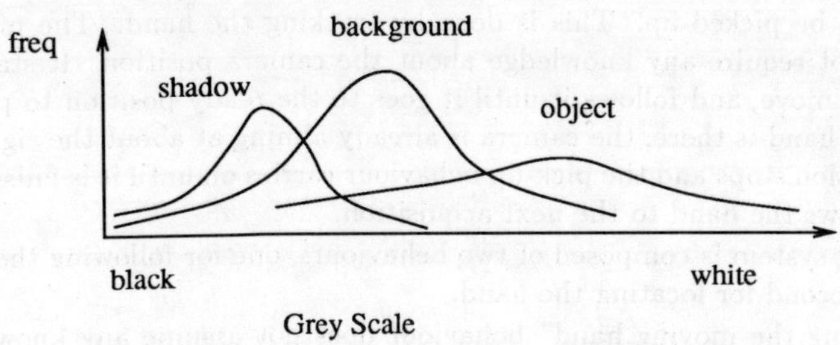

Figure 5: Components of the Histogram

4.5 Vision processing

The information needed from the image is the boundary of the Soma part, and the pixel coordinates of the fingertips.

In order to find in the first instance where the fingertips are, the difference of two pictures is calculated with only the fingers in different positions. Thus the initial finger positions can be located and two small windows enclosing the tips selected. Each fingertip is tracked throughout the hand motion by moving the windows to keep the finger centroids in the centres of the windows. The fingertips are painted white, so they are the brightest objects in the scene. This makes their identification more robust against the background.

The background is the easiest and most invariant thing to identify in the image. Use is made of knowledge about the image histogram (see Fig. 5), in particular that it is composed of three distributions; from the part, which has low peak and large variance, from the background which has high peak and small variance and from the shadow which is darker than the background. The histogram is first smoothed. To separate the background the maximum peak, which is the mean of the background distribution, is found. The variance of this distribution is analysed and the threshold selected. This algorithm works well even if shadows are present.

The system doesn't recognise the part in the usual sense. It follows the hand, using it as a pointer. A binary representation of the area of the image below the hand is made using thresholding as described above. The outline of the part is traced and the polygon approximation of that shape derived [10]. Then the most horizontal top edge is selected. It is the least obscured against the background and other objects, and the most reliable feature to look for.

4.6 Second Experiment : Tracking the robot hand

The manual positioning of the camera in the first experiment is replaced by automatic positioning using tracking. To establish the correct preconditions for the pick-up behaviour a second robot (the RTX) is used to hold the camera and to aim it correctly for each part to be picked up. This is done by tracking the hand. The method of tracking does not require any knowledge about the camera position. It starts with seeing the hand move, and follows it until it goes to the ready position to pick up a part. When the hand is there, the camera is already aiming at about the right place. The camera motion stops and the pick-up behaviour carries on until it is finished then the camera follows the hand to the next acquisition.

The tracking system is composed of two behaviours, one for following the moving hand, and the second for locating the hand.

The "following the moving hand" behaviour does not assume any knowledge of the shape of the hand. It is based on finding the difference between two successive images. The moving camera is instructed to centre its view on the part of the image that has changed. This in itself does not give accurately the location of the hand in motion and cannot be used when the hand does not move. In conjunction with this behaviour, a second behaviour was used to locate the hand.

The behaviour to locate the hand looks for a special mark on the robot hand. This mark is brightly coloured. A simple spectrum classification scheme based on intensity values of {Red,Green,Blue} is used to separate the mark from the background and calculate its centroid. Finally, the camera is view-centred on this point. The two behaviours are simple and complement each other well, and are sufficient to establish the proper preconditions for the pick-up behaviour.

5 Test results

The system works well in practice and is tolerant to changes, over a wide range, of lighting conditions and part and camera positions. From the experiment, we ran a number of assemblies with the different initial conditions and camera parameters. The distance of the camera to the part is 540 to 600 mm., the height of camera above the part is 330 mm., the pitch angle of the camera is 20 to 25 degrees, the yaw angle is -10 to 20 degrees. The hand travelled downward 40 mm. The part varies in the position from -40 to +30 mm. from its nominal position. The mean error and the standard deviation in determining the position and orientation of the part are 0.80 mm., 2.23 mm., 1.31 degrees and 2.18 degrees respectively.

The system will fail, as would be expected, if certain conditions are not met. If the contrast is sufficiently low the thresholding will fail. Tracking the fingers will fail if they appear too small in the image, for example if a wide angle lens is used. The angle of the camera with respect to the plane of the table is not critical, but near the horizontal the strategy employed gives less accurate results and near the vertical (top view) the fingers of the hand are more likely to be obscured.

It was possible to arrange the environment so that all these failure modes were avoided without difficulty.

6 Conclusions

Behavioural modules can make use of vision sensing to achieve robust reliable operations in the assembly cell. Programming using behavioural modules is a good method of problem subdivision, and of dealing with uncertainty.

The system discussed in this paper works without a common coordinate system for the components of the assembly system. Vision processing is simpler and robots can cooperate in a simpler way than would result from using such a coordinate system. In addition, the system is generally robust because it does not depend on calibration of subsystems.

By observing rather than reasoning or predicting we have been able to tightly couple vision sensing and action within the system. This is achieved without having to represent the world and its changes explicitly, thereby avoiding the danger of constructing Dennett's imaginary robot [3], which, as he describes it, is

> ... sitting, Hamlet-like, outside the room containing the ticking bomb, ... "Do something!" they yelled at it. "I am," it retorted. "I'm busily ignoring some thousands of implications I have determined to be irrelevant. Just as soon as I find an irrelevant implication, I put it on the list of those I must ignore, and..." the bomb went off.

References

[1] Andersson, R.L., "A Robot Ping-Pong Player". MIT Press, 1988.

[2] Brown, C.M., Ed., "Rochester Robot", *Technical Report 257*, University of Rochester, Computer Science, 1988.

[3] Dennett, D.C., "Cognitive Wheels: The Frame Problem of AI", in The Robot's Dilemma, Pylyshyn, Z.W., ed., Ablex Publishing, 1988.

[4] Gardner M., "Pleasurable Problems with Polycubes", *Scientific American*, Sept. 1972, p176.

[5] Hayes, G.M., "A Real Time Kinematic Depth System", *MSc Dissertation*, Edinburgh University, 1989.

[6] Heikkila, Matsushita and Sato, "Planning of Visual Feedback with Robot-Sensor Co-operation", *Proc. 1988 IEEE Inter. Workshop on Intelligent Robots and Systems* ,Tokyo, 1988.

[7] Inoue, I., "Hand Eye Coordination in Rope Handling", *1st Inter. Symp. on Robotics Research*, Bretten Woods USA, 1983.

[8] Jones, V.C., "Tracking: An Approach to Dynamic Vision and Hand-Eye Coordination", PhD Thesis in Electrical Engineering, University of Illinois, Urbana Champaign, 1974.

[9] Lozano-Perez et al, "Handeye : a robot system that recognises, plans, and manipulates", *Proc. 1987 IEEE Inter. Conf. on Robotics and Automation*, vol.2 ,1987, pp. 843-849.

[10] Malcolm, C.A., "The Outline Corner Filter", in Proc. of the 3rd *Inter. Conf. on Robot Vision and Sensory Controls*, Nov. 1983, pp 61-68.

[11] Malcolm, C.A., "Planning and Performing the Robotic Assembly of Soma Cube Constructions", *MSc Dissertation*, Edinburgh University, 1987.

[12] Malcolm, C.A., Smithers, T., "Programming Assembly Robots in terms of Task Achieving Behavioural Modules: First Experimental Results", In Proc. of the Inter. Advanced Robotics Program: *Second Workshop on Manipulators, Sensors and Steps Towards Mobility*, 1988. Also available as research report 410, Department of Artificial Intelligence, Edinburgh University.

[13] Malcolm, C.A., Smithers, T., "Symbol Grounding via a Hybrid Architecture in an Autonomous Assembly System", *Workshop on Knowledge Representation and Learning in an autonomous agent*, 1988. Also available as research report 420, Department of Artificial Intelligence, Edinburgh University.

[14] Popplestone, R.J., Ambler, A.P. and Bellos, I., "An Interpreter for a Language for Describing Assemblies", *Artificial Intelligence* 14, 1 (1980), pp. 79-107.

[15] Yin, B., "Using Vision Data in an Object-Level Robot Language RAPT", *Inter. Jour. Robotics Research*, vol. 6 no. 1, 1987, pp. 43-58.

A Multi-Operator Approach for the Segmentation of 3-D Images of Dental Imprints

Jean Côté, Denis Laurendeau and Denis Poussart
Computer Vision & Digital System
Department of Electrical Engineering
Laval University
Quebec, Province of Quebec, Canada, G1K 7P4

Abstract

This paper presents a computer vision system for the acquisition and processing of 3-D images of wax dental imprints recorded on human subjects. The system is dedicated to orthodontic diagnosis and epidemiological studies. The acquisition system is described in detail and experimental results are presented. The accuracy of the system is tested on a reference imprint.

A segmentation algorithm is used for the detection of the interstices between teeth on the 3-D images of the imprint. A multi-operator approach is developed to deal with teeth of different shape and size. Experiments performed on a large number of imprints show that the algorithm for interstice detection and position estimation is very reliable and accurate.

1 Introduction

The detection and correction of development of malocclusions and other dental abnormalities is one of the most important and critical phases of orthodontic diagnosis. The Department of Dentistry and the Computer Vision and Digital Systems Laboratory of Laval University are jointly working on the development of an expert system for orthodontic diagnosis and epidemiological studies. The expert system accepts a set of 10 orthodontic parameters as an input and produces a diagnosis based on inference rules considering the relations between the parameters [2] [3]. The diagnosis consists of one out of three choices: *i)* orthodontic treatment required, *ii)* elective orthodontic treatment, and *iii)* no treatment required. Actually, the orthodontic parameters are measured with great difficulty on costly plaster models. These models are obtained manually by specialized personnel and require a rather complex and lengthy acquisition procedure. A mold of the inside of the mouth must first be obtained. The acquisition of the mold is a very unpleasant experience for the patient. The plaster model is then cast into the mold. The plaster model must be manually improved to remove beards. The team at Laval University is actually working on the design of a computer vision-based 3-D data acquisition and analysis system for the measurement of orthodontic parameters. Instead of using costly plaster models, low-cost wax wafers are used to obtain an imprint of the teeth on both maxillaries.

This paper presents a computer vision system for the acquisition of 3-D profiles of both sides of the imprint recorded on the wax wafer. The acquisition system uses a specially designed 3-D camera based on the differential absorption of light by a dispersive medium. The 3-D profiles are used to detect the interstices between teeth on the maxillaries. The detection of these interstices is a very important step towards automatic measurement of orthodontic parameters. Since incisors, canines, pre-molars, and molars have very different shapes, several operators are combined for the segmentation of the 3-D images and the detection of the interstices.

Figure 1 shows the various steps that were designed for the detection of the interstices between teeth. Each step is explained in detail in the following sections.

Section 2 describes the acquisition setup that was designed for the acquisition of the 3-D profiles of wax dental imprints, and Section 3 presents experimental results of the acquisition of profiles. Accuracy of the 3-D data is also analyzed in this section. Section 4 describes the operators that were designed for the detection of interstices between teeth. Three operators are discussed: the case of upper maxillary incisors, the case of molars on both maxillaries, and the case of lower maxillary incisors. Section 5 presents experimental results of the segmentation technique for the detection of interstices between teeth and the robustness of the technique is discussed. Finally, Section 6 presents the user interface that was designed to facilitate the acquisition and processing of 3-D dental imprint images.

2 Acquisition

The dental imprints are first recorded by having the patient bite a low cost wax wafer similar to the one shown in Figure 2. Using these imprints a three-dimensional (3-D) image of both sides of the wafer is obtained with a special data acquisition system based on the differential absorption of light by a dispersive medium. The acquisition system is composed of two cameras, two optical bandpass filters of different central wave-lengths, two light sources, and a glass container filled with a colored liquid which acts as a dispersive medium. These components of the system are located on both sides of the container as shown in Figure 3.

The technique that is used to measure the 3D profile (e.g. 3-D image) of the wax wafer is very simple. The wafer is mounted on a support which is immersed in the

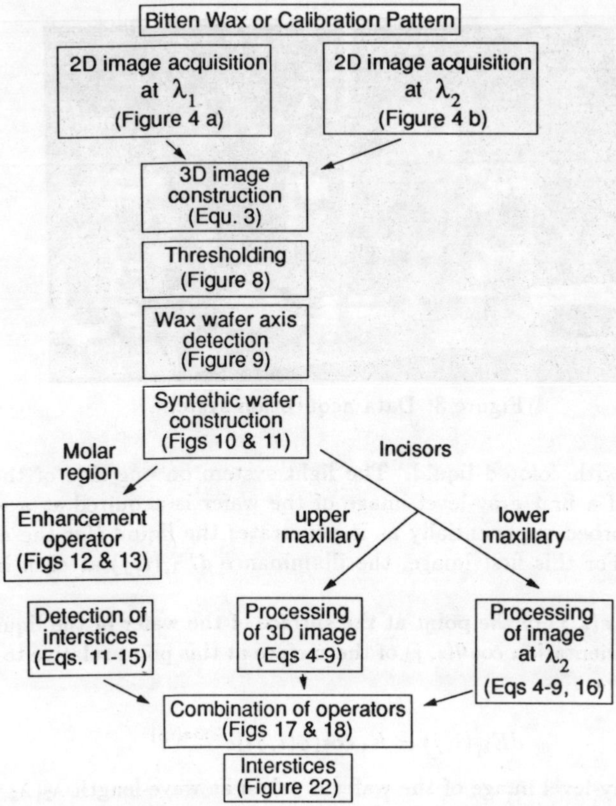

Figure 1: A schematic diagram of the process to detect the interstices between teeth.

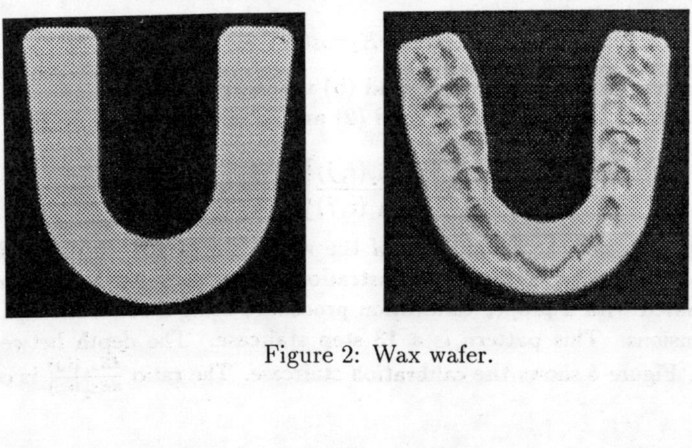

Figure 2: Wax wafer.

Figure 3: Data acquisition system.

container filled with colored liquid. The light system on one side of the container is turned on and a first gray-level image of the wafer is acquired at a wave-length λ_1 which is absorbed exponentially as it penetrates the liquid (for the experiments $\lambda_1 = 600nm$). For this first image, the illuminance $dE_{\lambda_1}(i,j)$ at each image point depends on

1) the depth $z(i,j)$ of the point at the surface of the wafer in the liquid and

2) the local orientation $\cos\theta(i,j)$ of the surface at this point relative to the optical axis of the camera:

$$dE_{\lambda_1}(i,j) = K_1 \cos(\phi(i,j))e^{-2\alpha z(i,j)} \tag{1}$$

A second gray-level image of the wafer is taken at wave-length $\lambda_2(\lambda_2 = 750nm)$. The spectral characteristics of the liquid are such that wave-length λ_2 is not absorbed. The illuminance $dE_{\lambda_2}(i,j)$ at each point in the second image only depends on the local orientation of the surface $\cos\theta(i,j)$, see [9]:

$$dE_{\lambda_2}(i,j) = K_2 \cos(\phi(i,j)) \tag{2}$$

Figure 4 shows (a) the image at λ_1 and (b) the image at λ_2.

Taking the ratio between Eqs. (1) and (2) and solving for $z(i,j)$ we get:

$$z(i,j) = \frac{1}{2\alpha}\ln\left[\frac{dE_{\lambda_2}(i,j)}{dE_{\lambda_1}(i,j)}\right] + \frac{1}{2\alpha}\ln\left[\frac{K_1}{K_2}\right] \tag{3}$$

The range image of the second side of the wafer is obtained similarly. The 3-D images on both sides are in perfect registration. Constants K_1, K_2 and α in Eq. (3) are estimated with a proper calibration procedure using a calibration pattern of known dimensions. This pattern is a 13 step staircase. The depth between each step is 1mm. Figure 5 shows the calibration staircase. The ratio $\frac{dE_{\lambda_1}(i,j)}{dE_{\lambda_2}(i,j)}$ is obtained

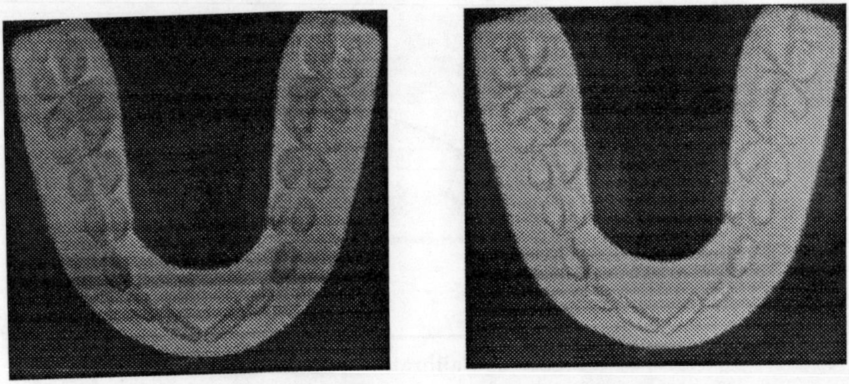

Figure 4: Image collected at (a) λ_1 and (b) λ_2.

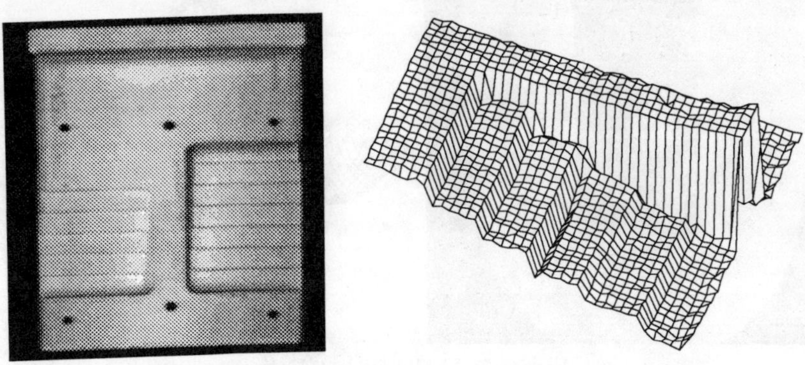

Figure 5: Staircase for calibration.

experimentally for each step of known depth . The parameters in Eq. (3) are then computed with a least squares method. Figure 6 shows a calibration curve. Cross symbols correspond to experimental values and the solid line is the regression curve.

3 Validation of 3-D Image Values

Three-dimensional image data was validated on a machine-made plexiglass model shown in Figure 7 and is similar to an actual wafer. The distance between adjacent teeth, and size and shape of each tooth on the model are known with accuracy. Table 1 compares the values measured with the 3-D acquisition setup on the image of the model to those measured manually on the model. Lateral measurements on the wafer

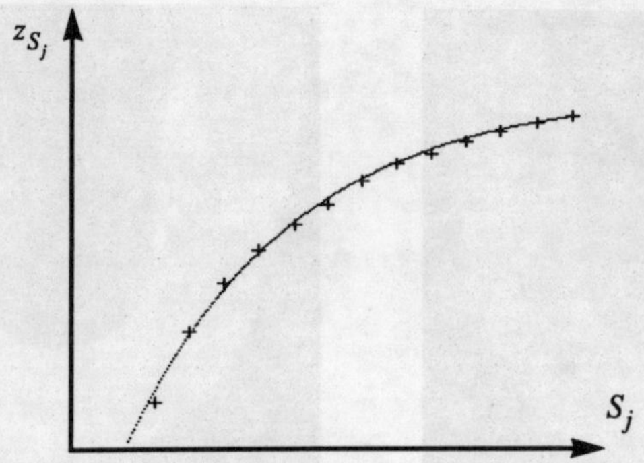

Figure 6: Calibration curve.

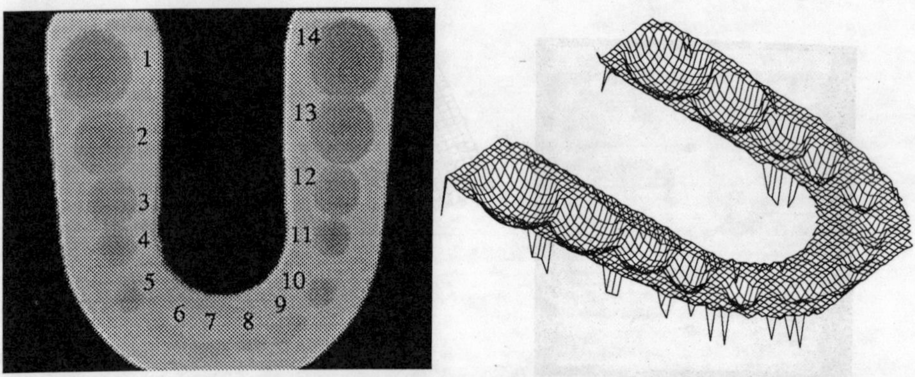

Figure 7: Model for validation.

(not shown in Table 1) are very accurate (within 5% accuracy). As shown in Table 1 the accuracy on depth measurements is also satisfactory with a relative error $\leq 10\%$. The error could be decreased by using a better lighting system and by cooling the cameras. The best concentration of the solution for the experiment was 0.02%. The colorant used was a blue organic dye [5].

4 Detection of the Interstices between Teeth

4.1 Pre-processing of the 3-D Image of the Wafers

The three-dimensional images (one for each side of the wafer) are thresholded to separate the wafer from the background. Figure 8 shows a thresholded image. Interstices between adjacent teeth are detected on this type of image.

Side 1	Concentration 0.01%	0.01%	0.02%	0.02%	0.03%	0.03%
1	5.71	1.20	5.71	5.11	5.11	5.11
2	3.97	5.63	2.32	4.97	0.00	3.31
3	7.87	9.45	11.4	2.36	9.44	0.39
4	20.1	20.7	13.1	9.76	19.2	12.5
5	17.0	19.8	9.09	6.32	9.09	9.88
6	7.47	12.1	4.67	7.48	12.1	0.93
7	0.87	5.22	2.61	4.35	9.57	13.0
8	0.93	5.61	7.48	9.35	8.41	1.87
9	2.70	3.60	10.8	6.31	12.6	3.60
10	21.8	22.2	15.6	17.8	8.89	5.78
11	25.6	24.1	6.64	6.64	6.64	3.80
12	0.43	8.70	2.17	4.35	2.17	4.35
13	3.47	2.52	4.10	7.57	7.89	10.4
14	2.07	0.89	5.03	3.55	6.51	5.03

Side 2	Concentration 0.01%	0.01%	0.02%	0.02%	0.03%	0.03%
1	6.61	3.58	4.96	3.31	4.68	3.58
2	0.71	1.78	7.14	3.57	6.42	3.57
3	7.69	6.92	1.92	6.92	3.07	7.69
4	17.8	19.2	7.74	14.1	6.40	15.8
5	15.9	13.9	7.56	16.0	4.20	7.56
6	0.00	2.17	6.52	7.61	8.70	0.00
7	4.67	8.41	3.74	7.48	7.48	4.67
8	13.9	0.00	12.0	0.93	15.7	10.2
9	3.36	0.84	6.72	0.00	4.20	0.84
10	24.4	24.4	7.56	11.8	1.26	0.84
11	18.3	20.0	6.67	11.7	6.67	12.7
12	8.61	5.61	4.87	1.50	4.87	2.62
13	0.95	1.90	3.17	0.95	6.35	3.17
14	3.57	1.79	3.93	3.57	5.71	1.43

Table 1: Accuracy of measurements. (% error)

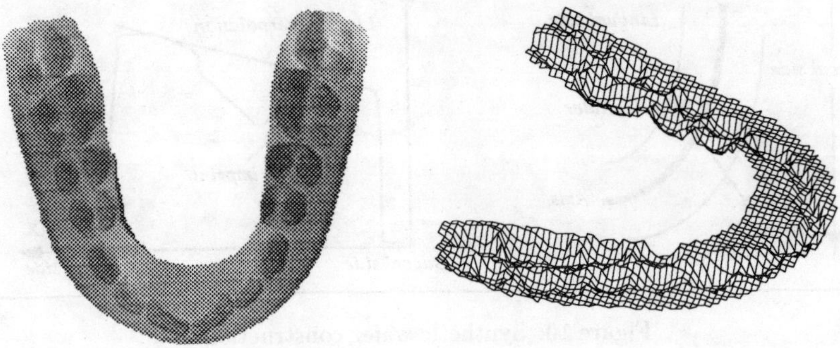

Figure 8: Thresholded dental imprint image.

All segmentation operations on a wax wafer image use the middle axis of the imprint as a reference. The image of the wafer is scanned from the upper right to the upper left along the scanlines shown in light grey in Figure 9. Points located at the midpoints between the extremities of the imprint along each scanline are detected. A cubic spline is fitted to these points. This spline curve is considered as the axis of the wax wafer. Figure 9 shows the axis as a dark solid line.

The wax wafer is often put out of shape after the bite. This deformation must be taken into account before the interstices between teeth are found. For this purpose, a synthetic wafer is built. It follows the general shape of the wafer without the teeth imprints. The method used to build this synthetic wafer is to interpolate values linearly between the lingual and the buccal sides of the image of the imprint on a ruler with a direction perpendicular to its axis as shown in Figure 10. Then a morphological opening followed by a closing operation [1],[4] is applied to the synthetic wafer in order to eliminate missing interpolated points. Figure 11 shows the synthetic wax for the

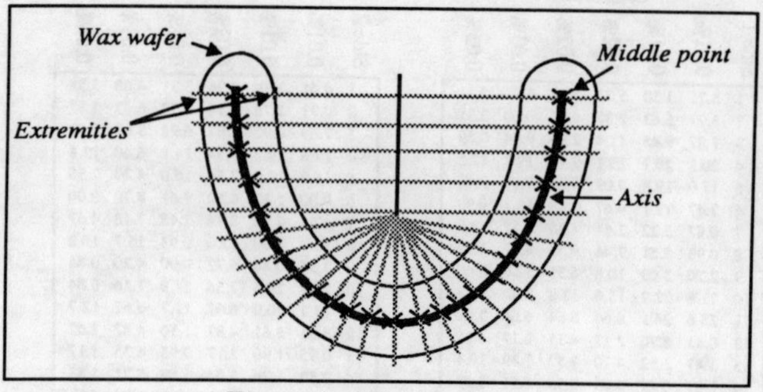

Figure 9: Wax wafer axis.

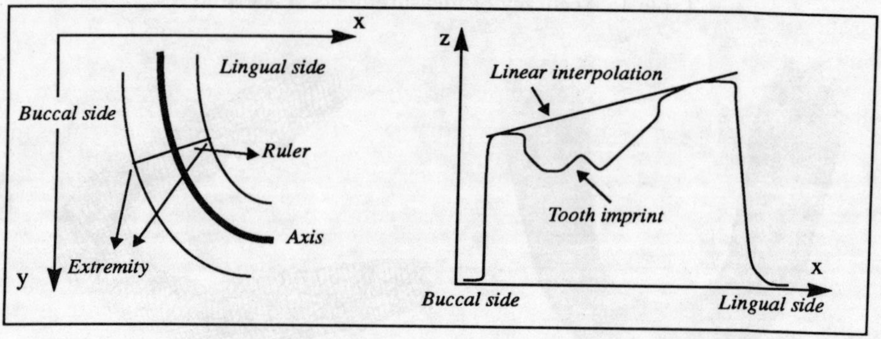

Figure 10: Synthetic wafer construction.

imprint of Figure 8.

An image enhancement operator is applied to the range image before the detection of the interstices between teeth, particularly molars. For each pixel of the image, the maximum value over the diamond shaped mask of Figure 12 is detected and is assigned to the central pixel of the mask. As shown in Figure 13, this operator amplifies the interstices between teeth and smoothes false interstices in the disto-buccal and distal fossae area [8] of molars. However, the operator has a spurious effect in the incisors area as demonstrated in the next section. Consequently, its output will be used only in the molar area.

4.2 Separation between Teeth

The axis drawn in Figure 9 is used to find the interstices between teeth. For each point along this axis, starting at the upper left end (s_o), a ruler is moved relative to the axis. For each position, it is oriented at different angles in the x,y plane relative to the axis. The function $\sigma(s_i)$ defined in Eq. (4) is computed for each value of s_i,

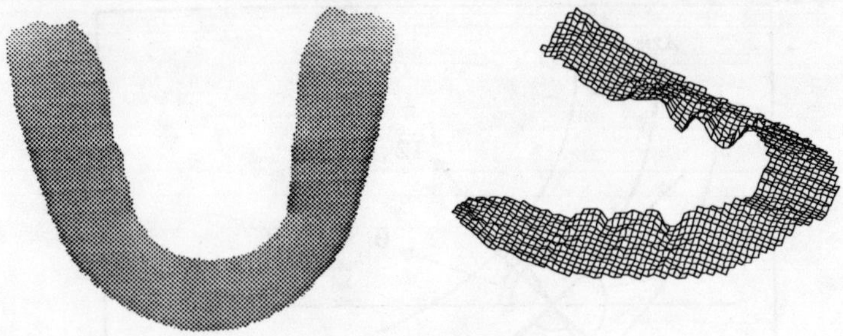

Figure 11: Synthetic wax.

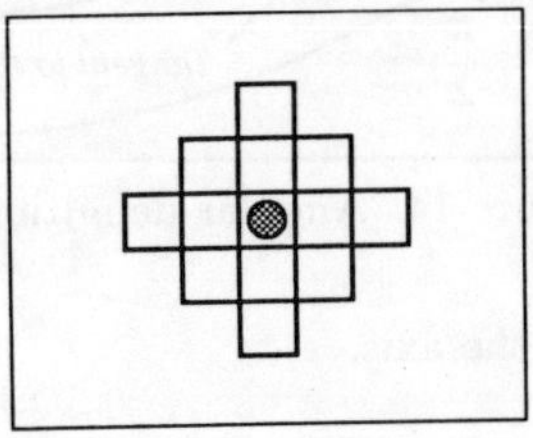

Figure 12: Diamond filter.

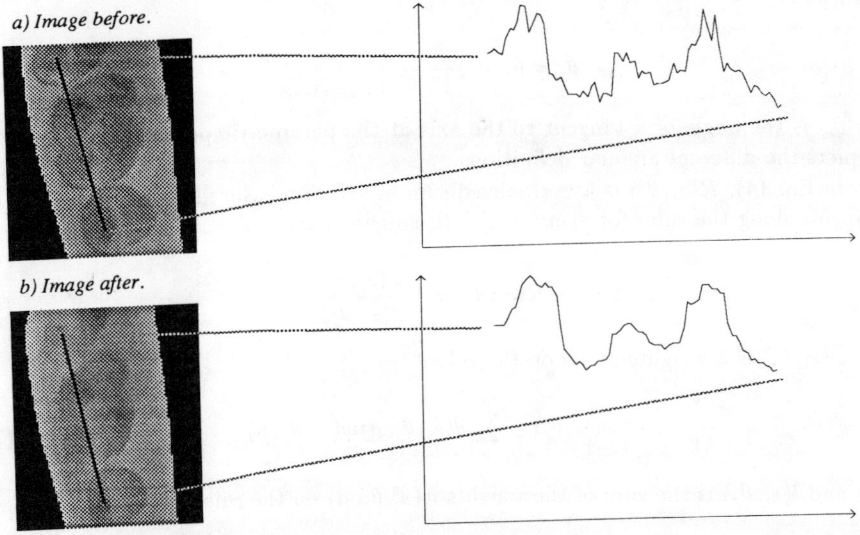

Figure 13: Image filtered with the diamond filter.

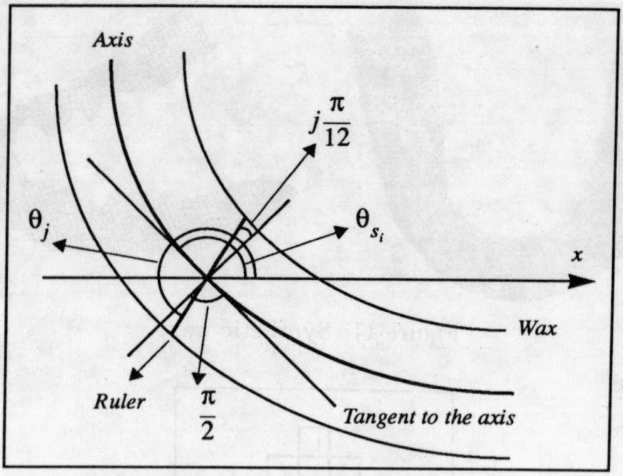

Figure 14: Angular definitions.

the parametric distance along the axis.

$$\sigma(s_i) = \max[\mathcal{R}(s_i, \theta_j)] \quad _{j=-4,\ldots,4} \tag{4}$$

where θ_j is the angle between the ruler and a perpendicular line to the axis at position s_i.

$$\theta_j = \theta_{s_i} + j\frac{\pi}{12} + \frac{\pi}{2} \quad _{j=-4,\ldots,4} \tag{5}$$

θ_{s_i} is the angle of a tangent to the axis at the parametric position s_i. Figure 14 depicts the different angular definitions.

In Eq. (4), $\mathcal{R}(s_i, \theta_j)$ is a normalized sum of values measured on the image of the imprint along the ruler for every angle. It is defined as:

$$\mathcal{R}(s_i, \theta_j) = -\frac{\omega(s_i, \theta_j)}{l(s_i, \theta_j)} \tag{6}$$

$\omega(s_i, \theta_j)$ is a weighted sum on the ruler:

$$\omega(s_i, \theta_j) = \sum_{\mu} d(s_i, \theta_j, \mu) w(s_i, \theta_j, \mu) \tag{7}$$

and $l(s_i, \theta_j)$ is the sum of the weights $w(s_i, \theta_j, \mu)$ on the ruler:

$$l(s_i, \theta_j) = \sum_{\mu} w(s_i, \theta_j, \mu) \tag{8}$$

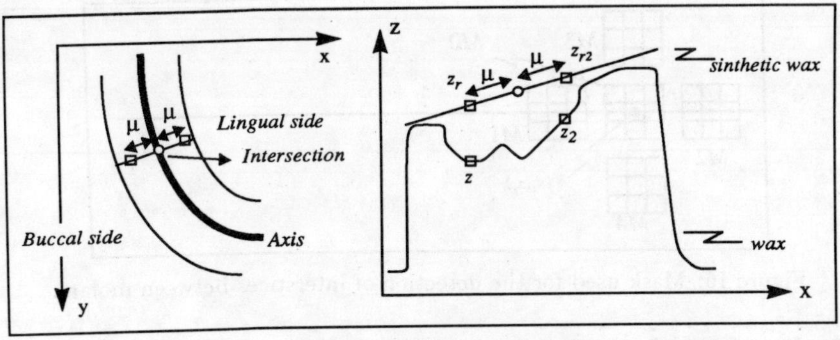

Figure 15: Mirror values.

$w(s_i, \theta_j, \mu)$ is a weigting factor applied to $d(s_i, \theta_j, \mu)$. μ is the parametric distance on the ruler and has the center point of the ruler for origin. $d(s_i, \theta_j, \mu)$ is defined differently for upper maxillary incisors, molars on both maxillaries, and lower maxillary incisors. $\omega(s_i, \theta_j, \mu)$ is also defined differently for incisors and molars. The design of the various operators used for the segmentation of the dental imprints was based on a thorough analysis of teeth morphology. Each operators seeks to amplify a given characteristic of each type of tooth [6].

4.2.1 Processing for upper maxillary incisors

Referring to Figure 15, the wafer is divided into two parts along the ruler; the buccal side and the lingual side. z and z_r are range values measured on the buccal side of the axis, and z_2 and z_{r_2} are values measured on the lingual side of the axis. z_r and z_{r_2} are measured on the synthetic wax, and z and z_2 are measured on the actual dental imprint image. z and z_2, as z_r and z_{r_2}, have opposite location and have the same distance μ to the intersection of the axis with the ruler. Note that the range values are measured on the 3-D image of the upper maxillary prior to filtering with the diamond-shaped filter described in Section 4.1.

Using values shown in Figure 15, $d(s_i, \theta_j, \mu)$ in Eq. (7) is defined as:

$$d(s_i, \theta_j, \mu) = \max[z(\mu) - z_r(\mu), z_2(\mu) - z_{r_2}(\mu)] \quad (9)$$

and $\omega(s_i, \theta_j, \mu)$ is set to $[d(s_i, \theta_j, \mu)]^2$. The curve labelled P_{IS_D} in Figure 18 is the result of the computation of the operator in Eq. (4)-(9) along the curved axis of the imprint. The operator is very reliable in the upper maxillary incisors area since large peaks appear on the curve between adjacent teeth. However, it does not perform as well in the molar area since peaks may appear that are not interstices between teeth but rather belong to the distal fossae or disto-buccal cusp of the molars. For this reason, a more robust operator has been defined for the detection of interstices between molars on both maxillaries.

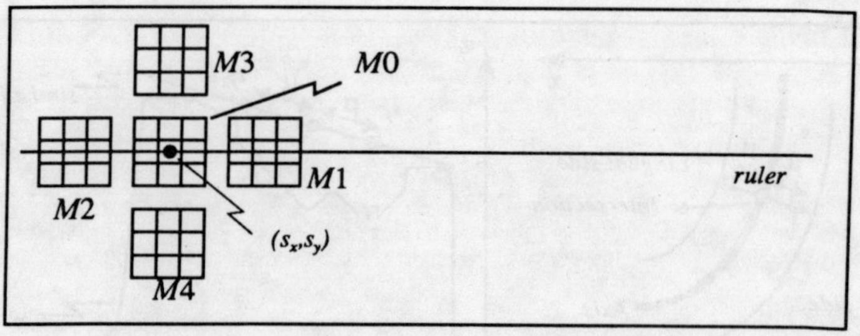

Figure 16: Mask used for the detection of interstices between molars.

4.2.2 Processing for molars on both maxillaries

Instead of using only range values along the ruler, as in Eq. (9), functions of the range values under the mask shown in Figure 16 are computed. The image filtered with the diamond-shaped filter is used for this purpose.

For each position of the mask along the ruler, the following parameters are computed:

$$M0 = \max[z(x + s_x, y + s_y)] \quad \begin{array}{l} x = -1, 0, 1 \\ y = -1, 0, 1 \end{array} \tag{10}$$

$$M1 = \sum_{x=3}^{5} \sum_{y=-1}^{1} \frac{z(x + s_x, y + s_y)}{9} \tag{11}$$

$$M2 = \sum_{x=-5}^{-3} \sum_{y=-1}^{1} \frac{z(x + s_x, y + s_y)}{9} \tag{12}$$

$$M3 = \sum_{x=-1}^{1} \sum_{y=3}^{5} \frac{z(x + s_x, y + s_y)}{9} \tag{13}$$

$$M4 = \sum_{x=-1}^{1} \sum_{y=-5}^{-3} \frac{z(x + s_x, y + s_y)}{9} \tag{14}$$

Then, these parameters are combined as follows:

$$z = M0 + M1 + M2 - M3 - M4 \tag{15}$$

This value is replaced in Eq. (9) for the computation of $d(s_i, \theta_j, \mu)$. For molars, $\omega(s_i, \theta_j, \mu)$ is set to $d(s_i, \theta_j, \mu)$.

This operator yields large values for on interstices between molar since $M0, M1$ and $M2$ are on the interstice and $M3$, $M4$, are on both sides of the interstice. On the other hand, the operator yields small values on areas occupied by a molar since $M1$ to $M4$ fall in the same range of small z values. On horizontal regions of the imprint (regions parallel to the x,y plane), z equals $M0$ since $M1 \approx M2 \approx M3 \approx M4$. The curve labeled P_{MS_F} in Figure 18 is the result of the processing step described above for the upper maxillary. Interstices between molars are clearly visible and correspond to large peaks on the curve. Peaks caused by the distal fossae and the disto-buccal cusps of molars are very small compared to those that were visible on the P_{IS_D} curve used for upper maxillary incisors. However, peaks that correspond to interstices between incisors are not as sharp on the P_{MS_F} curve as they were on the P_{IS_D}.

4.2.3 Processing for lower maxillary incisors

Lower maxillary incisors are very small and adjacent incisors are very close. Thus, the interstice between these incisors are very small and difficult to detect. More information is needed. The image at λ_2 provides this information. Referring to Figure 4 (b), it is observed that the bottom of lower maxillary incisors appears brighter then the sides of the imprint, while incisors appear very dark in the image at λ_1. For each pixel the following value is computed:

$$z(i,j) = z_{\lambda_1}(i,j) - z_{\lambda_2}(i,j) \tag{16}$$

This value is used in Eq. (9). Subtracting the image at λ_1 from the image at λ_2 yields an output image with significantly deeper values in the incisors area. A side effect is the amplification of the interstices between adjacent incisors.

4.2.4 Combination of the operators

The operators described above behave differently depending on the area of the imprint they are applied to. The output profiles of the operators are combined to yield a resulting profile that is used to find the interstices between teeth. For the upper maxillary the resulting profile P_S is computed as a weighted sum of the individual profiles

$$P_S = P_{IS_D} + F_S P_{IS_F} + (1 - F_S) P_{MS_F} \tag{17}$$

where F_S is defined as:

$$F_S = \frac{1 - \cos\left(\frac{2\pi s_i}{L}\right)}{2} \tag{18}$$

L is the total length of the axis of the imprint. Figure 17 shows the weighting function F_S and $(1 - F_S)$. This weighting function increases the contribution of P_{IS_F} in the incisors area and reduces its effect in the molar area (see Figure 18). The reverse process is observed for P_{MS_F}. For the lower maxillary, the weighted sum is

$$P_I = F_I P_{II_F} + F_I P_{II_{A-I}} + (1 - F_I) P_{MI_F} \tag{19}$$

with

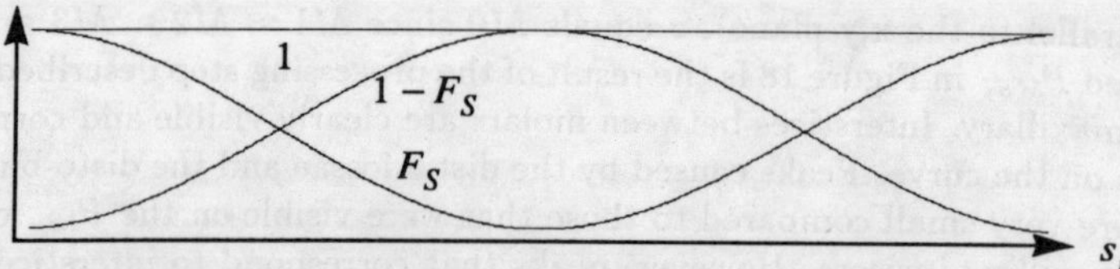

Figure 17: Weighting functions for upper maxillary.

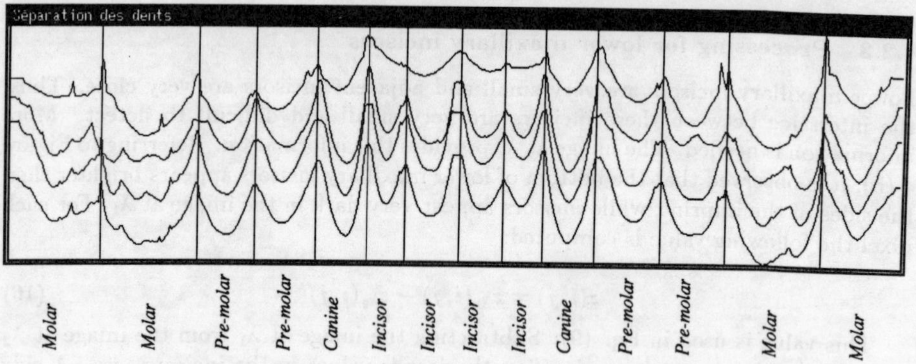

Figure 18: Detection of interstices. Upper maxillary.

$$F_I = 1 - \cos\left(\frac{2\pi s_i}{L}\right) \tag{20}$$

Figure 19 shows the weighting function for the lower maxillary. Figure 20 shows profiles for the lower maxillary. It is important to mention that the operators described above perform equally well for canines and pre-molars. No special operator was designed for these teeth

4.3 Peak detection

From the final profiles (P_S for upper maxillary and P_I for lower maxillary), the detection of interstices between teeth is reduced to a peak detection operation. The

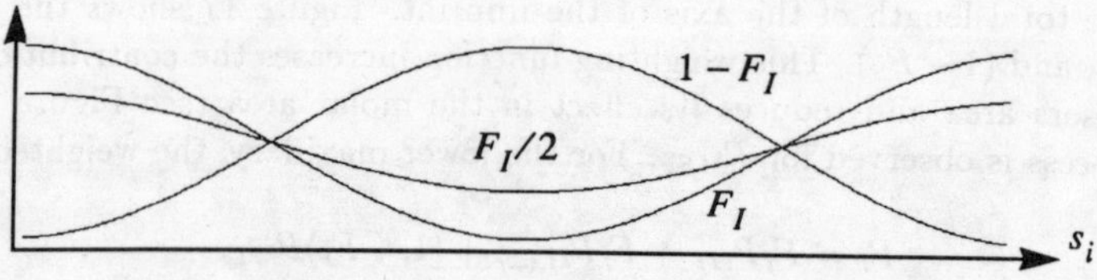

Figure 19: Weighting functions for lower maxillary.

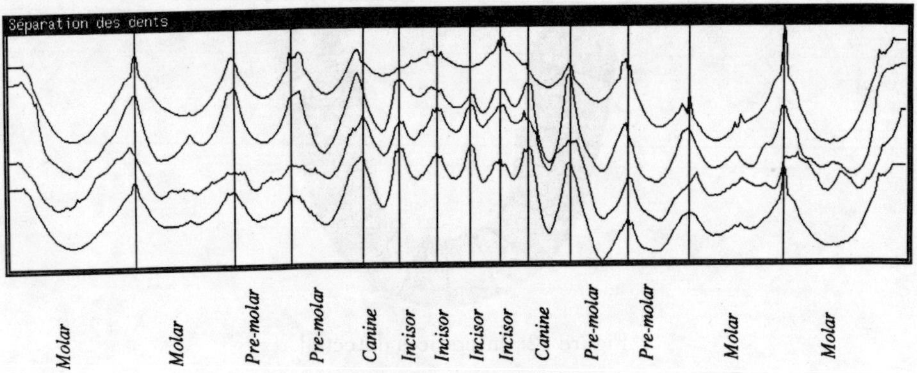

Figure 20: Detection of interstices. Lower maxillary.

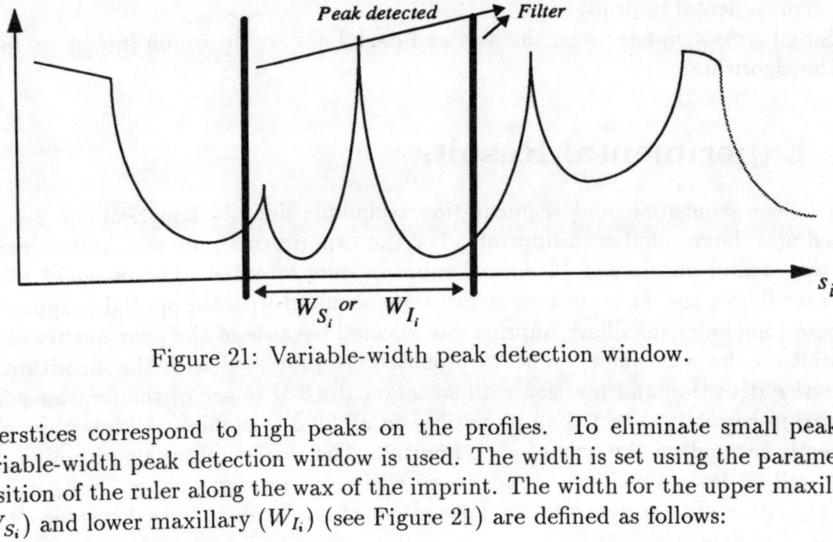

Figure 21: Variable-width peak detection window.

interstices correspond to high peaks on the profiles. To eliminate small peaks a variable-width peak detection window is used. The width is set using the parametric position of the ruler along the wax of the imprint. The width for the upper maxillary (W_{S_i}) and lower maxillary (W_{I_i}) (see Figure 21) are defined as follows:

$$W_{S_i} = \max\left[\frac{L}{45}\left(1+\cos\frac{2\pi(s_i - 2\frac{L}{45})}{L}\right)+\frac{L}{90}, 8\right] \tag{21}$$

$$W_{I_i} = \max\left[\frac{L}{45}\left(1+\cos\frac{2\pi(s_i - 2\frac{L}{45})}{L}\right)+\frac{L}{90}, 7\right] \tag{22}$$

L is the length of the axis. The 1/45 factor is the average half-width of a molar compared to the length of the axis along the wax. A threshold value $L/90$ is added because the width of the window is not allowed to take the value "zero".

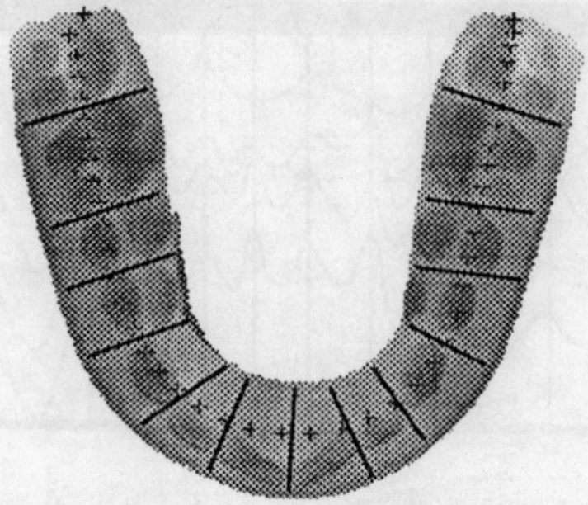

Figure 22: Interstices detected.

The results of interstices detection are shown on Figure 22 for the upper maxillary of a typical dental imprint.

Small cross symbols lie on the axis and dark lines are drawn on interstices found by the algorithm.

5 Experimental Results

The image acquisition and segmentation technique described in Sections 2-4 was tested on a large number of imprints. For the experiments presented in this paper, the wax dental imprints of 13 normal subjects were recorded. The image of 15 upper maxillaries and 14 lower maxillaries were acquired with the special purpose 3-D camera (one lower maxillary imprint was rejected because of the poor quality of the imprint on the wax wafer). The 29 profiles were processed with the algorithm for interstice detection and position estimation on the 3-D image of the imprints. The algorithm correctly detected all interstices on all 29 3-D profiles. All intertices were correctly located on the axis of each imprint. The segmentation algorithm worked very well on lower maxillary incisors even with poor imprint quality. Furthermore, the algorithm did not generate any false interstice in the disto-buccal cusp or buccal fossae of molars.

6 Conclusion

This paper has presented a computer vision system for the acquisition and the segmentation of 3-D profiles of wax dental imprints. The system is part of an expert system for automatic orthodontic diagnosis. The acquisition system is based on the absorption of light by a dispersive medium. The 3-D profiles of wax imprints measured with this special purpose camera showed an accuracy better than 10% for depth measurements, and better than 5% for lateral distance measurements on the imprint.

The segmentation algorithm searches for the interstice between teeth on both maxillaries. The algorithm is based on three different operators, each operator being tuned for a specific type of tooth. The algorithm was tested on many 3-D profiles and all interstices were correctly detected. No false interstices were produced by the operators.

The next step in the development of the system will be focused on the processing of profiles with missing teeth and on profiles with mixed dentition. A heuristic algorithm is currently under development for the identification of missing or false interstices that may appear on poorly recorded wax imprints. Tests on a larger number of imprints (> 100) will also be performed. Finally, a set of 10 orthodontic parameters will be measured on the 3-D profiles and fed to the expert system for automatic diagnosis.

Acknowledgements

The research work was supported by Grant OGPIN 011 of the Natural Sciences and Engineering Research Council of Canada (NSERC). J. Côté was supported by a NSERC postgraduate scholarship. The authors would like to thank R. Houde for his help in the design of the 3-D acquisition system, L. Guimond for comments on the segmentation technique, and MM. J. Lévesque and Y. Chalifour for their work on the experimental setup. J. Côté and D. Laurendeau are members of the Institute for Robotics and Intelligent Systems (IRIS) and wish to acknowledge the support of the Networks of Centres of Excellence program of the Government of Canada, the National Sciences and Engineering Research Council and the participation of PRECARN Associates Inc.

References

[1] Colin C. Archibald, Stanley R. Sternberg, Mathematical Morphology Applied to Range Image Processing. Proc. of Vision Interface '86, Vancouver B.C., May 1986, pp 293-299.

[2] C. Bernard, J.M. Brodeur, A. Fournier, D. Poussart, D. Laurendeau, R. Guay, M. Olivier, and P. Retrouvey, Computerized Diagnosis in Orthodontics for Epidemiological Studies: A Progress Report, Proc. of the 66th General Session of the International Association for Dental Research, Montreal, P.Q., Canada, March 9-13, 1988, pp.169.

[3] L. Guimond, Acquisition et segmentation d'images d'empreintes dentaires, Master's thesis, Laval University, Quebec, Canada, (in French).

[4] Robert M. Haralick, Stanley R. Sternberg and Xinhua Zhuang, Image Analysis Using Mathematical Morphology. IEEE Transaction on Pattern Analysis and Machine Intelligence, Vol. PAMI-9, #4, July 1987.

[5] R.Houde, D.Laurendeau and D.Poussart, 3-D camera based on differential optical absorbance, Proc. SPIE, International Symposium on Optical & Optoelectronic Applied Science & Engineering. San Diego, Cal. July 1990 (under press).

[6] B.S. Kraus, R.E. Jordan and L. Abrams, Dental Anatomy and Occlusion, B.C. Decker Inc. Toronto, 1988.

[7] D.Laurendeau, L.Guimond and D.Poussart, A Computer-Vision Technique for the Acquisition of 3D Profiles of Dental Imprints and the Detection of Teeth in the Profiles: An Application in Orthodontics, to appear in IEEE Transactions on Medical Imaging.

[8] J.H. Scott and N.B.B. Symons, Introduction to Dental Anatomy. 9th Edition, Churchil Livignstone, New York 1982.

[9] John W.T. Walsh, Photometry. Constable, London 3rd edition 1958.

Modelling Sonar Range Sensors

David Wilkes
Department of Computer Science, University of Toronto
Toronto, Canada M5S 1A4.

Gregory Dudek
McGill Research Center for Intelligent Machines
McGill University, 3480 University St.
Montreal, Canada H3A 2A7

Michael Jenkin
Department of Computer Science, York University
Downsview, Canada

Evangelos Milios
Department of Computer Science, York University
Downsview, Canada

Abstract

We present a simulation-based model of sonar range sensing for robot navigation that accounts for multiple reflections of the sonar signal between transmission and reception. This gives more realistic results than previous models. The approach is based on simulation of the reflection and diffraction of sonar rays from reflecting surfaces until they are attenuated beyond detectability or return to the receiver. Parameters of the model include frequency, minimum and maximum range, and signal detection threshold (relative to emitted signal strength, after linear gain compensation). Finally, the usefulness of the model to the development of more effective algorithms for the interpretation of sonar data is discussed.

1 Introduction

Sonar range sensing refers to technologies for emitting sound signals and measuring the characteristics of the echo that returns. The most common form of sonar range sensing – the one addressed here – is based on emitting a sonar signal (a "chirp") and measuring the time delay until an echo is detected. This provides a very simple and economical method for distance measurement.

Unfortunately, distance measurement with a sonar device is not as straightforward as one might expect. In a typical sonar scan taken around the compass from a single position, very few of the distance readings returned are accurate ranges to the nearest object in the aiming direction. Several authors [2, 3] have produced clever interpretation methods that attempt to cope with this unreliable data. The methods essentially throw away all measurements that do not have sustained support from multiple viewpoints. They do not incorporate sophisticated models of the measurement process. The model discussed herein accounts for much of the structure in the data that is typically thrown away. As we shall see, it is inadequate to model the ranging errors as the result of some simple stochastic noise process.

An encouraging conclusion that has been reinforced by this work is that many classes of ranging errors are in fact *repeatable, predictable,* features of the sonar device. The hope is that we may develop better sonar interpretation algorithms, given a deeper understanding of these features.

In work on algorithms for robotic exploration of unknown environments [4, 5] a crucial problem is that of recovering the shape of the open areas of an environment. The generation of realistic synthetic data facilitates the analysis of the algorithms in specialized scenarios. The performance of sonar-based methods can be studied in a controlled, repeatable framework. Furthermore, alternative environments can be simulated quickly and easily.

The class of sonar sensor dealt with here operates by measuring the time required for a pulse of ultrasound to travel from a transmitter, bounce off an object, and return to approximately the same point. The transmitter and receiver are often the same transducer. From knowledge of the time taken for the sound to return and of the speed of sound, the distance to the object reflecting the pulse can be inferred. A typical and widely used ultrasonic ranging device is that produced by the Polaroid Corporation. Its use is described in detail in references [6] and [7]. It operates using sound signals with frequencies of roughly 50 kHz.

There are several reasons why false range values may be inferred by such systems. The return signal detected by the system may be the result of multiple reflections of the sound pulse, or diffraction from edges of objects, or reflection of part of the wavefront off the line along which the transducer is aimed. Also, there are minimum and maximum useful ranges of the device imposed by the combination of transmitter and receiver in a single package, and receiving circuitry limitations, respectively.

Kuc and Siegel [1] have done a careful analysis of the detailed physics of signal transmission, reflection, diffraction and reception. They give results on the behavior of sonar signals at right angled corners and walls. Their simulation accurately matches actual sonar measurements from such features, provided that the signal returns to the receiver without multiple intervening reflections. In real environments, however, phenomena such as multiple reflections, amplifier characteristics, and the room shape play a significant role. In order to generate a realistic simulation of sonar sensing, these phenomena have been incorporated in our simulation. The result is a model that uses an approach somewhat akin to the "ray-tracing" models used in the computer graphics community. Sample signal paths are followed from transmitter, through bounces and diffraction, to the receiver.

Details of the approach follow in the next section, followed by a discussion of the implementation. The third section summarizes the validation of the model. Finally, the practical consequences of the phenomena that have been modelled are discussed.

2 Model and Implementation

The model is premised on several assumptions about the behavior of sonar signals in normal circumstances. First, we assume predominantly specular reflection of the sonar signal from surfaces, and that floors and ceilings can safely be ignored. The degree to which the former assumption is valid depends on the surface texture and the wavelength of the sonar. For the Polaroid unit's typical frequencies, reflection from most wall surfaces is highly specular. The latter assumption regarding floors and ceilings may not always be valid. In this case, the two-dimensional simulation may be run on different cross-sections of the environment to recover reliable range readings. For example, the simulation could be run on one horizontal and one vertical cross-section intersecting the transducer, or (with slight modification) on several parallel cross-sections of the environment. Although this increases the running time of the simulation, running time on a single cross-section is sufficiently small that this would not pose significant difficulties. As is commonly the case, it is assumed that the sonar system responds only to the first occurrence of the reception of a signal over the threshold strength, in a given trial. Also, it is assumed that the rangefinder incorporates an amplifier whose gain increases linearly with time to compensate for the dispersion of the signal into space (this is also consistent with commonly-used technology [6]).

The results from Kuc and Siegel [1] that are used to set parameters of the model are related to the transducer impulse response and the ratio of the strengths of a specularly reflected signal and a signal diffracted from a straight edge. Under the assumption that the objects reflecting the sonar pulse of wavelength λ are at distances z much larger than $\frac{a^2}{\lambda}$ from a circular transducer with radius a, the impulse response of a transducer at angle α to the sonar wavefront is given by

$$h_R(t, z, a, \alpha) = \frac{2c * \cos\alpha}{\pi a * \sin\alpha}\sqrt{1 - \frac{c^2\,(t - 2z/c)^2}{a^2 \sin^2\alpha}} \tag{1}$$

for c the speed of sound in the environment, $t \epsilon \left[\frac{2z - a*\sin\alpha}{c}, \frac{2z + a*\sin\alpha}{c}\right]$, and $0 < |\alpha| \le \frac{\pi}{2}$.

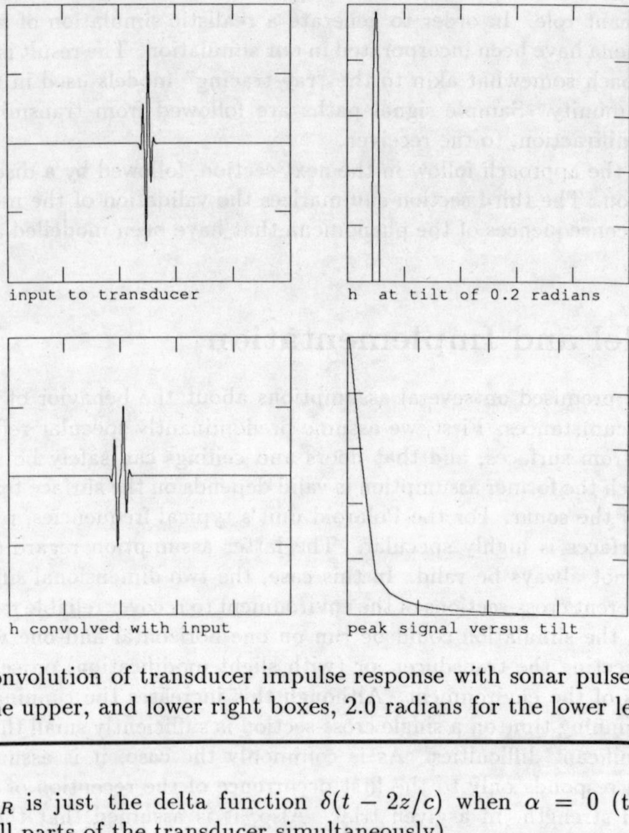

Figure 1: Convolution of transducer impulse response with sonar pulse. Box widths 1.0 ms for the upper, and lower right boxes, 2.0 radians for the lower left box.

Note that h_R is just the delta function $\delta(t - 2z/c)$ when $\alpha = 0$ (the wavefront leaves/hits all parts of the transducer simultaneously).

When h_R is convolved with the sonar pulse waveform (e.g. a Gaussian-modulated sine wave), the degree to which the sonar signal is attenuated is determined as a function of the angle α at which the resulting signal leaves (or arrives at) the transducer. Figure 1 shows the result of the convolution, and a plot of signal attenuation as a function of transducer angle.

The second result of Kuc and Siegel that is used to parameterize the model concerns the relative strength of the signal due to diffraction from corners. Modelling corners as line sources of echoes with cylindrical wavefronts, the relative strength is $\left(2\pi\sqrt{\frac{z}{\lambda}}\right)^{-1}$.

The model parameters are the sonar frequency, minimum and maximum measurable ranges, and the signal detection threshold relative to the strength of the outgoing signal (after linear-gain amplification). The environment is modelled as a set of line segments with specified reflectance coefficients from which incident sonar signals are reflected specularly or diffracted. Minimum range accounts for the recovery time of the transducer from the transmission transient. This time corresponds to a minimum distance consistent with the assumption that $z >> \frac{a^2}{\lambda}$. Maximum range accounts for limits on amplifier gain.

The simulation proceeds by generating a finely spaced fan of rays leaving the front of the transducer, and following their progress in the environment until they again hit the front face of the transducer or until they are sufficiently attenuated. Figure 2 illustrates this. For a single range measurement, the following steps occur. First, the fan of rays is generated from the transducer with an signal strength distribution that is a function of emittance angle. The intersections of this initial fan of rays with walls are computed. These intersections are added to a heap with the closest at the top. Subsequently the closest (and hence earliest) intersection is examined. If it is within the ray-spacing of a corner, a diffracted ray is generated in the direction of the receiver (since it propagates in all directions). Otherwise, the direction of the reflected ray is computed. In both cases the current intersection is replaced in the heap with the new ray's next intersection with a model line, and the signal strength is attenuated appropriately. The new closest intersection is then examined as the process repeats. This terminates when an intersection with a small line segment representing the receiver is encountered as the earliest intersection in the heap. If the arriving signal has a sufficiently small angle of incidence (so that the further attenuation of the signal does not render it undetectable) then half the cumulative distance is returned as the range measurement. If no reception occurs within the maximum distance, then the maximum distance is returned as the range value. In computing the receiver line length and distance from model corners at which to generate a diffracted ray, allowance is made for the spacing of the model rays, so that features are not missed as a result of being between model rays.

Figure 3 is a very high-level statement of the algorithm.

3 Experimental Results

Figure 4 shows the results of running the simulation on a model of the room used by Kuc and Siegel to validate their predictions. The data in Kuc and Siegel includes range measurements as a function of transducer orientation for a small sample room. Kuc and Siegel accurately account for a portion of this data, but their model does not account for those readings which are due to interactions of the signal with multiple surfaces. The experiments reported herein include the single-bounce reflectance measurements obtained by Kuc and Siegel (which are consistent with the actual sonar data), but also reproduce the multiple-reflection readings from the actual sonar data in Kuc and Siegel.

The use of sonar data for robotic exploration is further complicated by the variable

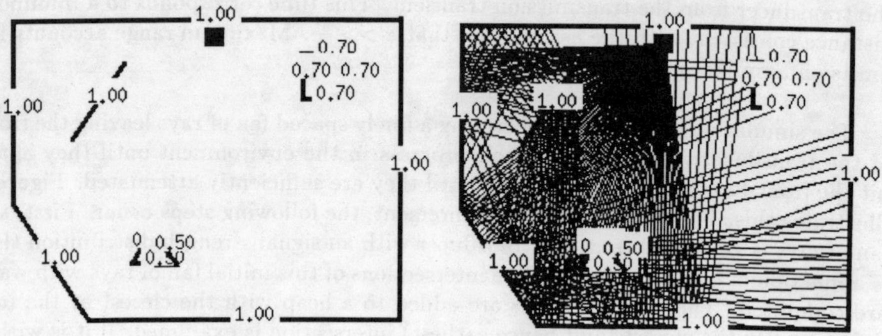

Figure 2: A single range measurement. Left: the room with robot. Right: some rays generated by a single range measurement. The numbers are relative wall reflectances.

reflectances of various objects in the environment. The simulation reported here for sonar range measurement readily allows for surfaces with various sonar reflectance properties. By modelling this aspect of real surfaces in addition to multi-bounce effects, it is possible to replicate much of the complex nature of sonar sensing. Figure 5 shows the results of running our model on more complicated data which includes multiple-reflection range readings, and surfaces having varying reflectance coefficients. The rays representing the range readings have partially obscured two small objects in the room in this figure. The upper right object is a small square. The lower left object is triangular.

More detailed experiments with an aimable sonar device have been initiated. Many of the peculiarities of the data demonstrated in the simulation show up with surprising robustness. Examples are multiple reflection artifacts as shown in the simulation of figure 5. Note especially the example of false high-range readings between the device and the adjacent wall, that occurs because the wall is closer than the minimum detectable distance.

The major remaining limitation of this model is that it uses only a planar model of the world. As such, it fails to capture bounces of the sonar signal that might come from interactions with objects outside this plane. Experimentation with the sonar device and the simulation suggests that, in many environments, sufficient accuracy may be obtained by projecting all objects near the horizontal plane scanned by the device onto this plane for modelling purposes.

```
range(θ, x, y): determine the range returned by a sonar measurement
        from position (x, y) of the sonar sensor facing direction θ

        initialize heap of intersections to be empty
        for each angle φ from θ − π
                to θ + π in steps of size δ
                determine: the first intersection I of a ray leaving (x, y)
                at angle φ with a model wall segment, the degree of
                attenuation of the signal as a function of |φ − θ|,
                the total distance travelled from the transducer, and angle
                of incidence at the intersection.
                add I to the heap (ordered by distance travelled
                        from transducer)
        while the heap is non-empty
                if the intersection I at the top of the heap is at a
                        distance d > maxdistance from the transducer
                        return(maxdistance)
                if the intersection is within a single ray-spacing
                        of the end of the model segment involved
                        generate a diffracted-ray intersection with the
                        receiver also, with appropriate signal strength and
                        angle of incidence
                generate the reflected ray r leaving the intersection,
                        based on the angle of incidence of the incoming ray
                        and reduce its amplitude as a function of the
                        characteristics of the segment
                if there exists an intersection of r with a model wall segment
                        if the intersection is with the model segment
                                representing the transducer
                                compute the further attenuation of the
                                incoming signal as a function of the angle
                                of incidence
                                if the remaining signal strength is above
                                        detection threshold
                                        return(total distance travelled / 2)
                else
                        add the closest intersection between r and a
                        model segment to the heap
         end while
```

Figure 3: A high-level statement of the algorithm

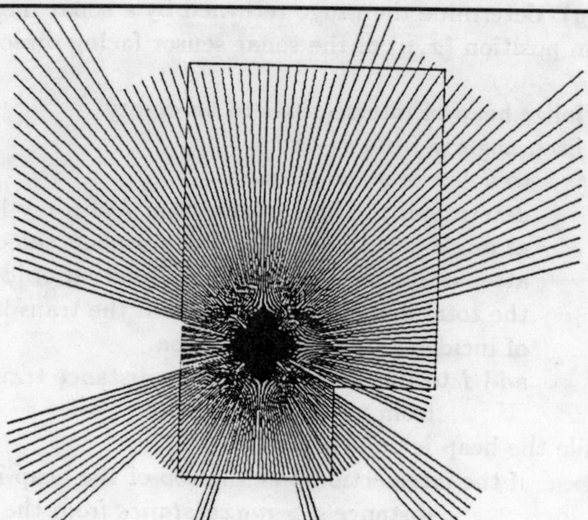

Figure 4: Validation on Kuc and Siegel's data. The horizontal and vertical lines indicate the positions of the simulated walls of a room. The radial lines indicate the range measurements predicted as a function of angle.

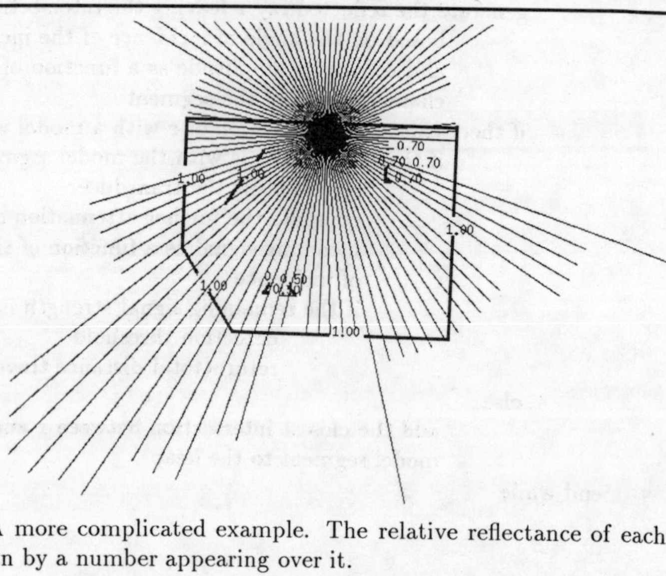

Figure 5: A more complicated example. The relative reflectance of each modelled wall is shown by a number appearing over it.

4 Conclusions

The model described here allows for the realistic simulation of sonar range sensing in real environments. As shown, it is consistent with real data collected from such devices.

The availability of realistic, effective and efficient simulation permits rapid and repeatable trials of experimental scenarios. Experimental work is essential to the enumeration of realistic problems in sensing. On the other hand, the ability to examine rapidly particular situations, and more importantly to repeat exactly a series of (simulated) measurements is critical to the rigorous, scientific development of sensing strategies. In addition, this allows a large suite of test cases to be examined rapidly. It is expected that this will facilitate the development and evaluation of robust sensing methodologies.

Methods for efficiently inferring the reflecting-surface geometry from sonar range measurements are currently under investigation. It is hoped that the deeper understanding of these measurements acquired through efforts to simulate them will allow conversion of aspects of the ranging data that have long been considered "errors" into useful features that can be part of the interpretation process.

Acknowledgements

The support of the Natural Sciences and Engineering Research Council of Canada and of the Government of Ontario through the Information Technology Research Centre is gratefully acknowledged. We also gratefully acknowledge the thoughtful comments of the anonymous reviewers.

References

[1] Kuc, R. and Siegel, M.W., "Physically Based Simulation Model for Acoustic Sensor Robot Navigation," *IEEE Transactions on Pattern Analysis and Machine Intelligence,* Vol. PAMI-9, No. 6, pp.766-78. 1987.

[2] Crowley, J.L., "Dynamic World Modeling for an Intelligent Mobile Robot Using a Rotating Ultrasonic Ranging Device," *IEEE Conference on Robotics and Automation*, St. Louis, March 1985.

[3] Moravec, H.P., and Elfes, A., "High-Resolution Maps from Wide-Angle Sonar," *IEEE Conference on Robotics and Automation*, St. Louis, March 1985.

[4] Dudek, Gregory, Jenkin, Michael, Milios, Evangelos, Wilkes, David, "Robotic Exploration as Graph Construction," *IEEE Transactions on Robotics and Automation* (to appear).

[5] Dudek, Gregory, Jenkin, Michael, Milios, Evangelos, Wilkes, David, "Using Multiple Markers in Graph Exploration," *Proceedings of the International Society for Optical Engineering Symposium on Advances in Intelligent Robotics Systems: Conference on Mobile Robotics*, Philadelphia, PA, November 1989.

[6] Polaroid Corporation, a: Untitled product spec sheet on Ultrasonic Ranging System. b: "Ultrasonic Ranging System," Users manual.

[7] Biber, C., Ellin, S., Shenk, E. and Stempeck, J., "The Polaroid Ultrasonic Ranging System," *The 67th Convention of the Audio Engineering Society*, New York, 1980.